VOLUME SEVENTY TWO

ADVANCES IN MICROBIAL PHYSIOLOGY

Nitric Oxide and Other Small Signalling Molecules

Edited by

ROBERT K. POOLE

West Riding Professor of Microbiology
Department of Molecular Biology and Biotechnology
The University of Sheffield
Firth Court, Western Bank
Sheffield, UK

ELSEVIER

ACADEMIC PRESS

An imprint of Elsevier

Academic Press is an imprint of Elsevier
The Boulevard, Langford Lane, Kidlington, Oxford OX5 1GB, United Kingdom
125 London Wall, London, EC2Y 5AS, United Kingdom
50 Hampshire Street, 5th Floor, Cambridge, MA 02139, United States
525 B Street, Suite 1800, San Diego, CA 92101-4495, United States

First edition 2018

ISBN: 978-0-12-814413-8
ISSN: 0065-2911

For information on all Academic Press publications
visit our website at https://www.elsevier.com/books-and-journals

Publisher: Zoe Kruze
Acquisition Editor: Ashlie Jackman
Editorial Project Manager: Joanna Collett
Production Project Manager: Abdulla Sait
Cover Designer: Matthew Limbert

Typeset by SPi Global, India

CONTENTS

CONTRIBUTORS

Takaaki Akaike
Tohoku University Graduate School of Medicine, Sendai, Japan

Rika I. Astuti
Faculty of Mathematics and Natural Sciences, Bogor Agricultural University, Bogor, Indonesia

Sujata S. Chaudhari
University of Nebraska Medical Center, Omaha, NE, United States

Jeffrey A. Cole
School of Biosciences and Institute of Microbiology and Infection, University of Birmingham, Birmingham, United Kingdom

Elizabeth H. Hutfless
University of Nebraska Medical Center, Omaha, NE, United States

Tomoaki Ida
Tohoku University Graduate School of Medicine, Sendai, Japan

Christopher Lupfer
Missouri State University, Springfield, MO, United States

Ryo Nasuno
Graduate School of Biological Sciences, Nara Institute of Science and Technology, Ikoma, Nara, Japan

Motohiro Nishida
Okazaki Institute for Integrative Bioscience, National Institute for Physiological Sciences, National Institutes of Natural Sciences, Okazaki; Graduate School of Pharmaceutical Sciences, Kyushu University, Fukuoka, Japan

Katsuhiko Ono
Graduate School of Medical Sciences, Kumamoto University, Kumamoto, Japan

Rajalaksmy A. Ramachandran
UT Southwestern Medical Center, Dallas, TX, United States

Tomohiro Sawa
Graduate School of Medical Sciences, Kumamoto University, Kumamoto, Japan

Hiroshi Takagi
Graduate School of Biological Sciences, Nara Institute of Science and Technology, Ikoma, Nara, Japan

Vinai C. Thomas
University of Nebraska Medical Center, Omaha, NE, United States

Masanori Toyofuku
Graduate School of Life and Environmental Sciences, University of Tsukuba, Tsukuba, Japan

Hiroyasu Tsutsuki
Graduate School of Medical Sciences, Kumamoto University, Kumamoto, Japan

Sang-Sun Yoon
Brain Korea 21 PLUS Project for Medical Science, Institute for Immunology and Immunological Diseases, Yonsei University College of Medicine, Seoul, Korea

Hasan Zaki
UT Southwestern Medical Center, Dallas, TX, United States

Tianli Zhang
Graduate School of Medical Sciences, Kumamoto University, Kumamoto, Japan

PREFACE

The chemistry, biochemistry and microbiology of several small bioactive gases have been covered several times in recent volumes of *Advances in Microbial Physiology*, especially oxygen, nitric oxide (NO) and carbon monoxide (CO) in the chapters, for example, by Sousa et al., Forte et al., Walker et al. and Southam et al. in Volume 71 and Bacon et al. in Volume 70. However, nitric oxide (NO) maintains a special place in this family of small signalling molecules, being the first to be called a 'gasotransmitter' (Wang, 2004). NO is one of several small molecules, formally known primarily as toxic gases, that have been shown to be endogenously generated, membrane-permeable and potent signalling molecules. The biological signalling activities associated with NO, CO, and hydrogen sulphide and their derived species have become topics of great interest to chemists, molecular biologists, clinicians and certainly microbiologists. Although these molecules have been termed 'gasotransmitters', the term is misleading, since all these species are completely soluble at concentrations of physiological relevance and therefore cannot be considered as gases in these biological contexts (Fukuto et al., 2012). Nevertheless, these small molecule signalling agents constitute complex and critical signalling webs that affect numerous physiological processes in animals, plants and microbes. This volume focuses on NO and its diverse roles and synthesis in yeast and bacteria, in pathogenesis and numerous aspects of bacterial physiology. Furthermore, because NO is generally toxic to bacteria, aspects of NO detoxification are covered. The impact of sulphides is covered in an opening comprehensive review of persulphides in microbiology and their importance as antioxidants and signalling agents.

Robert K. Poole

REFERENCES

Fukuto, J. M., Carrington, S. J., Tantillo, D. J., Harrison, J. G., Ignarro, L. J., Freeman, B. A., et al. (2012). Small molecule signaling agents: The integrated chemistry and biochemistry of nitrogen oxides, oxides of carbon, dioxygen, hydrogen sulfide, and their derived species. *Chemical Research in Toxicology*, *25*, 769–793.

Wang, R. (2004). The evolution of gasotransmitter biology and medicine. In R. Wang (Ed.), *Signal transduction and the gasotransmitters: NO, CO and H_2S in biology and medicine* (pp. 3–31). Totowa, NJ: Humana Press.

CHAPTER ONE

Reactive Cysteine Persulphides: Occurrence, Biosynthesis, Antioxidant Activity, Methodologies, and Bacterial Persulphide Signalling

Tomohiro Sawa*[,1], Katsuhiko Ono*, Hiroyasu Tsutsuki*, Tianli Zhang*, Tomoaki Ida[†], Motohiro Nishida[‡,§], Takaaki Akaike[†]

*Graduate School of Medical Sciences, Kumamoto University, Kumamoto, Japan
[†]Tohoku University Graduate School of Medicine, Sendai, Japan
[‡]Okazaki Institute for Integrative Bioscience, National Institute for Physiological Sciences, National Institutes of Natural Sciences, Okazaki, Japan
[§]Graduate School of Pharmaceutical Sciences, Kyushu University, Fukuoka, Japan
[1]Corresponding author: e-mail address: sawat@kumamoto-u.ac.jp

Contents

Abstract

Cysteine hydropersulphide (CysSSH) is a cysteine derivative having one additional sulphur atom bound to a cysteinyl thiol group. Recent advances in the development of analytical methods for detection and quantification of persulphides and polysulphides have revealed the biological presence, in both prokaryotes and eukaryotes, of hydropersulphides in diverse forms such as CysSSH, homocysteine hydropersulphide, glutathione hydropersulphide, bacillithiol hydropersulphide, coenzyme A hydropersulphide, and protein hydropersulphides. Owing to the chemical reactivity of the persulphide moiety, biological systems utilize persulphides as important intermediates in the

Advances in Microbial Physiology, Volume 72
ISSN 0065-2911
https://doi.org/10.1016/bs.ampbs.2018.01.002

synthesis of various sulphur-containing biomolecules. Accumulating evidence has revealed another important feature of persulphides: their potent reducing activity, which implies that they are implicated in the regulation of redox signalling and antioxidant functions. In this chapter, we discuss the biological occurrence and possible biosynthetic mechanisms of CysSSH and related persulphides, and we include descriptions of recent advances in the analytical methods that have been used to detect and quantitate persulphide species. We also discuss the antioxidant activity of persulphide species that contributes to protecting cells from reactive oxygen species-associated damage, and we examine the signalling roles of CysSSH in bacteria.

ABBREVIATIONS

biotin-PEG-MAL biotin-polyethylene glycol-conjugated maleimide
CARS2 mitochondrial cysteinyl-tRNA synthetase
CARSs cysteinyl-tRNA synthetases
CBS cystathionine β-synthase
COPD chronic obstructive pulmonary disease
CSE cystathionine γ-lyase
CstR CsoR-like sulphurtransferase repressor
CysSSH cysteine hydropersulphide
DM diabetes mellitus
EcCARS CARS from *Escherichia coli*
ETHE1 ethylmalonic encephalopathy protein 1
FRET fluorescence resonance energy transfer
GSH glutathione
GSSH glutathione hydropersulphide
HCysSSH homocysteine hydropersulphide
H_2O_2 hydrogen peroxide
HPE-IAM β-(4-hydroxyphenyl)ethyl iodoacetamide
H_2S hydrogen sulphide
LC-MS/MS liquid chromatography-tandem mass spectrometry
MBB monobromobimane
MSBT methylsulphonyl benzothiazole
8-nitro-cGMP 8-nitroguanosine 3′,5′-cyclic GMP
PCMB *p*-chloromercuribenzoic acid
PDO persulphide dioxygenase
PEG polyethylene glycol
PLP pyridoxal phosphate
PMSA PEG-conjugated maleimide-labelling gel shift assay
Rhod rhodanese
ROS reactive oxygen species
RSS reactive sulphur species
8-SH-cGMP 8-mercapto-cGMP
SQR sulphide:quinone oxidoreductase
TRP14 thioredoxin-related protein of 14 kDa
Trx1 thioredoxin1
TrxR1 thioredoxin reductase-1

1. INTRODUCTION

Cysteine hydropersulphide (CysSSH) is a cysteine derivative having one additional sulphur atom bound to a cysteinyl thiol group (Fukuto et al., 2012; Ono et al., 2014). CysSSH has been thought to serve as an important biosynthetic intermediate by donating its sulphur atom during biosynthesis of various sulphur-containing biofactors, such as iron–sulphur clusters, biotin, and lipoic acid (Hidese, Mihara, & Esaki, 2011; Zhang et al., 2012). CysSSH is also involved in the regulation of tRNA methylthiolation and insulin secretion (Takahashi et al., 2017). Recent advances in the development of analytical methods for detection and quantification of persulphides and polysulphides have revealed the biological presence of hydropersulphides such as CysSSH, homocysteine hydropersulphide (HCysSSH), glutathione hydropersulphide (GSSH), and protein hydropersulphides (see later for details). In addition to persulphides possessing the sulphur-donating functions mentioned earlier, recent evidence has suggested that persulphide species can regulate physiological signalling, at least in part via their redox-active property, as Toohey previously hypothesized (Toohey, 1986, 1989, 2011).

In this chapter, we discuss the biological occurrence and possible biosynthetic mechanisms of CysSSH and related persulphides, and we describe recent advances in the analytical methods that have been used to detect and quantitate persulphide species. We also discuss the antioxidant activity of persulphide species that contributes to protecting cells from reactive oxygen species (ROS)-associated damage, and we describe the signalling roles of CysSSH in bacteria. Readers are also referred to several useful review articles on the sulphur-donating functions of CysSSH (Cavuzic & Liu, 2017; Dahl, 2015; Fontecave, Ollagnier-de-Choudens, & Mulliez, 2003; Mueller, 2006; Py & Barras, 2010), which we do not explore in this chapter.

2. OCCURRENCE

Production of CysSSH and related low-molecular-weight persulphide species such as GSSH has long been suggested to occur in various biological systems. Recent advances in liquid chromatography-tandem mass spectrometry (LC-MS/MS) (see later for details) have led to identification of a number of low-molecular-weight persulphides in prokaryotes and eukaryotes (Fig. 1). Table 1 presents data for the endogenous occurrence of persulphide species determined in various cultured mammalian cells. CysSSH and GSSH

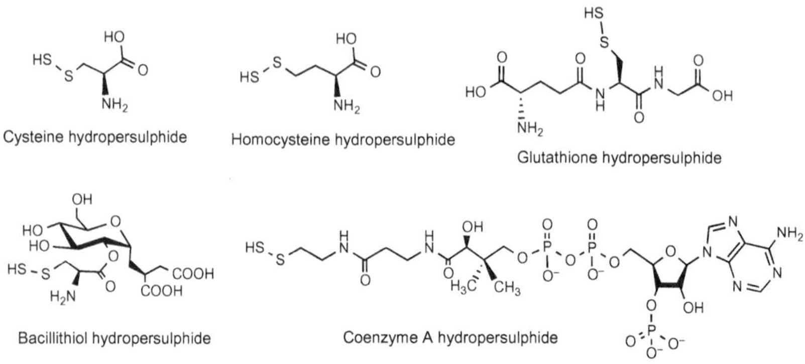

Fig. 1 Chemical structures of representative persulphide species (reduced forms) identified in biological systems.

Table 1 Endogenous Occurrence of Persulphide Species in Cultured Mammalian Cells

Cells (Species)	Persulphide Species Reported	Notes	References
A549 (human)	CysSSH, CysSSSSCys, CysSSSSSCys, GSSH, GSSSH, GSSSG	LC-MS/MS with MBB derivatization. Overexpression of CBS or CSE increased persulphide levels. Sulphasalazine (xCT inhibitor) treatment reduced CysSSH levels	Ida et al. (2014)
SH-SY5Y (human)	CysSSH, GSSH	LC-MS/MS with MBB derivatization. Overexpression of CBS or CSE increased persulphide levels	Ida et al. (2014)
C6 (rat)	CysSSH, GSSH	LC-MS/MS with MBB derivatization. Overexpression of CBS or CSE increased persulphide levels	Ida et al. (2014)
HeLa (human)	CysSSH, GSSH, GSSSH	LC-MS/MS with MBB derivatization. Preculture in cystine/methionine-free DMEM decreased persulphide levels. Supplementation with extracellular cysteine or *N*-acetylcysteine increased persulphide levels	Ono et al. (2017)
Bronchial epithelial cells (human)	CysSSH, GSSH	LC-MS/MS with HPE-IAM derivatization. Primary cells derived from non-COPD subjects showed higher persulphide levels than those from COPD patients	Numakura et al. (2017)
HEK293T cells (human)	CysSSH, GSSH	LC-MS/MS with HPE-IAM derivatization. Knock-down of CARS2 markedly reduced persulphide levels	Akaike et al. (2017)

were detected in all cell types examined. Cellular levels of those persulphide and polysulphide species were influenced by culture conditions such as the overexpression of cysteine-metabolizing enzymes and the availability of sulphur-containing amino acids in the cultures, as well as the origins of the cells.

Studies of animals revealed the presence of persulphide and polysulphide species in all organs and tissues examined, including plasma, liver, brain, lung, and heart (Akaike et al., 2017; Ida et al., 2014). Concentrations of GSSH were >100 μM in the brain and about 50 μM in other major organs, including the heart and liver (Ida et al., 2014) (Table 2). Because the same or higher amounts of HCysSSH compared with amounts of parental homocysteine were observed, HCysSSH seems to be the major homocysteine derivative in such organs (Ida et al., 2014). Similar to the cell culture

Table 2 Tissue and Organ Levels of Persulphide Species in Mice In Vivo

	Concentration (μM)				
Persulphide	**Heart**	**Liver**	**Brain**	**Lung**	**Plasma**
GSH	1378 ± 92	4418 ± 123	3036 ± 45	2089 ± 232	2.21 ± 0.74
GSSH	54 ± 2	60 ± 19	154 ± 4	4 ± 1	0.26 ± 0.06
GSSSH	0.67 ± 0.02	0.20 ± 0.05	0.32 ± 0.03	0.08 ± 0.01	N.D.
Cys	13.3 ± 0.4	68 ± 17	90 ± 13	48 ± 3	1.67 ± 0.59
CysSSH	3.9 ± 0.2	1.0 ± 0.4	2.2 ± 0.2	0.2 ± 0.0	0.45 ± 0.19
HCys	0.15 ± 0.05	1.13 ± 0.27	0.07 ± 0.02	0.29 ± 0.03	0.09 ± 0.02
HCysSSH	0.18 ± 0.07	0.56 ± 0.30	0.19 ± 0.03	0.02 ± 0.00	0.02 ± 0.00
GSSG	203 ± 24	507 ± 33	157 ± 29	394 ± 37	0.31 ± 0.04
GSSSG	0.10 ± 0.01	0.75 ± 0.15	0.07 ± 0.02	0.247 ± 0.03	0.03 ± 0.02
GSSSSG	0.15 ± 0.06	0.06 ± 0.01	0.59 ± 0.19	0.03 ± 0.01	N.D.
CysSSCys	N.D.	0.20 ± 0.01	N.D.	0.05 ± 0.01	6.2 ± 1.1
CysSSSCys	N.D.	N.D.	N.D.	N.D.	N.D.
CysSSSSCys	0.43 ± 0.02	0.60 ± 0.11	N.D.	0.45 ± 0.14	N.D.
CysSSSSSCys	0.35 ± 0.14	N.D.	N.D.	0.46 ± 0.17	1.7 ± 0.1

N.D., not detected.
Data are means ± SD ($n = 3$).
Adapted from Ida, T., Sawa, T., Ihara, H., Tsuchiya, Y., Watanabe, Y., Kumagai, Y., et al. (2014). Reactive cysteine persulfides and S-polythiolation regulate oxidative stress and redox signaling. *Proceedings of the National Academy of Sciences of the United States of America*, *111*, 7606–7611.

experiments, studies of tissue levels of persulphides showed that these levels were mainly affected by the availability of sulphur-containing amino acids (cysteine and methionine) in the diet (Ida et al., 2014). A more recent study clearly demonstrated that the genetic status of mitochondrial cysteinyl-tRNA synthetase (CARS2) markedly affected tissue and organ levels of CysSSH and GSSH (Akaike et al., 2017) (see later for details). *Cars2* heterozygous mice showed 50% or more reduced levels of CysSSH and GSSH in the liver and lung compared with wild-type littermates (Akaike et al., 2017).

The endogenous occurrence of CysSSH and GSSH has been reported in human tissues. Akaike and colleagues demonstrated the presence of various persulphide species including CysSSH, HCysSSH, GSSH, CysSSSSCys, CysSSSSSCys, and GSSSG in human plasma (Ida et al., 2014). A noteworthy finding was that endogenous levels of persulphide species in humans were associated with certain diseases, such as chronic obstructive pulmonary disease (COPD). An inflammatory disease, COPD, mainly affects small airways and lung parenchyma and leads to airway obstruction that is progressive (Barnes, 2013). Oxidative stress in the lung has been suggested as the major etiological factor for COPD (Barnes, 2013). Ichinose and colleagues analysed the endogenous levels of persulphide species in the lung and found that levels of persulphide species including CysSSH, GSSH, and GSSSH were reduced in lung-resident cells and in epithelial lining fluid obtained from airways of patients with COPD (Numakura et al., 2017). They also found that the levels of these persulphide species in lung cells were positively correlated with the extent of airflow limitation (Numakura et al., 2017). As mentioned later, persulphide species act as strong antioxidants. These data thus suggest that a decrease in these persulphide species would be associated with a redox imbalance in the lungs of patients with COPD.

Nakazawa and colleagues reported the metabolomic profiles of persulphide species in aqueous and vitreous humour (Kunikata et al., 2017). They found that patients with diabetes mellitus (DM) had elevated levels of CysSSH, cystine, and GSSSG in the aqueous humour compared with control subjects. Similarly, patients with DM had increased levels of CysSSH, CysSH, and cystine in the vitreous humour compared with control subjects. In contrast, plasma levels of those persulphide species did not differ in patients with DM and control subjects. Thus, various persulphide species appear to be present in the eye, and some persulphide species are upregulated in the aqueous and vitreous humour in DM as antioxidant protectors against oxidative stress.

Staphylococcus aureus, a commensal pathogen of humans, is a major cause of nosocomial infections (Conrady et al., 2008). Giedroc and colleagues investigated the occurrence of persulphide species in this bacterium by using LC-MS/MS with monobromobimane (MBB) derivatization (Peng, Shen, et al., 2017). They demonstrated the presence of persulphides of bacillithiol, cysteine, and coenzyme A in *S. aureus* (Peng, Shen, et al., 2017). Intracellular levels of these low-molecular-weight persulphides were affected by both genetic background and environmental factors (Peng, Shen, et al., 2017). Deficiency of the copper-sensing operon repressor (CsoR)-like sulphurtransferase repressor (CstR) gene, which is a transcriptional repressor of sulphide oxidation enzymes including CstA, CstB, and sulphide:quinone oxidoreductase (SQR), resulted in a reduction in persulphide species in *S. aureus* (Peng, Shen, et al., 2017). We discuss the roles of CstR in persulphide regulation in detail later in this chapter. Exposure of *S. aureus* to biologically relevant oxidants such as nitroxyl and peroxynitrite transiently increased the intracellular levels of low-molecular-weight persulphide species (Peng, Shen, et al., 2017). Takagi and Ohtsu (2017) reportedly detected persulphide species including CysSSH and GSSH in the Gram-negative bacterium *Escherichia coli* via LC-MS/MS. They identified those reactive species by using mass spectrometry, although they reported no quantitative data.

3. BIOSYNTHESIS

Three enzymes that have been identified so far generate CysSSH as a product: cystathionine β-synthase (CBS), cystathionine γ-lyase (CSE), and cysteinyl-tRNA synthetases (CARSs). These enzymes differ from each other in terms of substrate specificity and specific activity.

CBS, a haem-containing pyridoxal phosphate (PLP)-dependent enzyme, catalyses the condensation of serine and homocysteine to produce cystathionine (Miles & Kraus, 2004). Cystathionine is then converted to cysteine, α-ketobutyrate, and ammonia by the catalytic action of CSE, a PLP-dependent enzyme, via γ-elimination (Stipanuk & Ueki, 2011). Thus, CBS and CSE participate in a transsulphuration pathway in sulphur metabolism (Stipanuk & Ueki, 2011). In vitro experiments with recombinant enzymes showed that both enzymes catalyse CysSSH formation from cystine, an oxidized form of cysteine, as a substrate (Ida et al., 2014). During the enzymatic reaction, negligible hydrogen sulphide (H_2S) formed (Ida et al., 2014). Thus, CysSSH is directly produced from cystine via CS bond cleavage by catalytic action of these enzymes. Cell culture experiments revealed the CBS- and

CSE-dependent formation of low-molecular-weight persulphides and protein persulphides (Ida et al., 2014). Overexpression of these enzymes significantly increased the cellular levels of CysSSH, GSSH, and protein persulphides (Table 1) (Ida et al., 2014).

Most bacteria synthesize cysteine from serine via a two-step pathway that is independent of the transsulphuration pathway (CBS and CSE) (Takagi & Ohtsu, 2017) (Fig. 2). In the first step, the hydroxyl group of serine is activated in the form of *O*-acetylserine by a serine transacetylase in the presence of acetyl-coenzyme A. In the second step, the methyl ester is displaced in the presence of a sulphur donor (sulphide or thiosulphate) by the PLP-dependent *O*-acetylserine sulfhydrylase to yield cysteine.

Akaike and colleagues recently demonstrated that CARSs, in addition to having a canonical role in protein translation, act as the principal CysSSH synthase in vivo (Akaike et al., 2017). CARSs are enzymes that catalyse cysteinyl-tRNA production via a two-step mechanism, in which cysteine

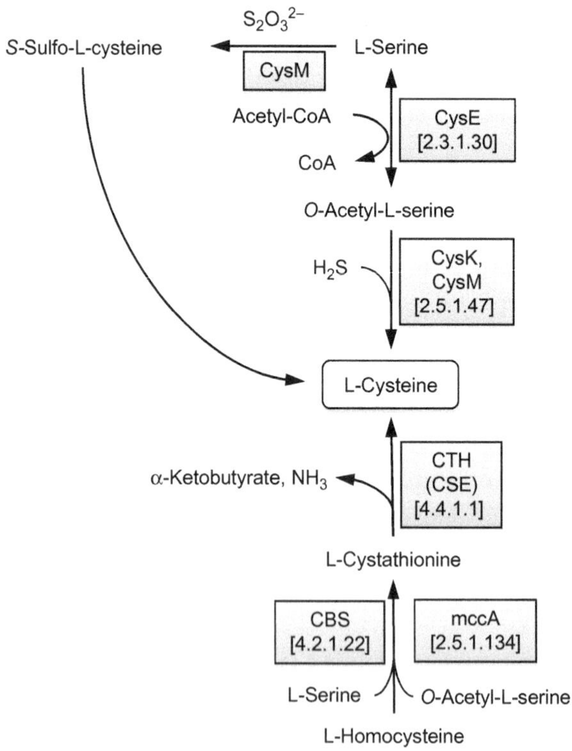

Fig. 2 L-Cysteine biosynthetic pathways. Corresponding enzyme commission (EC) numbers are shown in *square brackets*.

is first activated in the presence of ATP to form an enzyme-bound cysteinyl adenylate (Fig. 3). The activated cysteine is subsequently transferred to the ribose hydroxyl groups at the 3′-terminus of the cysteinyl-tRNA molecule, with the release of AMP. CARSs are class I aminoacyl-tRNA synthetases (Avalos, Corrochano, & Brenner, 1991). By means of LC-MS/MS, Akaike and colleagues determined that recombinant CARS from *E. coli* (EcCARS) effectively catalyses CysSSH formation from cysteine, used as a substrate. Enzymatic kinetic analyses showed that EcCARS is extremely efficient in producing CysSSH (sufficiently low K_m and high catalytic rate constant k_{cat}) (Akaike et al., 2017).

Biochemical and structural investigations revealed PLP-dependent CARS-mediated CysSSH formation (Akaike et al., 2017). Several lysine residues, including the well-conserved ^{73}KIIK^{76} and ^{266}KMSK^{269}, were determined to be PLP-binding sites (Akaike et al., 2017). Enzymes that had lysine-to-alanine mutations introduced at K73A, K76A, K266A, and K269A showed markedly less activity in terms of CysSSH formation, whereas these mutants possessed almost intact protein synthesis activity (Akaike et al., 2017). Introduction of cysteine-to-aspartate mutations at C28D and C209D, with both cysteine residues being critical for translation, did not affect persulphide production efficiency, so that it was similar to that of the wild type, although the mutants had strongly attenuated protein synthesis and translational activity (Akaike et al., 2017). In mammals, two different CARSs exist—cytosolic CARS1 and mitochondrial CARS2 (Coughlin et al., 2015; Hallmann et al., 2014). Both CARS1 and CARS2 showed strong CysSSH-producing activity (Akaike et al., 2017). Genetic knockdown analyses with cultured human HEK293T cells suggested that CARS2 is a major persulphide producer (Akaike et al., 2017). This idea is supported by the finding that endogenous persulphide levels in CARS2 heterozygous knockout mice (*Cars2*$^{+/-}$) were markedly lower compared with the levels in wild-type mice (Akaike et al., 2017). Thus, the mitochondrion is a key organelle for the formation of cysteine persulphides in mammalian cells.

Another important and novel function of CARSs is their catalysis of the incorporation of CysSSH directly into nascent polypeptides during translation (Fig. 3) (Akaike et al., 2017). CARSs are thus enzymes that catalyse both CysSSH formation from cysteine and cotranslational protein polysulphidation. Akaike and colleagues also demonstrated the critical involvement of mitochondrial CARS2 in the regulation of mitochondrial biogenesis and bioenergetics (Akaike et al., 2017). Knockout of CARS2,

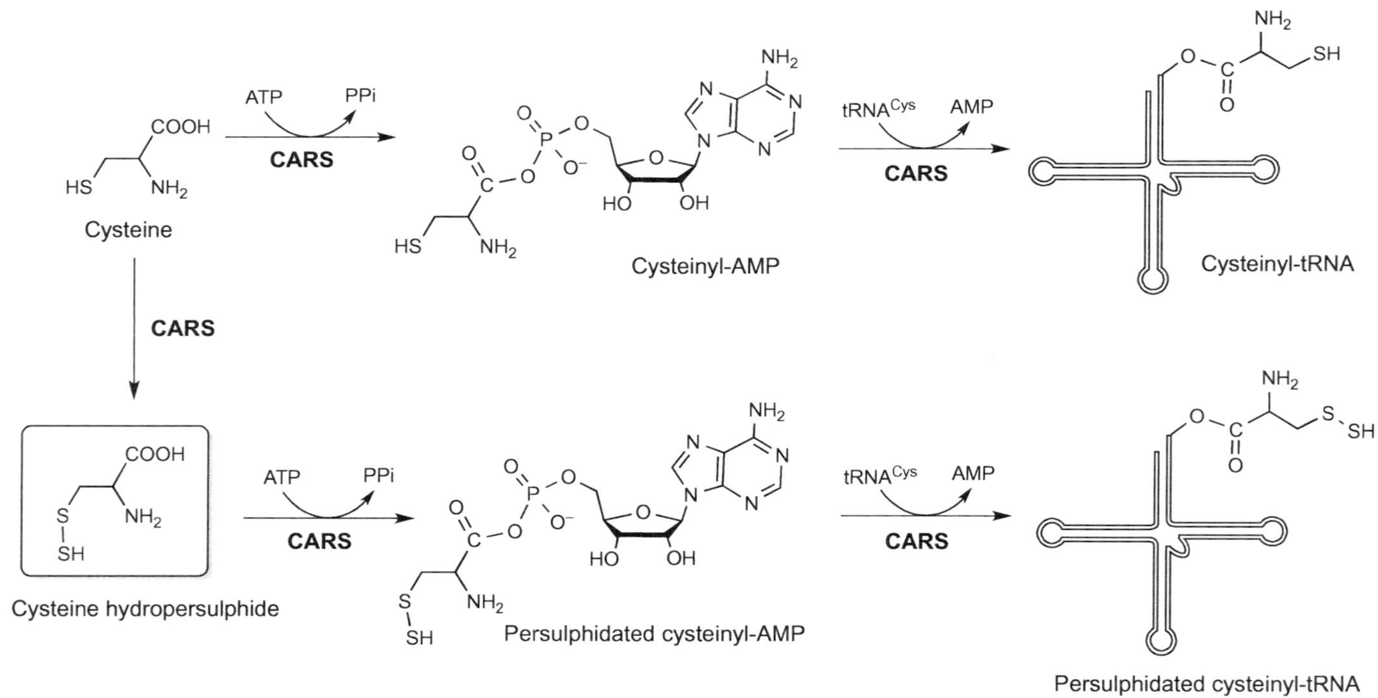

Fig. 3 CARSs catalyse both CysSSH formation and aminoacylation of CysSSH onto cysteinyl-tRNA. *PPi*, pyrophosphate.

and hence reduced CysSSH levels, induced markedly altered mitochondrial morphology (i.e. shrunken or fragmented appearance), possibly through desulphuration-mediated activation of dynamin-related protein (Drp1), a major mediator of mitochondrial fission (Akaike et al., 2017; Akhtar et al., 2016). CARS2-dependent CysSSH production also played a role in mitochondrial biogenesis (Akaike et al., 2017). A notable finding was that CysSSH produced by CARS2 in mitochondria can function as an electron acceptor at complex III in the electron transport chain (Akaike et al., 2017).

4. ANTIOXIDANT ACTIVITY

Persulphide and polysulphide species have been proposed to be more nucleophilic and superior reducing agents compared with their corresponding parental thiols (Francoleon, Carrington, & Fukuto, 2011; Fukuto et al., 2012). They therefore act as reactive sulphur species (RSS) (Ono et al., 2014). This unique characteristic of reactive CysSSH is due to the adjacent electron pairs, with the increased reactivity resulting from these electron pairs being known as the α-effect (Fukuto et al., 2012; Ono et al., 2014). In vitro experiments clearly showed that GSSH can directly scavenge ROS such as hydrogen peroxide (H_2O_2) in aqueous media at neutral pH (Ida et al., 2014). Simple glutathione (GSH) and sodium hydrogen sulphide did not demonstrate such direct scavenging of H_2O_2 (Ida et al., 2014). Overexpression of CSE in human adenocarcinoma A549 cells resulted in an increase in cellular persulphide levels (Table 1), and more important, protection against H_2O_2-induced cell death (Ida et al., 2014). Kunikata and colleagues recently reported that exogenous addition of GSSSG significantly suppressed the death of immortalized retinal ganglion cells (RGC-5) induced by H_2O_2 exposure (Kunikata et al., 2017). Reactive persulphide species can scavenge not only H_2O_2 but also various electrophiles (Abiko et al., 2015; Ida et al., 2014; Ihara et al., 2017). 8-Nitroguanosine 3′,5′-cyclic GMP (8-nitro-cGMP) is an endogenously occurring weak electrophile and acts as a second messenger in redox signalling (Akaike, Fujii, Sawa, & Ihara, 2010; Akashi et al., 2016; Fujii et al., 2010; Nishida et al., 2012; Sawa, Arimoto, & Akaike, 2010; Sawa et al., 2013, 2007; Tsutsuki et al., 2016; Zaki et al., 2009). Ida et al. showed that reactive persulphides efficiently react with 8-nitro-cGMP and convert it to 8-mercapto-cGMP (8-SH-cGMP) with a concomitant release of nitrite (Ida et al., 2014) (Fig. 4). Taken together, these findings demonstrate the unique

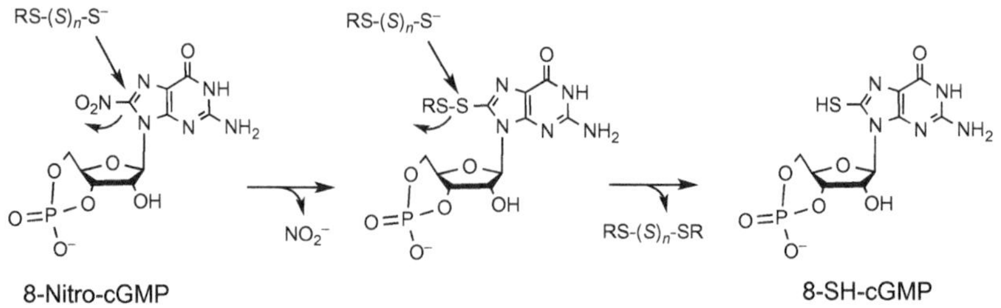

Fig. 4 Polysulphide-mediated metabolism of 8-nitro-cGMP to 8-SH-cGMP. Reactive polysulphides (RS-(S)$_n$-S$^-$) attack the C8 carbon of 8-nitro-cGMP to form an 8-perthio-cGMP intermediate with concomitant formation of nitrite. In a second step, the disulphide bond of 8-perthio-cGMP is cleaved by the reactive polysulphides to form 8-SH-cGMP.

potency of reactive cysteine persulphides as reductants and nucleophiles, so that they possess greater reactivity than thiols and can scavenge oxidants and cellular electrophiles, thereby serving as important antioxidant molecules (Fujii et al., 2016; Nishida et al., 2016, 2017).

5. METHODOLOGIES

Methodologies that allow specific detection or identification of persulphide species have been developed. These methodologies include mass spectrometry-based assays, fluorescent probes, and proteomics.

5.1 LC-MS/MS

Mass spectrometry is one of the most effective techniques for identification and quantification of biological molecules. A critical issue for mass spectrometry-based quantification is the need to have a set of authentic standards labelled with isotopes. Intensities of mass signals vary for each run because of the copresence of ionic substances. Thus, the use of isotope-labelled standards is indispensable for precise quantification and identification of persulphide species in complex biological milieu. The isotope-labelled standards should be spiked into the samples before mass spectrometry, and the peaks of the analytes are compared with those of spiked standards; this method is called isotope dilution mass spectrometry (Harwood, Kettle, Brennan, & Winterbourn, 2009; Lehmann, 2017; Zhu, Oe, & Blair, 2008).

Stabilization of oxidation-labile sulfhydryls is another important issue that should be kept in mind. Alkylation of sulfhydryls with thiol-specific

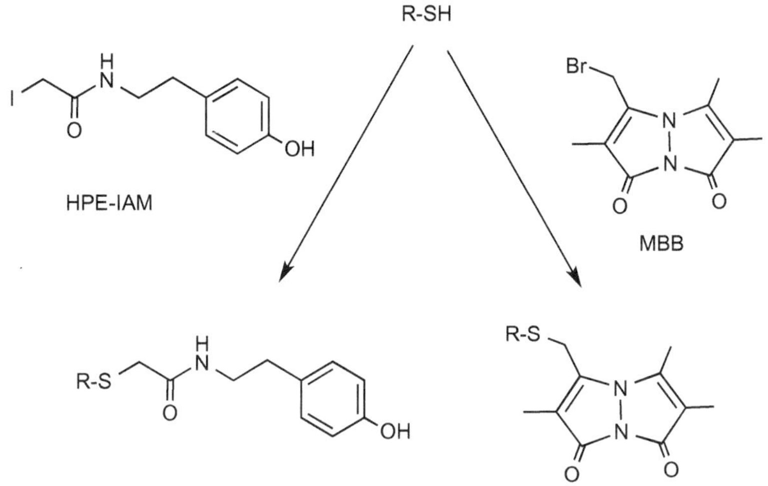

Fig. 5 Reaction between alkylating agents and sulfhydryls for persulphide derivatization in mass spectrometric analysis.

reagents such as MBB (Kosower & Kosower, 1987) (Fig. 5) has been utilized for such a purpose (Ida et al., 2014; Kunikata et al., 2017; Ono et al., 2017). Sulfhydryl alkylation also led to increased hydrophobicity of the analytes and therefore sufficient retention during reversed-phase liquid chromatography separation. β-(4-Hydroxyphenyl)ethyl iodoacetamide (HPE-IAM) is a thiol-alkylating agent that was used for successful labelling of persulphides and polysulphides to obtain LC-MS/MS measurements (Akaike et al., 2017; Numakura et al., 2017). Because of the mild electrophilicity of HPE-IAM, specific labelling of hydropolysulphides can be achieved by using HPE-IAM as an alkylating agent to form stable adducts without causing appreciable artefactual decay related to the dual nucleophilic and electrophilic attributes of polysulphides (Akaike et al., 2017).

Ono et al. developed a simple and efficient method for incorporating the stable isotope sulphur atom (^{34}S) into cysteine thiols (Ono et al., 2017). As mentioned earlier, bacterial *O*-acetylserine sulfhydrylase catalyses cysteine formation from *O*-acetylserine and H_2S (Cook & Wedding, 1976; Leu & Cook, 1994). Ono et al. carried out such enzyme reactions by using recombinant enzymes cloned from *Salmonella typhimurium* in the presence of *O*-acetylserine and isotope-labelled sulphide ($^{34}S^{2-}$) to yield ^{34}S-labelled cysteine (Ono et al., 2017). Isotope-labelled cysteine thus obtained can be used for in vivo sulphur transfer experiments involving mass spectrometry (Ono et al., 2017).

5.2 Fluorescent Probes

Several types of fluorescent probes for persulphide and polysulphide species have been developed (Lin, Chen, Xian, & Chang, 2015; Takano, Echizen, & Hanaoka, 2017). Sulphane sulphur refers to a sulphur atom with six valence electrons but no charge (known as S^0). Sulphane sulphur atoms are present in biologically relevant molecules such as cysteine-based hydropersulphides (R-S-SH), polysulphides (R-S-S_n-S-R, $n > 1$), and hydrogen polysulphides (H_2S_n, $n > 1$) (Mueller, 2006; Pan & Carroll, 2013; Toohey, 1989; Westley & Westley, 1991). Sulphane sulphur has electrophilic characteristics and hence reacts with nucleophiles such as cyanide and thiol groups (Wood, 1987). The traditional method for sulphane sulphur detection, i.e., cyanolysis, is based on a reaction with the cyanide ion to form thiocyanate, which can be measured as ferric thiocyanate by using spectrometry (Wood, 1987) (Fig. 6A). On the basis of this characteristic, fluorescent probes were developed that react with sulphane sulphur to become fluorescent (Fig. 6B). These probes include the SSP series (Chen et al., 2013; Ida et al., 2014) and the SSip series (Takano et al., 2017) (Fig. 6C), which have been used to detect intracellular formation of persulphide species (Ihara et al., 2017; Numakura et al., 2017).

Xian and colleagues developed the DSP series of fluorescent probes (Fig. 7) to detect hydrogen polysulphides (Liu et al., 2014). DSP probes such as DSP-3 show good selectivity and sensitivity for hydrogen polysulphides vs other thiol compounds such as cysteine, homocysteine, GSH, and H_2S and sulphur-containing species such as thiosulphate, sulphate, sulphonate, cystine trisulphide, and elemental sulphur (S8) (Liu et al., 2014). A drawback of this type of probe is that 2-fluoro-5-nitrobenzoic esters can also react with biothiols. Although such reactions do not switch on fluorescence, consumption of the probes is a problem. To overcome this drawback, Xian and colleagues designed the PSP series probes (Chen et al., 2015) (Fig. 7). The PSP type of probe showed a desirable off/on response to hydrogen polysulphides and high specificity, and these probes were successfully applied to visualization of intracellular hydrogen polysulphides (Chen et al., 2015). Xian's group more recently developed a single fluorescent probe, DDP-1 (Fig. 7), that detected both H_2S and hydrogen polysulphides as different fluorescence signals (Chen et al., 2016). This probe may facilitate greater understanding of cellular cross talk between H_2S and hydrogen polysulphides (Chen et al., 2016).

Ojida and colleagues described a reversible fluorescent probe for detection of hydropersulphides (Kawagoe, Takashima, Uchinomiya, & Ojida,

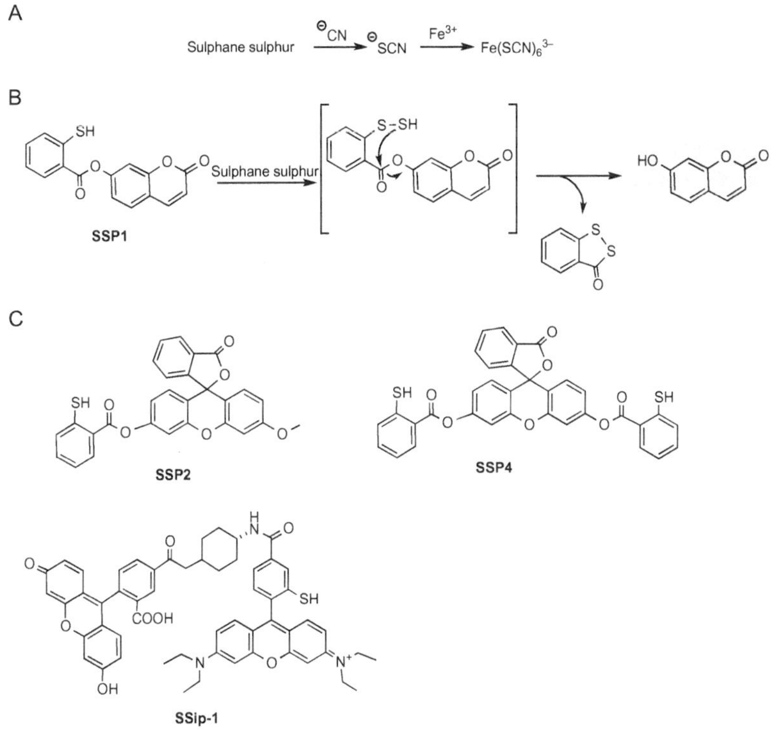

Fig. 6 Reaction between sulphane sulphur and cyanide (A), and reaction for the sulphane sulphur probe SSP1 (B) utilized for sulphane sulphur detection. *CN*, cyanide; *SCN*, thiocyanate. Structures of fluorescent sulphane sulphur probes are also shown (C).

2017) (Fig. 8). This probe uses a fluorescence resonance energy transfer (FRET) mechanism to control the fluorescence and modulates the intramolecular FRET efficiency to induce a dual-emission change. Because of this reversibility, real-time monitoring of hydropersulphides in living cells is possible (Kawagoe et al., 2017).

5.3 Proteomics

Cysteinyl side chains in proteins are endogenously polysulphidated. As mentioned earlier, during translation CysSSH is incorporated into nascent polypeptides by the catalytic action of CARSs (Akaike et al., 2017). Sulphur transfer from low-molecular-weight persulphides such as GSSH to protein thiols is another mechanism to form polysulphidated proteins in biological systems (Ida et al., 2014). Protein polysulphidation is thought to affect

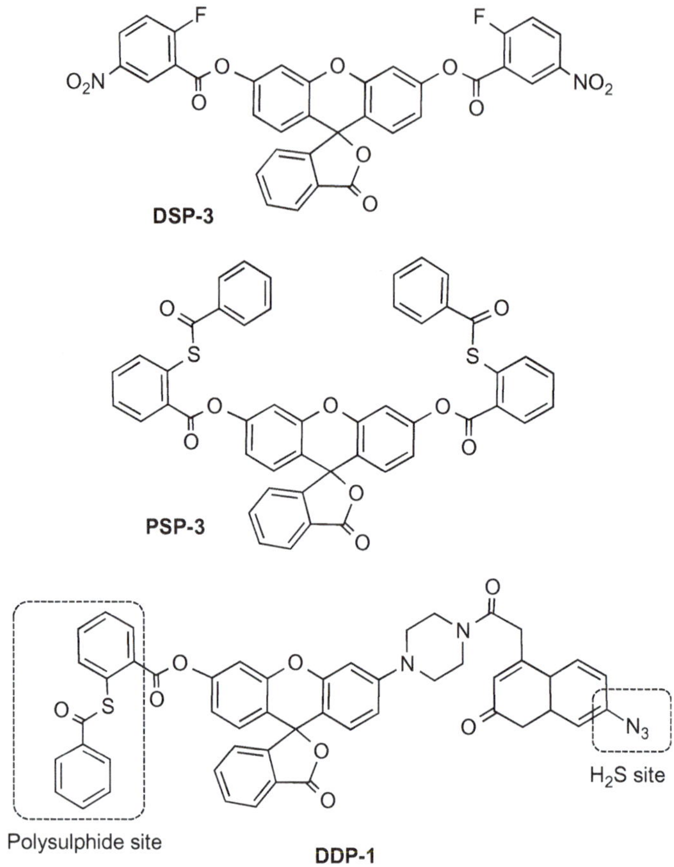

Fig. 7 Chemical structures of DSP series (DSP-3), PSP series (PSP-3), and DDP-1 probes.

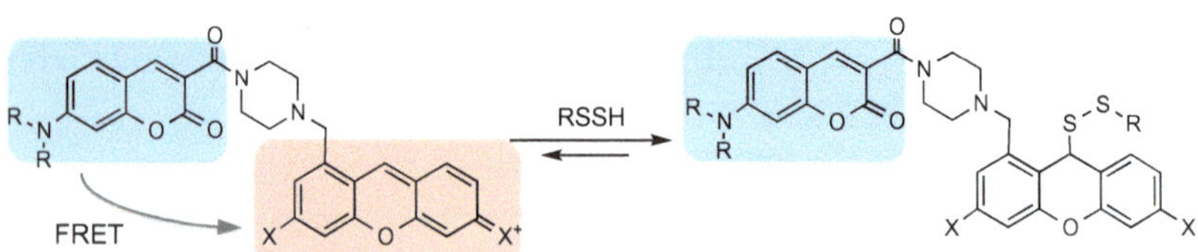

Fig. 8 Reaction of the reversible fluorescent probe for hydropersulphides. This mechanism is based on a nucleophilic reaction of a hydropersulphide with the probe's pyronine unit. *Modified from Kawagoe, R., Takashima, I., Uchinomiya, S., & Ojida, A. (2017). Reversible ratiometric detection of highly reactive hydropersulfides using a FRET-based dual emission fluorescent probe.* Chemical Science, 8, *1134–1140.*

protein functions, particularly regulation of redox signal transduction (Francoleon et al., 2011; Fukuto et al., 2012; Ono et al., 2014; Paul & Snyder, 2012).

The tag-switch assay was developed to detect persulphidated and polysulphidated proteins in biological systems (Ida et al., 2014) (Fig. 9). In this assay, nucleophilic thiol residues (thiol, hydropersulphide, and hydropolysulphide) are first tagged with the electrophile methylsulphonyl benzothiazole (MSBT) (Zhang, Devarie-Baez, Li, Lancaster, & Xian, 2012). Among all the tagged thiol residues, only those with a reactive sulphur residue present in polysulphides are subsequently modified by a biotin-cyano (CN)-labelled probe, which can be detected by conventional avidin-conjugated enzyme-linked chemiluminescence in Western blotting (Ida et al., 2014).

Nagy and colleagues reported an alternative method to detect persulphidated proteins (protein persulphide detection protocol, ProPerDP) (Doka et al., 2016). As Fig. 10 shows, proteinyl thiol and persulphide functional groups are alkylated via the biotin-labelled alkylating agent EZ-Link Iodoacetyl-PEG2-Biotin and can then be pulled down from a protein mixture by using streptavidin-coated magnetic beads (Doka et al., 2016). Other oxidized thiol derivatives, such as disulphides, are not alkylated under these conditions and remain in the supernatant. Derivatized persulphidated proteins are subsequently recovered by reductive cleavage of disulphide bonds and are then subjected to conventional protein assays such as sodium dodecyl sulphate-polyacrylamide gel electrophoresis. By using this method, Nagy and colleagues demonstrated that thioredoxin reductase-1 (TrxR1),

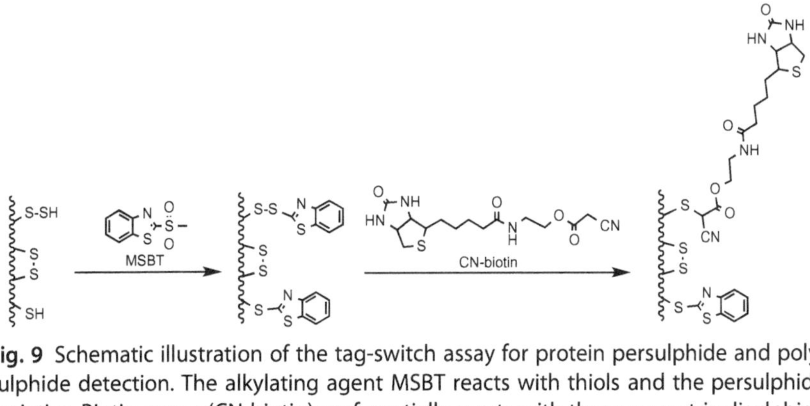

Fig. 9 Schematic illustration of the tag-switch assay for protein persulphide and polysulphide detection. The alkylating agent MSBT reacts with thiols and the persulphide moieties. Biotin-cyano (CN-biotin) preferentially reacts with the asymmetric disulphide bond to form CS bonding, which results in incorporation of the biotin tag at the sites of persulphide moieties.

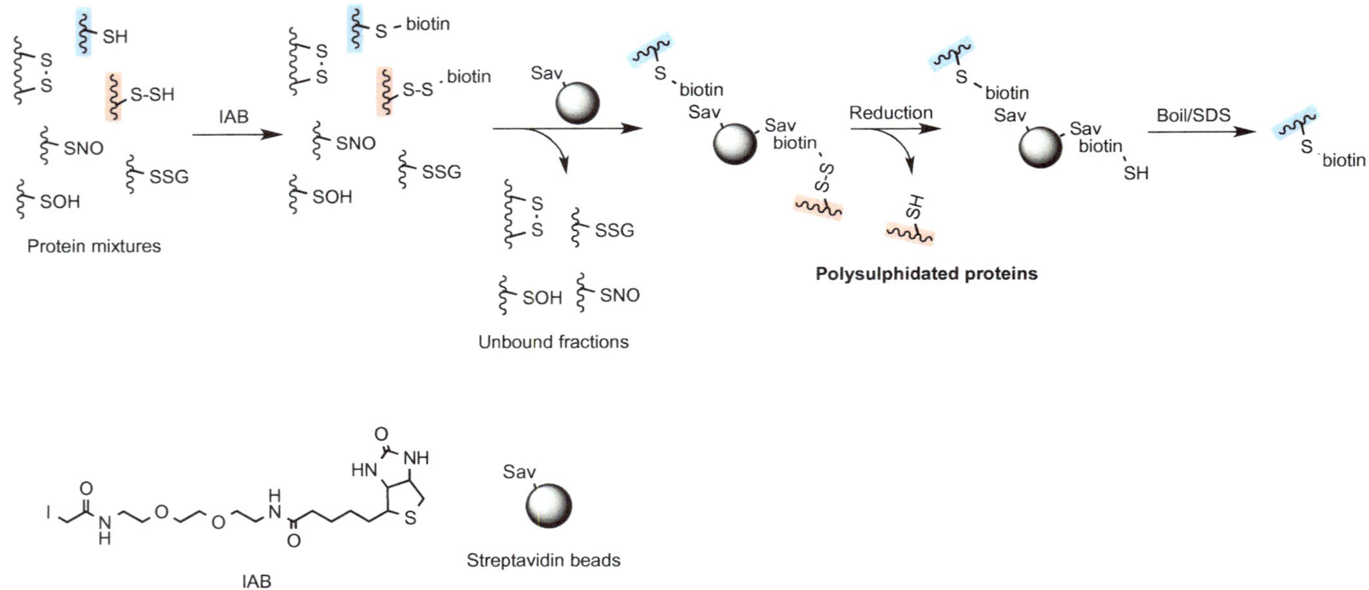

Fig. 10 ProPerDP for detection of persulphidated proteins in biological samples. *IAB*, biotin-labelled alkylating agent EZ-Link Iodoacetyl-PEG2-Biotin; *–SNO*, nitrosothiols; *–SOH*, sulphenic acids.

thioredoxin1 (Trx1), and the alternative TrxR1-dependent redox-active thioredoxin-related protein of 14 kDa (TRP14) (Jeong, Yoon, Lee, & Rhee, 2004) can reduce polysulphide and protein persulphide species (Doka et al., 2016). As mentioned earlier, a polysulphur linkage possesses both nucleophilic and electrophilic characteristics (Akaike et al., 2017). Akaike and colleagues developed unique protein labelling and extraction methods based on this dual electrophilic–nucleophilic attribute of polysulphidated proteins (Akaike et al., 2017). Protein thiols and polysulphides are labelled with biotin-polyethylene glycol (PEG)-conjugated maleimide (biotin-PEG-MAL). Polysulphur linkages labelled with biotin-PEG-MAL are susceptible to cleavage mediated by strong electrophiles such as *p*-chloromercuribenzoic acid (PCMB), which results in liberation of PEG moieties and hence protein size reduction. Such an alteration in protein size can be readily detected by means of electrophoresis. This method was named the PEG-conjugated maleimide-labelling gel shift assay or PMSA (Fig. 11). This PMSA allowed confirmation of several proteins as being polysulphidated during recombinant protein expression in *E. coli*, as well as

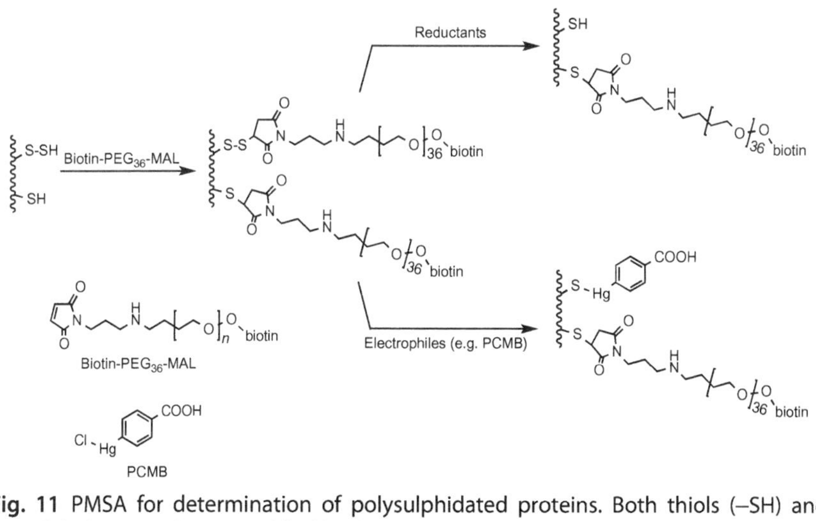

Fig. 11 PMSA for determination of polysulphidated proteins. Both thiols (–SH) and persulphides (–SSH) are modified by biotinylated PEG through maleimide-dependent alkylation, which leads to an increase in the apparent molecular size. Treatment with reductants such as dithiothreitol causes reductive cleavage of the disulphide bond occurring on the persulphide moiety. Certain electrophiles such as PCMB can also cleave polysulphide bonds. Reduced apparent molecular sizes can be detected by means of electrophoresis and are evidence of the presence of polysulphidated cysteine residues in proteins.

identification of endogenous polysulphidated proteins (Akaike et al., 2017). Jung et al. demonstrated an important finding by using this PMSA: ethylmalonic encephalopathy protein 1 (ETHE1) was extensively polysulphidated (Jung et al., 2016). ETHE1 is an enzyme that catalyses oxygen-dependent conversion of GSSH to GSH and sulphide and thus is critically involved in persulphide regulation (Kabil & Banerjee, 2012; Pettinati, Brem, Lee, McHugh, & Schofield, 2016; Tiranti et al., 2009). Biochemical analyses clearly demonstrated that polysulphidation at Cys247 was extremely important not only for overall polysulphidated status but also for persulphide dioxygenase activity (Jung et al., 2016). These data suggest that ETHE1 is a primary enzyme regulating endogenous GSSH and glutathione polysulphide and that its activity is controlled by polysulphidation of the Cys247 residue (Jung et al., 2016). Biotin-PEG-MAL was also used in the biotin-PEG-MAL capture method (Akaike et al., 2017; Takata et al., 2017). Watanabe and colleagues demonstrated that the activity of calcium/calmodulin-dependent protein kinase IV was regulated through its active-site polysulphidation (Takata et al., 2017).

6. CYSTEINE PERSULPHIDES IN BACTERIAL SIGNALLING

Accumulating evidence has suggested that RSS can serve as important mediators in the regulation of bacterial signal transduction. CstR is the transcriptional repressor controlling the operon that is involved in the H_2S oxidation system in *S. aureus* (Grossoehme et al., 2011). CstR possesses redox-sensitive cysteine thiols (Cys31 and Cys60). CstR with reduced thiols at Cys31 and Cys60 has a high affinity for the DNA operator region (Luebke et al., 2014). As several groups suggested, H_2S cannot directly react with reduced thiols (Greiner et al., 2013; Zhang et al., 2014). However, Luebke et al. clearly demonstrated that RSS including polysulphides such as Na_2S_4 and GSSH directly react with thiols of CstR, which leads to DNA operator dissociation and transcriptional derepression of the Cst operon (Luebke et al., 2014). Mass spectrometry demonstrated that RSS treatment produces oxidized sulphide bonds between CstR dimers, including di-, tri-, and tetrasulphide bonds (Fig. 12) (Luebke et al., 2014). Formation of oxidized sulphide bonds in CstR may reduce the binding affinity of the protein to the operator region through an allosteric mechanism (Luebke et al., 2014). A similar mechanism may function for regulation of the sulphide-responsive transcriptional repressor SqrR in the purple photosynthetic bacterium *Rhodobacter capsulatus* (Shimizu et al., 2017). In *R. capsulatus*, SQR is the major

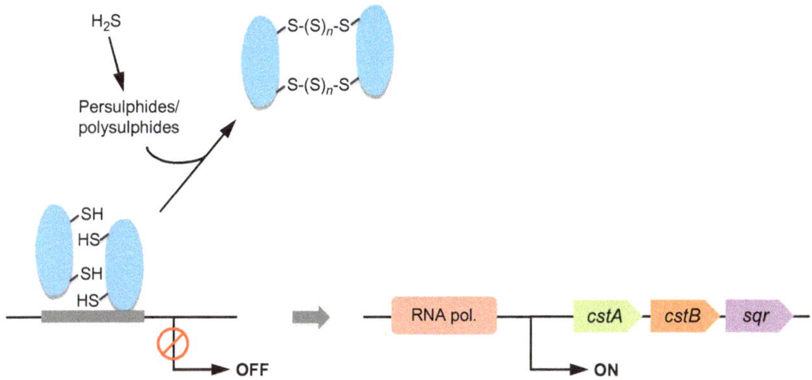

Fig. 12 Persulphide-mediated regulation of CstR function. RSS react with thiol residues of CstR to form di-, tri-, tetrasulphide bridges, which results in dissociation of CstR from DNA, possibly through an allosteric mechanism. RNA polymerase then binds to the operon to initiate gene expression to generate CstA, CstB, and SQR.

enzyme responsible for catalysing the first step in sulphide oxidation, and the level of SQR expression increases after sulphide treatment (Griesbeck et al., 2002). Masuda and colleagues reported that persulphide-dependent cysteine modification of SqrR regulates SqrR DNA-binding and transcriptional regulatory activities (Shimizu et al., 2017). Formation of a di-, tri-, or tetrasulphide bond between Cys41 and Cys107 of SqrR inhibits the ability of SqrR to bind to the promoter region (Shimizu et al., 2017). RNA polymerase binds to DNA and then induces gene expression.

Dissociation of CstR from the operator region results in the expression of three genes involved in H_2S detoxification, including CstA, CstB, and SQR (Fig. 12) (Luebke et al., 2014). Those three gene products cooperatively contribute to the protection of *S. aureus* against H_2S toxicity (Fig. 13) (Luebke et al., 2014). In *S. aureus*, Sqr encodes a type II SQR that catalyses two-electron oxidation of sulphide into sulphane sulphur in the presence of flavin adenine dinucleotide and a water-soluble quinone acceptor (Shen, Peng, Zhang, Trinidad, & Giedroc, 2016). The disulphide bond between C167 and C344 reacts with H_2S to form an intermediate persulphide moiety at C344, followed by donation of the sulphur atom to a nucleophilic acceptor such as coenzyme A to form persulphide (Fig. 13) (Shen et al., 2016). Low-molecular-weight persulphide thus formed can be further metabolized by CstB (Fig. 13) (Shen et al., 2015). CstB is a multidomain protein that contains an N-terminal persulphide dioxygenase (PDO) domain and a C-terminal rhodanese (Rhod) domain. The N-terminal PDO domain of

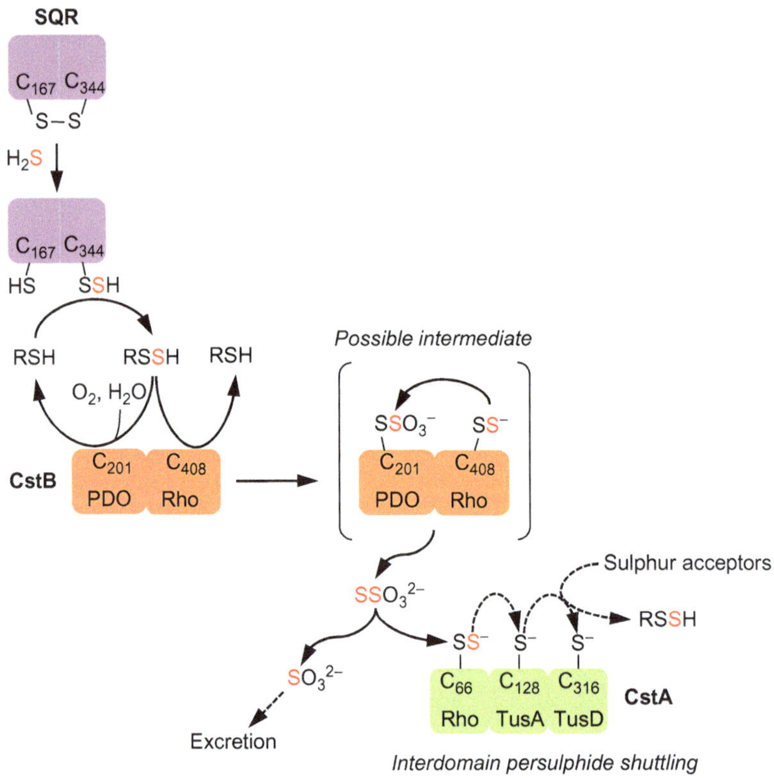

Fig. 13 Proposed mechanism for sulphur oxidation mediated by the Cst operon gene products SQR, CstB, and CstA.

CstB is homologous to full-length human ETHE1, with 21% homology, as well as to *Arabidopsis* ETHE1 (Holdorf et al., 2012; McCoy et al., 2006). In the presence of molecular oxygen, human ETHE1 oxidizes GSSH to generate sulphite and GSH (Kabil & Banerjee, 2012). In contrast, CstB oxidizes major cellular low-molecular-weight persulphide substrates directly from *S. aureus*, coenzyme A persulphide and bacillithiol persulphide, to generate thiosulphate and reduced thiols (Shen et al., 2015). Both Cys201 in the N-terminal PDO domain and Cys408 in the C-terminal Rho domain strongly enhance the thiosulphate-generating activity of CstB (Shen et al., 2015). Fig. 13 illustrates the possible reaction mechanism mediated by CstB. Thiosulphate is then further metabolized by CstA (Fig. 13). *S. aureus* CstA harbours three domains that function in sequential CysSSH transfer in other systems. These include an N-terminal Rho domain, a central TusA domain, and a C-terminal TusD domain. Higgins et al. showed that Cys66 in the

Rho domain receives a sulphur atom from thiosulphate and then transfers it to Cys128 in the TusA domain (Higgins, Peng, Luebke, Chang, & Giedroc, 2015). Sulphite formed during the reaction may be excreted from the cell for completion of H_2S detoxification. Additional persulphide transfer is hypothesized to occur from Cys128 to Cys316 and finally to a persulphide acceptor (Fig. 13) (Higgins et al., 2015).

Giedroc and colleagues recently reported that low-molecular-weight RSS including CoA persulphide, GSSH, and CysSSH play a role in regulating the global virulence of *S. aureus* (Peng, Zhang, et al., 2017). Those low-molecular-weight RSS affect bacterial virulence, at least in part, by modulating the activity of the global virulence regulator MgrA via protein persulphidation (Peng, Zhang, et al., 2017). They also found that two previously uncharacterized thioredoxin-like proteins, TrxP and TrxQ, can act as desulphurases and participate in the RSS-mediated regulation of virulence gene expression (Peng, Zhang, et al., 2017).

7. CONCLUSION

This chapter described the biological occurrence of CysSSH and related persulphides and polysulphides in both prokaryotes and eukaryotes. The chemical properties and abundance of these species suggest a pivotal role for reactive persulphides in cell regulatory processes. CysSSH and related persulphides can behave as potent antioxidants and cellular protectors and may function as redox signalling intermediates. Recent observations by Akaike and colleagues suggest that CysSSH regulates bioenergetics by acting as an electron acceptor in the mitochondrial electron transport chain (Akaike et al., 2017). CARSs are enzymes that efficiently catalyse CysSSH production from cysteine and are present in both bacteria and mammals. Additional studies are clearly needed to identify the molecular functions of CysSSH and related persulphides in bacterial physiology so that new and more effective strategies for treatment, prevention, and diagnosis of bacterial infections can be developed.

ACKNOWLEDGEMENTS

We thank J.B. Gandy for her editing of the manuscript. This work was supported in part by Grants-in-Aid for Scientific Research [(A), (B), Innovative Areas (Research in a Proposed Area), and Research Activity Start-up] from the Ministry of Education, Sciences, Sports, and Technology (MEXT), Japan, to T.S. (15H03115, 17K19205), K.O. (15H06509), H.T. (17K10019), T.I. (15K20876, 17K08619), and T.A. (26111008, 26111001, 15K21759, 25253020, 16K15208).

REFERENCES

Abiko, Y., Yoshida, E., Ishii, I., Fukuto, J. M., Akaike, T., & Kumagai, Y. (2015). Involvement of reactive persulfides in biological bismethylmercury sulfide formation. *Chemical Research in Toxicology*, *28*, 1301–1306.

Akaike, T., Fujii, S., Sawa, T., & Ihara, H. (2010). Cell signaling mediated by nitrated cyclic guanine nucleotide. *Nitric Oxide*, *23*, 166–174.

Akaike, T., Ida, T., Wei, F. Y., Nishida, M., Kumagai, Y., Alam, M. M., et al. (2017). Cysteinyl-tRNA synthetase governs cysteine polysulfidation and mitochondrial bioenergetics. *Nature Communications*, *8*, 1177.

Akashi, S., Ahmed, K. A., Sawa, T., Ono, K., Tsutsuki, H., Burgoyne, J. R., et al. (2016). Persistent activation of cGMP-dependent protein kinase by a nitrated cyclic nucleotide via site specific protein S-guanylation. *Biochemistry*, *55*, 751–761.

Akhtar, M. W., Sanz-Blasco, S., Dolatabadi, N., Parker, J., Chon, K., Lee, M. S., et al. (2016). Elevated glucose and oligomeric beta-amyloid disrupt synapses via a common pathway of aberrant protein S-nitrosylation. *Nature Communications*, *7*, 10242.

Avalos, J., Corrochano, L. M., & Brenner, S. (1991). Cysteinyl-tRNA synthetase is a direct descendant of the first aminoacyl-tRNA synthetase. *FEBS Letters*, *286*, 176–180.

Barnes, P. J. (2013). New anti-inflammatory targets for chronic obstructive pulmonary disease. *Nature Reviews Drug Discovery*, *12*, 543–559.

Cavuzic, M., & Liu, Y. (2017). Biosynthesis of sulfur-containing tRNA modifications: A comparison of bacterial, archaeal, and eukaryotic pathways. *Biomolecules*, *7*, 27.

Chen, W., Liu, C., Peng, B., Zhao, Y., Pacheco, A., & Xian, M. (2013). New fluorescent probes for sulfane sulfurs and the application in bioimaging. *Chemical Science*, *4*, 2892–2896.

Chen, W., Pacheco, A., Takano, Y., Day, J. J., Hanaoka, K., & Xian, M. (2016). A single fluorescent probe to visualize hydrogen sulfide and hydrogen polysulfides with different fluorescence signals. *Angewandte Chemie International Edition English*, *55*, 9993–9996.

Chen, W., Rosser, E. W., Matsunaga, T., Pacheco, A., Akaike, T., & Xian, M. (2015). The development of fluorescent probes for visualizing intracellular hydrogen polysulfides. *Angewandte Chemie International Edition English*, *54*, 13961–13965.

Conrady, D. G., Brescia, C. C., Horii, K., Weiss, A. A., Hassett, D. J., & Herr, A. B. (2008). A zinc-dependent adhesion module is responsible for intercellular adhesion in staphylococcal biofilms. *Proceedings of the National Academy of Sciences of the United States of America*, *105*, 19456–19461.

Cook, P. F., & Wedding, R. T. (1976). A reaction mechanism from steady state kinetic studies for O-acetylserine sulfhydrylase from Salmonella typhimurium LT-2. *Journal of Biological Chemistry*, *251*, 2023–2029.

Coughlin, C. R., 2nd, Scharer, G. H., Friederich, M. W., Yu, H. C., Geiger, E. A., Creadon-Swindell, G., et al. (2015). Mutations in the mitochondrial cysteinyl-tRNA synthase gene, CARS2, lead to a severe epileptic encephalopathy and complex movement disorder. *Journal of Medical Genetics*, *52*, 532–540.

Dahl, C. (2015). Cytoplasmic sulfur trafficking in sulfur-oxidizing prokaryotes. *IUBMB Life*, *67*, 268–274.

Doka, E., Pader, I., Biro, A., Johansson, K., Cheng, Q., Ballago, K., et al. (2016). A novel persulfide detection method reveals protein persulfide- and polysulfide-reducing functions of thioredoxin and glutathione systems. *Science Advances*, *2*, e1500968.

Fontecave, M., Ollagnier-de-Choudens, S., & Mulliez, E. (2003). Biological radical sulfur insertion reactions. *Chemical Reviews*, *103*, 2149–2166.

Francoleon, N. E., Carrington, S. J., & Fukuto, J. M. (2011). The reaction of H(2)S with oxidized thiols: Generation of persulfides and implications to H(2)S biology. *Archives of Biochemistry and Biophysics*, *516*, 146–153.

Fujii, S., Sawa, T., Ihara, H., Tong, K. I., Ida, T., Okamoto, T., et al. (2010). The critical role of nitric oxide signaling, via protein S-guanylation and nitrated cyclic GMP, in the antioxidant adaptive response. *Journal of Biological Chemistry, 285*, 23970–23984.

Fujii, S., Sawa, T., Nishida, M., Ihara, H., Ida, T., Motohashi, H., et al. (2016). Redox signaling regulated by an electrophilic cyclic nucleotide and reactive cysteine persulfides. *Archives of Biochemistry and Biophysics, 595*, 140–146.

Fukuto, J. M., Carrington, S. J., Tantillo, D. J., Harrison, J. G., Ignarro, L. J., Freeman, B. A., et al. (2012). Small molecule signaling agents: The integrated chemistry and biochemistry of nitrogen oxides, oxides of carbon, dioxygen, hydrogen sulfide, and their derived species. *Chemical Research in Toxicology, 25*, 769–793.

Greiner, R., Palinkas, Z., Basell, K., Becher, D., Antelmann, H., Nagy, P., et al. (2013). Polysulfides link H2S to protein thiol oxidation. *Antioxidants & Redox Signaling, 19*, 1749–1765.

Griesbeck, C., Schutz, M., Schodl, T., Bathe, S., Nausch, L., Mederer, N., et al. (2002). Mechanism of sulfide-quinone reductase investigated using site-directed mutagenesis and sulfur analysis. *Biochemistry, 41*, 11552–11565.

Grossoehme, N., Kehl-Fie, T. E., Ma, Z., Adams, K. W., Cowart, D. M., Scott, R. A., et al. (2011). Control of copper resistance and inorganic sulfur metabolism by paralogous regulators in *Staphylococcus aureus*. *Journal of Biological Chemistry, 286*, 13522–13531.

Hallmann, K., Zsurka, G., Moskau-Hartmann, S., Kirschner, J., Korinthenberg, R., Ruppert, A. K., et al. (2014). A homozygous splice-site mutation in CARS2 is associated with progressive myoclonic epilepsy. *Neurology, 83*, 2183–2187.

Harwood, D. T., Kettle, A. J., Brennan, S., & Winterbourn, C. C. (2009). Simultaneous determination of reduced glutathione, glutathione disulphide and glutathione sulphonamide in cells and physiological fluids by isotope dilution liquid chromatography-tandem mass spectrometry. *Journal of Chromatography B: Analytical Technologies in the Biomedical and Life Sciences, 877*, 3393–3399.

Hidese, R., Mihara, H., & Esaki, N. (2011). Bacterial cysteine desulfurases: Versatile key players in biosynthetic pathways of sulfur-containing biofactors. *Applied Microbiology and Biotechnology, 91*, 47–61.

Higgins, K. A., Peng, H., Luebke, J. L., Chang, F. M., & Giedroc, D. P. (2015). Conformational analysis and chemical reactivity of the multidomain sulfurtransferase, *Staphylococcus aureus* CstA. *Biochemistry, 54*, 2385–2398.

Holdorf, M. M., Owen, H. A., Lieber, S. R., Yuan, L., Adams, N., Dabney-Smith, C., et al. (2012). Arabidopsis ETHE1 encodes a sulfur dioxygenase that is essential for embryo and endosperm development. *Plant Physiology, 160*, 226–236.

Ida, T., Sawa, T., Ihara, H., Tsuchiya, Y., Watanabe, Y., Kumagai, Y., et al. (2014). Reactive cysteine persulfides and S-polythiolation regulate oxidative stress and redox signaling. *Proceedings of the National Academy of Sciences of the United States of America, 111*, 7606–7611.

Ihara, H., Kasamatsu, S., Kitamura, A., Nishimura, A., Tsutsuki, H., Ida, T., et al. (2017). Exposure to electrophiles impairs reactive persulfide-dependent redox signaling in neuronal cells. *Chemical Research in Toxicology, 30*, 1673–1684.

Jeong, W., Yoon, H. W., Lee, S. R., & Rhee, S. G. (2004). Identification and characterization of TRP14, a thioredoxin-related protein of 14 kDa. New insights into the specificity of thioredoxin function. *Journal of Biological Chemistry, 279*, 3142–3150.

Jung, M., Kasamatsu, S., Matsunaga, T., Akashi, S., Ono, K., Nishimura, A., et al. (2016). Protein polysulfidation-dependent persulfide dioxygenase activity of ethylmalonic encephalopathy protein 1. *Biochemical and Biophysical Research Communications, 480*, 180–186.

Kabil, O., & Banerjee, R. (2012). Characterization of patient mutations in human persulfide dioxygenase (ETHE1) involved in H2S catabolism. *Journal of Biological Chemistry, 287*, 44561–44567.

Kawagoe, R., Takashima, I., Uchinomiya, S., & Ojida, A. (2017). Reversible ratiometric detection of highly reactive hydropersulfides using a FRET-based dual emission fluorescent probe. *Chemical Science*, *8*, 1134–1140.
Kosower, N. S., & Kosower, E. M. (1987). Thiol labeling with bromobimanes. *Methods in Enzymology*, *143*, 76–84.
Kunikata, H., Ida, T., Sato, K., Aizawa, N., Sawa, T., Tawarayama, H., et al. (2017). Metabolomic profiling of reactive persulfides and polysulfides in the aqueous and vitreous humors. *Scientific Reports*, *7*, 41984.
Lehmann, W. D. (2017). A timeline of stable isotopes and mass spectrometry in the life sciences. *Mass Spectrometry Reviews*, *36*, 58–85.
Leu, L. S., & Cook, P. F. (1994). Kinetic mechanism of serine transacetylase from Salmonella typhimurium. *Biochemistry*, *33*, 2667–2671.
Lin, V. S., Chen, W., Xian, M., & Chang, C. J. (2015). Chemical probes for molecular imaging and detection of hydrogen sulfide and reactive sulfur species in biological systems. *Chemical Society Reviews*, *44*, 4596–4618.
Liu, C., Chen, W., Shi, W., Peng, B., Zhao, Y., Ma, H., et al. (2014). Rational design and bioimaging applications of highly selective fluorescence probes for hydrogen polysulfides. *Journal of the American Chemical Society*, *136*, 7257–7260.
Luebke, J. L., Shen, J., Bruce, K. E., Kehl-Fie, T. E., Peng, H., Skaar, E. P., et al. (2014). The CsoR-like sulfurtransferase repressor (CstR) is a persulfide sensor in *Staphylococcus aureus*. *Molecular Microbiology*, *94*, 1343–1360.
McCoy, J. G., Bingman, C. A., Bitto, E., Holdorf, M. M., Makaroff, C. A., & Phillips, G. N., Jr. (2006). Structure of an ETHE1-like protein from Arabidopsis thaliana. *Acta Crystallographica D Biological Crystallography*, *62*, 964–970.
Miles, E. W., & Kraus, J. P. (2004). Cystathionine beta-synthase: Structure, function, regulation, and location of homocystinuria-causing mutations. *Journal of Biological Chemistry*, *279*, 29871–29874.
Mueller, E. G. (2006). Trafficking in persulfides: Delivering sulfur in biosynthetic pathways. *Nature Chemical Biology*, *2*, 185–194.
Nishida, M., Kumagai, Y., Ihara, H., Fujii, S., Motohashi, H., & Akaike, T. (2016). Redox signaling regulated by electrophiles and reactive sulfur species. *Journal of Clinical Biochemistry and Nutrition*, *58*, 91–98.
Nishida, M., Nishimura, A., Matsunaga, T., Motohashi, H., Kasamatsu, S., & Akaike, T. (2017). Redox regulation of electrophilic signaling by reactive persulfides in cardiac cells. *Free Radical Biology and Medicine*, *109*, 132–140.
Nishida, M., Sawa, T., Kitajima, N., Ono, K., Inoue, H., Ihara, H., et al. (2012). Hydrogen sulfide anion regulates redox signaling via electrophile sulfhydration. *Nature Chemical Biology*, *8*, 714–724.
Numakura, T., Sugiura, H., Akaike, T., Ida, T., Fujii, S., Koarai, A., et al. (2017). Production of reactive persulfide species in chronic obstructive pulmonary disease. *Thorax*, *72*, 1074–1083.
Ono, K., Akaike, T., Sawa, T., Kumagai, Y., Wink, D. A., Tantillo, D. J., et al. (2014). Redox chemistry and chemical biology of H2S, hydropersulfides, and derived species: Implications of their possible biological activity and utility. *Free Radical Biology and Medicine*, *77*, 82–94.
Ono, K., Jung, M., Zhang, T., Tsutsuki, H., Sezaki, H., Ihara, H., et al. (2017). Synthesis of L-cysteine derivatives containing stable sulfur isotopes and application of this synthesis to reactive sulfur metabolome. *Free Radical Biology and Medicine*, *106*, 69–79.
Pan, J., & Carroll, K. S. (2013). Persulfide reactivity in the detection of protein s-sulfhydration. *ACS Chemical Biology*, *8*, 1110–1116.
Paul, B. D., & Snyder, S. H. (2012). H(2)S signalling through protein sulfhydration and beyond. *Nature Reviews Molecular Cell Biology*, *13*, 499–507.

Peng, H., Shen, J., Edmonds, K. A., Luebke, J. L., Hickey, A. K., Palmer, L. D., et al. (2017). Sulfide homeostasis and nitroxyl intersect via formation of reactive sulfur species in *Staphylococcus aureus*. *mSphere*, *2*, e00082-17.

Peng, H., Zhang, Y., Palmer, L. D., Kehl-Fie, T. E., Skaar, E. P., Trinidad, J. C., et al. (2017). Hydrogen sulfide and reactive sulfur species impact proteome S-sulfhydration and global virulence regulation in *Staphylococcus aureus*. *ACS Infectious Diseases*, *3*, 744–755.

Pettinati, I., Brem, J., Lee, S. Y., McHugh, P. J., & Schofield, C. J. (2016). The chemical biology of human metallo-beta-lactamase fold proteins. *Trends in Biochemical Sciences*, *41*, 338–355.

Py, B., & Barras, F. (2010). Building Fe-S proteins: Bacterial strategies. *Nature Reviews Microbiology*, *8*, 436–446.

Sawa, T., Arimoto, H., & Akaike, T. (2010). Regulation of redox signaling involving chemical conjugation of protein thiols by nitric oxide and electrophiles. *Bioconjugate Chemistry*, *21*, 1121–1129.

Sawa, T., Ihara, H., Ida, T., Fujii, S., Nishida, M., & Akaike, T. (2013). Formation, signaling functions, and metabolisms of nitrated cyclic nucleotide. *Nitric Oxide*, *34*, 10–18.

Sawa, T., Zaki, M. H., Okamoto, T., Akuta, T., Tokutomi, Y., Kim-Mitsuyama, S., et al. (2007). Protein S-guanylation by the biological signal 8-nitroguanosine 3′,5′-cyclic monophosphate. *Nature Chemical Biology*, *3*, 727–735.

Shen, J., Keithly, M. E., Armstrong, R. N., Higgins, K. A., Edmonds, K. A., & Giedroc, D. P. (2015). *Staphylococcus aureus* CstB is a novel multidomain persulfide dioxygenase-sulfurtransferase involved in hydrogen sulfide detoxification. *Biochemistry*, *54*, 4542–4554.

Shen, J., Peng, H., Zhang, Y., Trinidad, J. C., & Giedroc, D. P. (2016). *Staphylococcus aureus* sqr encodes a type II sulfide:quinone oxidoreductase and impacts reactive sulfur speciation in cells. *Biochemistry*, *55*, 6524–6534.

Shimizu, T., Shen, J., Fang, M., Zhang, Y., Hori, K., Trinidad, J. C., et al. (2017). Sulfide-responsive transcriptional repressor SqrR functions as a master regulator of sulfide-dependent photosynthesis. *Proceedings of the National Academy of Sciences of the United States of America*, *114*, 2355–2360.

Stipanuk, M. H., & Ueki, I. (2011). Dealing with methionine/homocysteine sulfur: Cysteine metabolism to taurine and inorganic sulfur. *Journal of Inherited Metabolic Disease*, *34*, 17–32.

Takagi, H., & Ohtsu, I. (2017). L-Cysteine metabolism and fermentation in microorganisms. *Advances in Biochemical Engineering/Biotechnology*, *159*, 129–151.

Takahashi, N., Wei, F. Y., Watanabe, S., Hirayama, M., Ohuchi, Y., Fujimura, A., et al. (2017). Reactive sulfur species regulate tRNA methylthiolation and contribute to insulin secretion. *Nucleic Acids Research*, *45*, 435–445.

Takano, Y., Echizen, H., & Hanaoka, K. (2017). Fluorescent probes and selective inhibitors for biological studies of hydrogen sulfide- and polysulfide-mediated signaling. *Antioxidants & Redox Signaling*, *27*, 669–683.

Takano, Y., Hanaoka, K., Shimamoto, K., Miyamoto, R., Komatsu, T., Ueno, T., et al. (2017). Development of a reversible fluorescent probe for reactive sulfur species, sulfane sulfur, and its biological application. *Chemical Communications*, *53*, 1064–1067.

Takata, T., Ihara, H., Hatano, N., Tsuchiya, Y., Akaike, T., & Watanabe, Y. (2017). Reactive sulfur species inactivate Ca2+/calmodulin-dependent protein kinase IV via S-polysulfidation of its active-site cysteine residue. *Biochemical Journal*, *474*, 2547–2562.

Tiranti, V., Viscomi, C., Hildebrandt, T., Di Meo, I., Mineri, R., Tiveron, C., et al. (2009). Loss of ETHE1, a mitochondrial dioxygenase, causes fatal sulfide toxicity in ethylmalonic encephalopathy. *Nature Medicine*, *15*, 200–205.

Toohey, J. I. (1986). Persulfide sulfur is a growth factor for cells defective in sulfur metabolism. *Biochemistry and Cell Biology*, *64*, 758–765.

Toohey, J. I. (1989). Sulphane sulphur in biological systems: A possible regulatory role. *Biochemical Journal, 264*, 625–632.

Toohey, J. I. (2011). Sulfur signaling: Is the agent sulfide or sulfane? *Analytical Biochemistry, 413*, 1–7.

Tsutsuki, H., Jung, M., Zhang, T., Ono, K., Ida, T., Kunieda, K., et al. (2016). Endogenous occurrence of protein S-guanylation in Escherichia coli: Target identification and genetic regulation. *Biochemical Biophysical Research Communications, 478*, 7–11.

Westley, A. M., & Westley, J. (1991). Biological sulfane sulfur. *Analytical Biochemistry, 195*, 63–67.

Wood, J. L. (1987). Sulfane sulfur. *Methods in Enzymology, 143*, 25–29.

Zaki, M. H., Fujii, S., Okamoto, T., Islam, S., Khan, S., Ahmed, K. A., et al. (2009). Cytoprotective function of heme oxygenase 1 induced by a nitrated cyclic nucleotide formed during murine salmonellosis. *Journal of Immunology, 182*, 3746–3756.

Zhang, B., Crack, J. C., Subramanian, S., Green, J., Thomson, A. J., Le Brun, N. E., et al. (2012). Reversible cycling between cysteine persulfide-ligated [2Fe-2S] and cysteine-ligated [4Fe-4S] clusters in the FNR regulatory protein. *Proceedings of the National Academy of Sciences of the United States of America, 109*, 15734–15739.

Zhang, D., Devarie-Baez, N. O., Li, Q., Lancaster, J. R., Jr., & Xian, M. (2012). Methylsulfonyl benzothiazole (MSBT): A selective protein thiol blocking reagent. *Organic Letters, 14*, 3396–3399.

Zhang, D., Macinkovic, I., Devarie-Baez, N. O., Pan, J., Park, C. M., Carroll, K. S., et al. (2014). Detection of protein S-sulfhydration by a tag-switch technique. *Angewandte Chemie International Edition English, 53*, 575–581.

Zhu, P., Oe, T., & Blair, I. A. (2008). Determination of cellular redox status by stable isotope dilution liquid chromatography/mass spectrometry analysis of glutathione and glutathione disulfide. *Rapid Communications in Mass Spectrometry, 22*, 432–440.

CHAPTER TWO

Nitric Oxide Signalling in Yeast

Rika I. Astuti*[,2], Ryo Nasuno†[,2], Hiroshi Takagi†[,1]

*Faculty of Mathematics and Natural Sciences, Bogor Agricultural University, Bogor, Indonesia
†Graduate School of Biological Sciences, Nara Institute of Science and Technology, Ikoma, Nara, Japan
[1]Corresponding author: e-mail address: hiro@bs.naist.jp

Contents

Abstract

Nitric oxide (NO) is a cellular signalling molecule widely conserved among organisms, including microorganisms such as bacteria, yeasts, and fungi, and higher eukaryotes such as plants and mammals. NO is mainly produced by the activities of NO synthase (NOS) or nitrite reductase (NIR). There are several NO detoxification systems, including NO dioxygenase (NOD) and *S*-nitrosoglutathione reductase (GSNOR). NO homeostasis, based on the balance between NO synthesis and degradation, is important for regulating its physiological functions, since an excess of NO causes nitrosative stress due to the high reactivity of NO and NO-derived compounds. In yeast, NO may be involved in stress responses, but the role of NO and the mechanism underlying NO signalling are poorly understood due to the lack of mammalian NOS orthologs in the yeast genome. NOS and NIR activities have been observed in yeast cells, but the gene-encoding NOS

[2] The first two authors contributed equally to this work.

Advances in Microbial Physiology, Volume 72
ISSN 0065-2911
https://doi.org/10.1016/bs.ampbs.2018.01.003

and the mechanism by which NO production is catalysed by NIR remain unclear. On the other hand, yeast cells employ NOD and GSNOR to maintain intracellular redox balance following endogenous NO production, treatment with exogenous NO, or exposure to environmental stresses. This article reviews NO metabolism (synthesis, degradation) and its regulation in yeast. The physiological roles of NO in yeast, including the oxidative stress response, are also discussed. Such investigations into NO signalling are essential for understanding how NO modulates the genetics and physiology of yeast. In addition to being responsible for the pathology and pharmacology of various degenerative diseases, NO signalling may be a potential target for the construction and engineering of industrial yeast strains.

1. INTRODUCTION

The systematic name for NO is nitrogen monoxide, but 'nitric oxide' is more commonly used. About 25 years ago, NO was identified as a unique cellular signalling molecule by Dr Furchgott of the State University of New York Downstate Medical Center. Shortly thereafter, the pharmacological effects of NO on the contraction of blood vessels were investigated. Subsequently, endothelial cells were found to generate an unknown signal molecule for the relaxation of vascular smooth muscle cells, termed an endothelium-derived relaxing factor (EDRF) (Furchgott & Zawadzki, 1980). Independently, Dr Murad from the University of Virginia found that NO is released from NO_3^- and relaxes smooth muscle tissue, thereby initiating vasodilation (Murad, Mittal, Arnold, Katsuki, & Kimura, 1978). In 1987, Dr Ignarro from the UCLA School of Medicine proved that EDRF was actually NO (Ignarro, Buga, Wood, Byrns, & Chaudhuri, 1987). In recognition of these fundamental and important findings on NO, the Nobel Prize in Physiology or Medicine was awarded to these three scientists in 1998.

NO is a radical gas molecule with one unpaired electron, which leads to its high reactivity with other radicals and the formation of derivatives with higher reactivity and toxicity (Hughes, 2008). NO is chemically reactive, but it is unlikely that NO forms a dimer either in the gas phase or in solution (Fukuto et al., 2012). Instead, NO modifies cellular compounds via *S*-nitrosylation or the coordination of iron atoms at the haem centre. There are two distinct mechanisms for these modifications: *S*-nitrosation and *S*-nitrosylation. Most strikingly, NO-based protein modifications mediate cellular responses involving the complex regulatory network underlying NO signalling. Studies of NO have been widely conducted from a variety of viewpoints, with particular focus paid to the biological significance of this gaseous signalling molecule. Indeed, the importance of NO signalling has been investigated not only in mammals but also in other organisms, including plants, yeast, and bacteria.

Numerous studies have demonstrated the physiological functions of NO in a variety of cellular responses in yeast cells, including those performed in our laboratory (Almeida et al., 2007; Astuti, Watanabe, & Takagi, 2016; Nasuno et al., 2014; Nishimura, Kawahara, & Takagi, 2013; Sasano et al., 2012). Although most yeast genera are unlikely to possess ortholog genes for the typically classical mammalian NO synthase (NOS), several NO synthetic mechanisms have been proposed. NO has been reported to elicit both beneficial and deleterious effects in yeast cells, much like NO signalling in higher eukaryotic systems, including cardiovascular systems. Studies on NO signalling in yeast are extremely important, since yeast is a crucial model system for analysing NO-mediated cellular responses in higher eukaryotes, and because NO signalling has potential application in the fermentation industry.

This review article describes recent findings regarding the metabolic regulation and physiological functions of NO in yeast. In addition, these mechanisms and roles are compared between yeast and other organisms, and the potential applications of NO in medical biology and biotechnology are discussed.

2. NITRIC OXIDE AS A BIOMOLECULE MODIFIER

Three modes of action of NO have been reported in association with three different NO species, i.e., the NO radical ($NO^{\cdot}$), the nitrosonium cation (NO^{+}), and the nitroxyl anion (NO^{-}) (Arnelle & Stamler, 1995) (Fig. 1). Each of these NO species exhibits unique properties and modifies biomolecules by nitrosylation or nitrosation. Although these two reactions are often confused, they have distinct reaction mechanisms that are reflected in their names: nitrosylation is the direct conjugation of $NO^{\cdot}$ to its target, while nitrosation is the reaction between a nucleophile and NO^{+} or similar molecular species.

NO^{+} directly modifies the nucleophilic centre, often the sulphur of a thiol group of the target molecule, generating an *S*-nitrosothiol (RSNO) (Heinrich et al., 2013). Collectively, such modifications are called the *S*-nitrosation reaction. RSNO compounds can also be generated through $NO^{\cdot}$. However, due to its negative electromagnetic charge, $NO^{\cdot}$ cannot bind directly to the nucleophilic sulphur atom of a biomolecule (Heinrich et al., 2013). Instead, the NO radical reacts with thiyl radicals ($RS^{\cdot}$) arising from the oxidation of the thiol

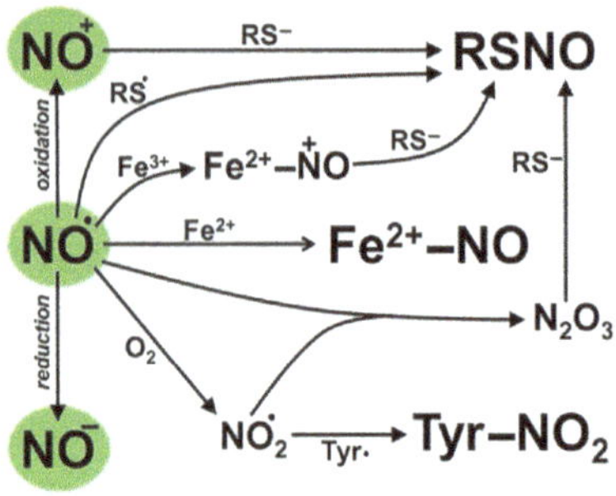

Fig. 1 Three reactions of nitric oxide (NO) with biomolecules. The NO radical (NO·) is rapidly reduced by one electron to form nitroxyl anion (NO^-) or is oxidized by one electron to form nitrosonium cation (NO^+). NO^+ reacts with thiolate anion, yielding RSNO. The NO radical reacts with thiyl radical (RS·), Fe^{3+}, or Fe^{2+} to form RSNO, or unstable (Fe^{2+}-NO^+) or stable (Fe^{2+}-NO) nitrosyl ferrous complex, respectively. Furthermore, the NO radical reacts with molecular oxygen, generating the NO_2 radical. The reaction of NO_2 radical with NO radical produces the N_2O_3 nitrosating thiolate anion. The NO_2 radical, which is also generated from peroxynitrous acid or nitrosoperoxycarbonate, reacts with the tyrosine radical, forming 3-nitrotyrosine (Tyr-NO_2).

compound, yielding RSNO compounds in a reaction commonly referred to as oxidative nitrosylation (Forman, Fukuto, & Torres, 2004; Heinrich et al., 2013). On the other hand, the NO radical can also react with oxygen or a transition metal such as iron, generating an NO_2 radical or metal–NO complex, respectively (Ford, Fernandez, & Lim, 2005; Heinrich et al., 2013). In the reaction with iron, the resultant products are dependent on the iron redox state: the reaction of NO radical with ferric ion produces the unstable complex Fe^{2+}-NO^+, which reacts with RS– to yield RSNO, whereas the reaction with ferrous ion generates the stable complex Fe^{2+}-NO (Heinrich et al., 2013). The formation of Fe^{2+}-NO triggers the activation of soluble guanylate cyclase (sGC) (Hunt & Lehnert, 2015). NO and NO_2 radical react with each other, yielding N_2O_3, which converts RS– into RSNO. It is worth mentioning that both *S*-nitrosation and -nitrosylation are prominent reactions in the development of intracellular RSNOs. Additionally, the NO_2 radical generates 3-nitrotyrosine via reaction with a tyrosine radical (Gunaydin & Houk, 2009). The reaction product of $NO^{\cdot}$ with the superoxide anion, peroxynitrite (PN), also mediates tyrosine nitration (Gunaydin & Houk, 2009). Quantum chemical calculations using *p*-methylphenol as a model compound for tyrosine have been performed to investigate the reaction mechanism of PN-mediated tyrosine nitration (Gunaydin & Houk, 2009). The peroxide bond of peroxynitrous acid, the protonated form of PN, undergoes hemolysis to produce the hydroxyl radical and NO_2 radical, whereas PN reacts with carbon dioxide to yield the NO_2 radical and carbonate radical ($CO_3^{\cdot -}$) through the formation of an intermediate, nitrosoperoxycarbonate. These reactions occur because the hydroxyl radical or $CO_3^{\cdot -}$ oxidize tyrosine to a tyrosine radical, PN, or peroxynitrous acid nitrate via radical chemistry using the NO_2 radical as a direct nitration reagent. Depending on the reaction rate for nitrosoperoxycarbonate-dependent tyrosine radical formation and peroxynitrous acid-dependent radical formation, reaction through the formation of $CO_3^{\cdot -}$ is clearly preferable, a conclusion also supported by experimental analyses. The nitration of tyrosine on the protein can alter the protein's function. For example, nitration at position 3 decreases the pK_a value of the phenolic hydroxyl group of tyrosine from 10.3 to 7.3, leading to a 50% yield of the corresponding phenolate (Radi, 2013).

The mechanism by which NO attacks a specific residue or residues of a target molecule has also been examined. In the case of the *S*-nitrosation of a protein, thiolate (RS–), the deprotonated form of cysteine thiol, is preferred because the thiolate anion is a better nucleophile than thiol, leading to the specificity of the *S*-nitrosation reaction (Kovacs & Lindermayr, 2013). Several sequence motifs suitable for protein *S*-nitrosation have been identified. It has been reported that acid–base sequences in proximity to the targeted cysteine residue regulate nitrosation (Hess, Matsumoto, Nudelman, & Stamler, 2001; Jia et al., 2014). Kovacs and Lindermayr (2013) revealed that these acid–base motifs provide reactive cysteine thiol sites favourable for *S*-nitrosation. One particular motif comprises a cysteine residue flanked by acidic (aspartate and glutamate) and basic (arginine, histidine, and lysine) residues (the [KRHDE]-C-[DE] motif), which ensures the formation of nucleophilic RS– through electrostatic interactions (Kovacs & Lindermayr, 2013). A comparable motif, denoted as I/L-X-C-X_2-D/E, has also been observed at *S*-nitrosation sites (Jia et al., 2014) and differs from the former motif in terms of its hydrophobic amino acid residues (isoleucine or leucine). In addition, the targeted cysteine residues in the primary structure exhibit high nucleophilicity (Hess et al., 2001). Although the consensus motifs of nitration-based modifications are unknown, the generation of 3-nitrotyrosine may occur at the accessible tyrosine residue (Ferrer-Sueta & Radi, 2009; Radi, 2004).

3. NO BIOSYNTHESIS

3.1 NO Synthase

Little is currently known about the NO synthetic mechanism in yeast, whereas a family of enzymes known as NOSs (EC 1.14.13.39) has been well studied in mammalian cells. NOSs catalyse the conversion of L-arginine

(Arg) into NO and L-citrulline (Cit) in the presence of molecular oxygen (O_2) and NADPH (Bredt & Snyder, 1990). NOSs are active in the homodimeric form, and each of the subunits comprises two distinctive catalytic domains: the N-terminal oxygenase and the C-terminal reductase domains (Stuehr, 1999). There are three main isoforms of NOSs, namely, inducible NOS (iNOS), endothelial NOS (eNOS), and neuronal NOS (nNOS), and each exhibits unique features in terms of structure, function, and regulation. iNOS generates a high level of NO following induction of its expression by inflammatory mediators such as lipopolysaccharide or IFN-γ in macrophages (Bergeron & Olivier, 2006; Zamora, Vodovotz, & Billiar, 2000). It is thus likely that iNOS-derived NO is correlated with the critical role of NO in the immune system response against pathogens. In contrast, eNOS and nNOS are constitutively expressed as NOSs with low NO-releasing activity and function in a Ca^{2+}-dependent manner (Campbell, Smith, Potter, Carragher, & Marletta, 2014; McMurry et al., 2011). These features of eNOS and nNOS are associated with the homeostatic functions of NO in the cardiovascular and nervous systems, respectively (Esplugues, 2002; Liu & Huang, 2008; Versari, Daghini, Virdis, Ghiadoni, & Taddei, 2009; Wang & Golledge, 2013).

In addition to Arg, O_2, and NADPH, several other molecules are required for NOS-mediated NO synthetic reactions—namely, the redox cofactor tetrahydrobiopterin (BH_4), flavins (flavin adenine dinucleotide (FAD) and flavin mononucleotide (FMD)), calmodulin (CaM), and haem (Bredt & Snyder, 1990). The reductase domain provides binding sites for one molecule each of NADPH, FAD, and FMN, which are essential for electron transfer, whereas the oxidase domain is important for binding Arg, BH_4, and haem and contributes to the O_2-mediated oxidation of Arg to Cit and NO. An electron from NADPH is sequentially transferred via flavin compounds to the final acceptor, O_2 bound to haem, thus providing the oxidizing power for Arg oxidation. In addition, haem is substantial enough for the dimerization of two NOS subunits (Chen, Panda, & Stuehr, 2002), thus allowing the binding of BH_4 to the oxygenase domains. BH_4 is crucial for dimer stabilization as well as electron flow in the catalytic reaction (Tejero & Stuehr, 2013), since absence of BH_4 leads to the uncoupled reduction of O_2, and thereby the production of O_2^-. In addition, the binding site for CaM between these two domains is essential for maintaining both the protein structure and its enzymatic function. The conversion of Arg to Cit and NO is conducted via a two-step oxidation reaction. Arg is converted into the intermediate compound N^G-

hydroxy-Arg (OH-Arg) to yield Cit and NO. The NOS-catalysed reaction consumes 2 mol of O_2 and 1.5 mol of NADPH to produce 1 mol of NO (Andrew & Mayer, 1999).

Bacteria have also been shown to generate Arg-dependent NO through NOS systems (Adak, Aulak, & Stuehr, 2002; Sudhamsu & Crane, 2009). Based on the analysis of genome sequences, NOS-like proteins have been identified in many prokaryotes, including *Nocardia* sp. (Chen & Rosazza, 1995), *Staphylococcus aureus* (Bird et al., 2009; Hong et al., 2003), *Bacillus subtilis* (Adak et al., 2002), *Sorangium cellulosum* (Agapie et al., 2009), *Streptomyces turgidiscabies* (Kers et al., 2004), and *Bacillus anthracis* (Gusarov et al., 2008). Unlike mammalian NOSs, bacterial NOSs (bNOSs), especially those from the genera *Bacillus* (Adak et al., 2002; Gusarov et al., 2008; Holden et al., 2013; Wang et al., 2007) and *Staphylococcus*, comprise only an oxygenase domain (Bird et al., 2009). The tertiary structures as well as the active sites of mammalian and bNOSs are nearly identical, with the exception that bacterial NOSs lack the N-terminal fragment of mammalian NOSs (Pant, Bilwes, Adak, Stuehr, & Crane, 2002). Although the N-terminal hook is absent in bNOSs, they can still form dimers, indicating that an extra compensatory alteration occurs during dimer formation in these molecules (Bird et al., 2009). It is believed that eukaryotic NOS originated evolutionarily earlier than bNOS, with the latter going on to employ a dedicated reductase domain (Gusarov et al., 2008). However, the bNOS of the bacterium *S. cellulosum* contains both an oxygenase domain and a reductase domain, providing a new perspective to the likelihood that NOS in both eukaryotes and prokaryotes was evolved from a common ancestor (Agapie et al., 2009). The lack of a devoted reductase domain in bNOS in certain genera of bacteria leads to the use of a cellular redox partner available within the corresponding bacterial cellular systems. Certain NADPH-utilizing proteins, including flavodoxin reductase/flavodoxin proteins (Wang et al., 2007) or proteins harbouring a FMN or FAD flavin group (Holden et al., 2013), support the production of NO as a redox partner of bNOS. These redox partners act as a supplier of NADPH-derived electrons to bNOS (Holden et al., 2013).

The identification of NO biosynthetic mechanisms in yeast presents a challenge, because the yeast genome is believed to lack mammalian NOS orthologs. It was reported, however, that crude extracts of the budding yeast *Saccharomyces cerevisiae* show positive immunoreactivity against an anti-nNOS antibody and that the protein responsible has a molecular weight of approximately 60 kDa (Kanadia, Kuo, Mcnabb, & Botchway, 1998;

Kuo et al., 1996). This finding suggests that yeast NOS activity is regulated in a manner similar to the constitutive NOS of mammalian cells. Indeed, addition of a NOS cofactor, CaM, and a NOS inhibitor, such as L-N^G-nitroarginine methyl ester (L-NAME), significantly affects the formation of NO. Addition of CaM plus Arg to a sample prepared by either histone-agarose chromatography or antineuronal NOS immunoprecipitation doubles the rate of Cit formation compared to the comparable enzymatic reaction using Arg alone. In contrast, treatment with L-NAME reduces NO formation by more than 60% (Kanadia et al., 1998).

In *S. cerevisiae* and the fission yeast *Schizosaccharomyces pombe*, the generation of NO via NOS-like activity has been detected by in vitro and in vivo analyses (Almeida et al., 2007; Kig & Temizkan, 2009; Yoshikawa et al., 2016), yet the corresponding NOS-encoding gene has not been identified. More recently, our laboratory demonstrated the generation of NO via an Arg-dependent mechanism mediated by the flavoprotein Tah18 in *S. cerevisiae* (Nishimura et al., 2013) (Fig. 2). This particular Tah18-dependent generation of NO is connected to the expression of *MPR1* and *PUT1* in the proline (Pro)-Arg metabolic pathway, since induction of these two genes leads to a significant increase in the level of Arg, which is the NOS substrate. This increase in Arg occurs particularly in response to environmental stresses, such as high temperature, air-drying, and freeze–thaw stress in baker's yeast strains of *S. cerevisiae* (Nishimura et al., 2013; Sasano et al., 2012). Although the expression of *TAH18* is required for NOS-like activity in *S. cerevisiae*, complete details regarding yeast Tah18-associated NOS activity have not been clearly elucidated. One confounding factor is that the primary structure of the Tah18 protein does not indicate an intact NOS oxygenase domain and is instead homologous to the reductase domain. A previous study indicated that Tah18 transfers electrons from NADPH to the Fe–S clusters of the Dre2 protein via the flavin molecules FAD and FMD, leading to the biogenesis of Fe–S cluster proteins (Netz et al., 2010). Thus, it is worth examining the potential contribution of the Tah18–Dre2 interaction in relation to NO production in *S. cerevisiae*. Recently, our biochemical analyses showed that Tah18 dissociates from Dre2 in response to oxidative stress, accompanied with an increase in the intracellular NO level. Additionally, enhancement of the Tah18–Dre2 interaction by the fusion of Tah18 with Dre2 suppressed NO production under oxidative stress conditions, indicating that the interaction of Tah18 with Dre2 inhibits Tah18-dependent NOS-like activity (Yoshikawa

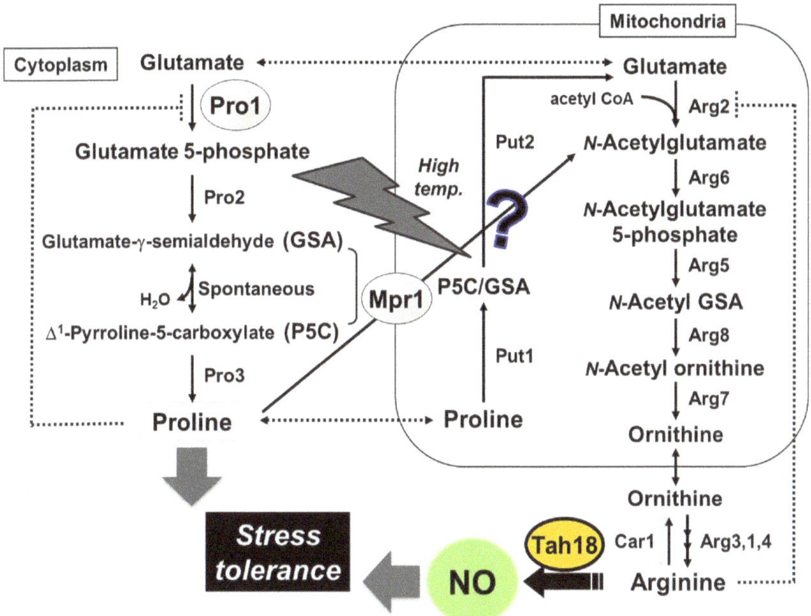

Fig. 2 Metabolic pathway of Pro and Arg in *S. cerevisiae*. In addition to our previous work showing that Pro protects yeast from various stresses (Morita, Nakamori, & Takagi, 2003; Takagi, 2008; Takagi, Taguchi, & Kaino, 2016), we recently found that, after exposure to high temperature that induced the generation of intracellular ROS, the Arg content of cells was significantly increased in the presence of a novel *N*-acetyltransferase, Mpr1 (Nishimura, Kotani, Sasano, & Takagi, 2010). More importantly, we discovered that NO is produced from the increased amounts of Arg through a Tah18-dependent activity, and this NO confers stress tolerance to yeast (Nasuno et al., 2014; Nishimura et al., 2013). Protein names: *Arg1*, arginosuccinate synthetase; *Arg2*, *N*-acetyl-glutamate synthase; *Arg3*, ornithine carbamoyltransferase; *Arg4*, arginosuccinate lyase; *Arg5*, *N*-acetyl-glutamyl-5-phospate reductase; *Arg6*, *N*-acetyl-glutamate kinase; *Arg7*, *N*-acetyl-ornithine acetyltransferase; *Arg8*, *N*-acetyl-ornithine aminotransferase; *Car1*, arginase; *Pro1*, γ-glutamyl kinase; *Pro2*, γ-glutamyl phosphate reductase; *Pro3*, P5C reductase; *Put1*, proline oxidase; *Put2*, P5C dehydrogenase.

et al., 2016). Our findings raise the possibility that Tah18 acts as a reductase domain to transfer electrons from NADPH to an unidentified oxygenase protein that oxidizes Arg to Cit and NO, and Dre2 inhibits Tah18-dependent NOS-like activity by removing electrons derived from NADPH exclusively. The Tah18–Dre2 complex might function as a molecular switch to control NO production in response to environmental oxidative conditions (Fig. 3).

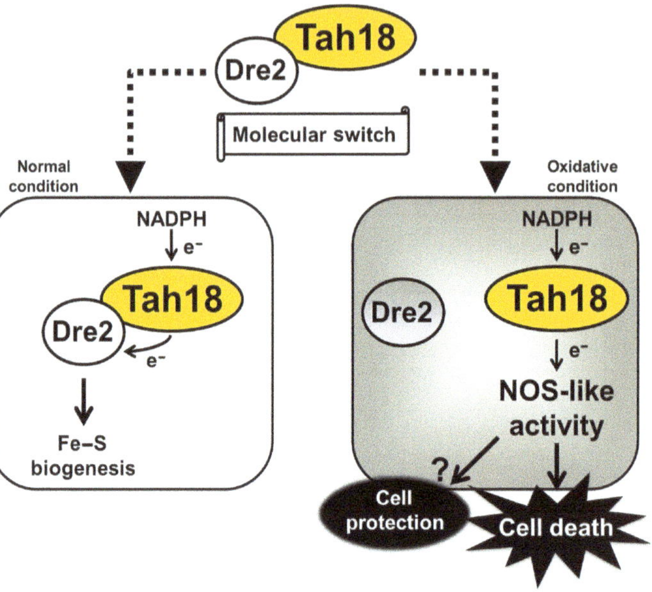

Fig. 3 Tah18–Dre2 complex regulates cell death through its NO-generating ability. Tah18 transfers electrons from NADPH to Dre2 for Fe–S cluster biogenesis under normal conditions. Oxidative stress induces the dissociation of Tah18 from Dre2, leading to a functional change of Tah18. Tah18 uses electrons from NADPH and exhibits NOS-like activity that is dependent on the availability of NADPH. NO production derived from Tah18-dependent NOS-like activity induces cell death and may be involved in cell protection. The Tah18–Dre2 complex functions as a molecular switch in *S. cerevisiae* and regulates cellular health via NO production.

3.2 Nitrite Reductase

Nitrite reductase (NIR) has been determined to be the major NO producer in mammalian cells under hypoxic conditions, such as those that commonly occur in hypoxic vasodilation of cardiovascular systems (Totzeck et al., 2012), muscle tissue during extensive exercise (Larsen, Weitzberg, Lundberg, & Ekblom, 2007), and ischemic tissues (Webb et al., 2004) (Fig. 4). Unlike NOS, NIR produces NO in an O_2-independent manner, suggesting the potential contribution of NO in mediating hypoxic signalling-dependent cellular responses (Dijkers & O'Farrell, 2009). Haem-binding proteins, such as haemoglobin (Gladwin & Kim-Saphiro, 2008) and myoglobin (Hendgen-Cotta, Kelm, & Rassaf, 2010; Totzeck et al., 2012), and cytochrome *c* (Basu et al., 2008), cytochrome *bc1* (Nohl et al., 2000; Tischner, Planchet, & Kaiser, 2004), and cytochrome *c* oxidase (Cco) (Castello, David, Mcclure, Crook, & Poyton, 2006; Tischner et al., 2004) of the mitochondrial electron

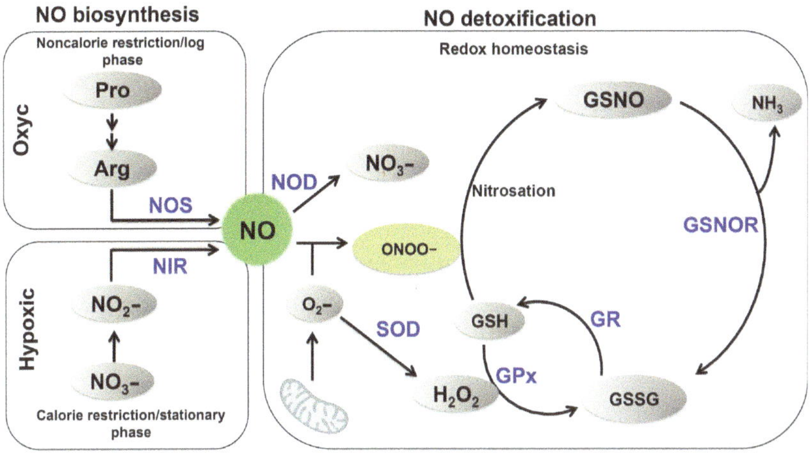

Fig. 4 Dual effects of NO in yeast cellular systems. NO causes both beneficial and deleterious effects, depending on the level and the target molecules of NO. Intracellular NO levels can be altered endogenously or exogenously. Endogenous NO production via NOS activity, mitochondrial respiratory activity, or NO-induced environmental stresses, as well as exogenous sources of NO, mediate many cellular responses through various pathways. The feedback regulatory roles of NO in mitochondria also determine the sequential NO-mediated cellular events in yeast cells. NO might interfere with intracellular redox homeostasis, modify biomolecules, and regulate stress response pathways in yeast cells. Fine-tuned NO signalling is required for the NO-mediated cytoprotective effects that are beneficial for yeast cells, leading to improved cellular metabolism, stress resistance, or fermentation ability. On the other hand, failure in NO signalling might impair essential physiological pathways, leading to cytotoxic effects such as nitrosative stress, defective metabolism, growth inhibition, or cell death.

transport chain, have been reported to generate NO via NIR-dependent catalysis. Most strikingly, hypoxic conditions markedly trigger the overproduction of NO within nodules formed by the symbiotic interaction between the legume *Medicago truncatula* and the bacterium *Sinorhizobium meliloti*, possibly via plant and bacterial nitrate reductases and electron transfer chains (Horchani et al., 2011).

Previous studies have demonstrated the ability of *S. cerevisiae* Cco to produce NO from NO_2^- through a reaction catalysed by NIR activity, particularly under low oxygen conditions (Castello et al., 2008; Li et al., 2011). It has been postulated that the mechanism by which NO_2^- is converted into NO via Cco in yeast cells mimics the NIR-mediated generation of NO by haemoglobin (Castello et al., 2008; Huang et al., 2005), as both are haem-containing proteins that are mainly expressed in low oxygen or acidic environments (Basu et al., 2008; Castello et al., 2008). During

NIR-dependent NO synthesis via the haemoglobin system, NO_2 is rapidly reduced to NO at low pH concomitantly with the oxidation of ferrous deoxyhaemoglobin ($HbFe^{2+}$) to methaemoglobin ($HbFe^{3+}$). The NO produced by the first reaction subsequently binds to vacant $HbFe^{2+}$ molecules to generate iron nitrosyl species ($HbFe^{2+}$-NO) (Shiva, 2013). In *S. cerevisiae*, NO is mainly produced through the activity of the Vb subunits of Cco, and Cco is highly expressed under hypoxic conditions (Castello et al., 2008). Mitochondrial NO production regulates the expression of the Vb-encoded gene *COX5b*, escalating NO synthesis via Cco, particularly under hypoxic or completely anoxic conditions (Castello et al., 2008). Similar to *S. cerevisiae*, two NO synthetic mechanisms via yet-unidentified NOS and NIR activities of mitochondrial complex III have been reported in *S. pombe* (Astuti et al., 2016; Kig & Temizkan, 2009). Treatment with L-NAME and antimycin A, which inhibit NOS and complex III of the mitochondria respiratory chain (MRC), respectively, or gene disruption of the complex III subunit Qcr8 (*Δqcr8*), significantly reduced NO production by *S. pombe* cells, as detected using the specific NO fluorescence probe 4-amino-5-methylamino-2′,7′-difluorofluorescein diacetate (DAF-FM DA) (Astuti et al., 2016). It is likely that these two distinctive NO synthetic mechanisms function in a growth phase-dependent manner when *S. pombe* cells are grown in rich medium. NOS activity predominantly produces NO during log phase growth and the complex III of MRC synthesizes NO through NIR activity during stationary phase (Astuti et al., 2016). It is worth noting that oxygen plays a key role in determining the NO synthetic mechanism, since the former system is O_2-dependent but the latter system is not (Andrew & Mayer, 1999; Bredt & Snyder, 1990; Dijkers & O'Farrell, 2009) (Fig. 4). Mitochondrial-NIR activity has been observed in highly respiring cells (Li et al., 2011). A high mitochondrial respiration rate may lead to hypoxic environments in individual mitochondria, as seen in the stationary phase of cells. In support of this possibility, the *S. pombe* hypoxia-inducible cytochrome *c* gene, *cyc1*$^+$, a homologue of *S. cerevisiae CYC7*, was significantly induced when cells entered stationary phase, but there was no alteration of *sre1*$^+$, which encodes an endoplasmic reticulum membrane-bound transcription factor that responds to changes in oxygen-dependent sterol synthesis (unpublished data). NO generation from the MRC may compete with oxygen for cytochrome *c* oxidase reactions (Benamar et al., 2008), which could further inhibit the consumption of oxygen, thereby alleviating local hypoxic environments. Thus, we speculate that such O_2 homeostasis is important to maintain efficient energy production via the MRC, as well as to prevent cellular hypoxia in yeast. In this context, it is worth noting that NO generated via the MRC has a potential role in the negative control of MRC activity for oxygen homeostasis.

In addition to oxygen, the level of the carbon source is likely to regulate NO production. A low glucose environment, such as seen during stationary phase or calorie restriction, might increase the respiratory activity of mitochondria (Astuti et al., 2016; Li et al., 2011; López-Lluch et al., 2006). This could further induce NO production through the MRC, as observed in yeast (Astuti et al., 2016; Guarente, 2005; Li et al., 2011) (Fig. 4) and mammalian HEK23 cells, which lack NOS isoforms (Ball, Nelson, Foster, & Poyton, 2012).

Our recent findings revealed a possible correlation between intracellular NO localization and the corresponding NO synthetic mechanism. In *S. pombe*, a significant amount of NO was observed within the vacuoles and in the vicinity of the mitochondria during log and stationary phases, respectively (Astuti et al., 2016). The physiological role of NO within vacuole compartments is unclear, but the presence of NO supports the existence of NOS activity in vacuoles, since cells in log phase growth store quite high levels of the NOS substrate Arg. Indeed, Arg accounts for more than 20% of the total vacuolar amino acids (Astuti et al., 2016; Chardwiriyapreecha et al., 2010). Relatively high mitochondrial activity during stationary phase may lead to MRC-dependent NO generation, as supported by the observation of intracellular NO accumulation in mitochondria (Astuti et al., 2016; Castello et al., 2008).

4. NO DETOXIFICATION

As discussed earlier, NO is a radical that easily reacts with other substances to yield NO-derived compounds. Various NO derivatives are produced inside cells that are either beneficial or deleterious to the cell. NO is subject to tight regulation, since an imbalance of NO could influence intracellular redox homeostasis, which is potentially harmful to cells. Redox imbalance due to the accumulation of NO derivatives, especially reactive nitrogen species (RNS), induces specific conditions termed nitrosative stress (Ridnour et al., 2004). Several RNS molecules, such as peroxynitrite, the higher oxides of nitrogen, *S*-nitrosothiols, and dinitrosyl iron complexes, have been reported to play roles in cellular systems (Gaston, Carver, Doctor, & Palmer, 2003; Heck, 2001). However, nitrosative stress may occur whenever cells are incapable of removing an excess of these RNS molecules. RNS accumulation is not inevitable following NO production, as NO may be easily oxidized to nitrate within cells via deoxygenases. Instead, the concurrent production of reactive oxygen species (ROS) such as O_2^- unavoidably leads to the production of RNS, particularly peroxynitrite (Pacher, Beckman, & Liaudet, 2007; Packer, Porteous, & Murphy, 1996). Indeed, a significant decrease in intracellular superoxide concentration due to high superoxide dismutase activity does not affect the viability of either wild-type cells or cells lacking NO detoxification systems (Li et al., 2011).

Intracellular NO detoxification mechanisms are important for ensuring that NO can exert its cytoprotective actions instead of having cytotoxic effects on the cell. Two well-known mechanisms, NO dioxygenase (NOD) and *S*-nitrosoglutathione reductase (GSNOR), are known to play roles in maintaining intracellular NO homeostasis (Astuti et al., 2016; Gardner, 2005) (Fig. 4). Together, these enzymes constitute the major mechanisms for NO detoxification, suggesting that their respective catalytic activities combat excessive intracellular NO levels and maintain NO at its physiological level to concomitantly protect against nitrosative stress. Interestingly, each of these NO detoxification systems plays a particular role under specific physiological conditions. Our latest study in *S. pombe* cells reveals that NOD and GSNOR might be induced by different mechanisms. The former is likely a NOS-dependent system, whereas the latter is a NOS-independent system that is potentially associated with mitochondrial-dependent NO production (Astuti et al., 2016). It is noteworthy that two enzymatic NO detoxification systems are conserved among eukaryotes, from yeast to mammals, indicating that NO is potentially involved in primary physiological activities that are conserved throughout evolution.

4.1 Nitric Oxide Dioxygenase

Nitric oxide dioxygenase (NOD) (or flavohaemoglobin (flavoHb)) mediates deoxygenation by using the reductive power of NAD(P)H to covalently bond the two oxygen atoms in O_2 with NO, thus yielding NO_3^- (Gardner, 2005). In higher eukaryotic organisms, haemoglobin exhibits NOD activity (for reviews, see Gardner, 2005, 2012). NOD consists of three distinct functional domains, namely, the N-terminal globin, and the FAD-binding and C-terminal NADPH-binding domains (Bonamore & Boffi, 2008). The haem pocket region of the globin domain is essential for NOD catalytic activity. The Fe^{3+}-haem prosthetic group is the final electron acceptor from the electron donor NADPH. The Fe^{2+}-haem pocket resulting from electron reduction binds O_2 with high affinity to yield a haem-Fe^{3+}-O_2^- intermediate that reacts rapidly with $NO^\cdot$ to produce NO_3^- (Gardner, Gardner, Martin, & Salzman, 1998).

Sequence alignment of the NOD-encoding genes from three diverse yeast species (*S. cerevisiae, Candida albicans*, and *S. pombe*) revealed a conserved domain that is regarded as a hallmark of both the globin-binding and cofactor-binding sites (Astuti et al., 2016). In addition, the amino acid sequence of the NOD of yeast is highly homologous to that of the NOD-encoding gene *hmp1* in *E. coli*, suggesting that NOD has been conserved during evolution, thus further underscoring the importance of NO signalling within cellular systems. Indeed, the globin structure of the Yhb1 protein conserves two groups of important amino acids residues: Tyr58 and Gln77, and His114, Tyr124, and Glu167 of *S. pombe* NOD. These two groups of amino acids are conserved among the genera *Saccharomyces, Candida*, and *Schizosaccharomyces* (Astuti et al., 2016). The former group of amino acids contributes to the distal stabilization of the iron-ligand during oxygen fixation, while the latter is responsible for the structural architecture of the haem pocket region inside the globin structure. In addition, the binding motifs of both the pyrophosphate (GQY) and isoalloxazine (RXYS) moieties of FAD molecules (Dym & Eisenberg, 2001), as well as the NADPH-binding domain (GXGXXP) motifs (Roma, Crowley, Davis, & Barber, 2005), are conserved among the Yhb1 protein structures of these three yeast genera.

Our recent findings confirmed the presence of NOD activity in *S. pombe*. This activity was likely associated with NOS activity, because the NOD expression levels weakened following treatment with NAME in a dose-dependent manner (Astuti et al., 2016). The manner, in which cells regulate intracellular NO via NOD, and the associative interaction of NOD with NO synthetic mechanisms remain unclear. However, functional NO detoxification systems (NOD and GSNOR) in *S. cerevisiae* are required to support NO production either via NOS activity or the mitochondria, and the role of cytochrome *c* oxidase is particularly important in cells subjected to calorie restriction (Li et al., 2011). Interestingly, NOD and GSNOR may function in a redundant or cooperative manner, depending on the particular yeast genera. In *S. pombe*, deletion of the *yhb1*$^+$ gene alone does not induce intracellular NO accumulation, except in cells lacking both the NOD- and GSNOR-encoding genes (Astuti et al., 2016), unlike in the case of *S. cerevisiae* (Lewinska & Bartosz, 2006; Liu et al., 2001) and *C. albicans*

(Hromatka, Noble, & Johnson, 2005). In addition, significant induction of GSNOR expression was observed in *Δyhb1* cells of *S. pombe*, suggesting a compensatory relation between these two NO detoxification enzymes (Astuti et al., 2016). Thus, it is likely that the NO signalling regulated by NOD and GSNOR overlaps substantially between *S. pombe* and other yeast species. However, NO detoxification systems exhibit distinct regulatory functions according to the physiology of the cells, since NOD is constitutively expressed throughout the growth cycle of *S. pombe* cells, whereas GSNOR is induced only once cells enter stationary phase (Astuti et al., 2016).

The induction of *YHB1* in *C. albicans* and *S. cerevisiae* cells depends on nonhomologous transcription factors, namely, the $Zn(II)_2$-Cys_6 transcription factor Cta4 and the zinc finger transcription factor Fzf1, respectively (Chiranand et al., 2008; Sarver & DeRisi, 2005). Interestingly, Cta4 from *C. albicans* has been shown to fully complement the function of Pdr1 and Pdr3, the *S. cerevisiae* transcription factors that regulate the genes for pleiotropic drug resistance, such as *PDR5* (Chiranand et al., 2008; Coste, Ramsdale, Ischer, & Sanglard, 2008). Thus, in the case of excessive intracellular NO concentration due to exogenous NO treatment, such as following exposure to a drug, it is likely that yeast cells sense the presence of NO at some level. The zinc finger transcription factor haem-activator protein 1 (Hap1) and the Hap2/3/4 complex, which respond to intracellular haem levels, have also been reported to regulate the expression of *YHB1* in *S. cerevisiae* (Crawford, Sherman, & Goldberg, 1995).

Currently, little is known regarding the regulation of NOD expression in *S. pombe*. However, Pap1, which is a transcription factor critical for inducing the protein expression of antioxidative genes, is required for the viability of *S. pombe* cells under nitrosative stress conditions, as cell survival under these conditions requires NOD activity (Calvo, Ayté, & Hidalgo, 2013; Kang, Jung, Hyun, Park, & Lim, 2011; Sahoo et al., 2006). In addition to the wide regulatory activities of the transcription factors Msn2/4 and Fzf1, the *S. cerevisiae* transcription factor Yap1, which is homologous to *S. pombe* Pap1, is also essential for protecting yeast cells from nitrosative stresses (Horan, Bourges, & Meunier, 2006).

In various yeast cells, including *C. albicans*, *S. cerevisiae*, and *S. pombe*, the *YHB1* gene is transcriptionally upregulated in response to exogenous NO (Astuti et al., 2016; Hromatka et al., 2005; Ullmann et al., 2004). The loss of Yhb1 activity in *S. cerevisiae* cells increases intracellular levels of RSNOs more than 10-fold compared to wild-type cells, creating the potential for nitrosative stress conditions (Lewinska & Bartosz, 2006). Further, loss of

Yhb1 activity increases sensitivity to NO-mediated growth inhibition of *C. albicans* (Ullmann et al., 2004), *S. cerevisiae* (Gardner et al., 2000; Helmick et al., 2005), *Blastomyces dermatitidis* (Rocco, Carmen, & Klein, 2011), and *S. pombe* (Astuti et al., 2016). Yhb1 is also essential for virulence, since macrophage-induced NO is particularly likely to harm pathogenic yeast cells when Yhb1 activity is low. Therefore, Yhb1 of pathogenic yeast is a potential target for treating patients with yeast infections such as *B. dermatitidis* (Rocco et al., 2011) and *C. albicans* (Hromatka et al., 2005).

Yhb1 has also been reported to play a role in antioxidation by consuming NO, which is a potent inhibitor of Cco activity in the MRC (Brown, 2007; Castello et al., 2006; Liu, Zeng, Hausladen, Heitman, & Stamler, 2000). The loss of Yhb1 activity compromises the function of Cco, leading to the generation of superoxide, which easily reacts with NO to form peroxinitrite, a protein nitration agent (Bhattacharjee et al., 2009; Brown & Borutaite, 2004; Castello et al., 2006). Thus, unlike in the mode of action of antioxidative superoxide dismutase, which dismutates superoxide molecules, Yhb1 reduces the intracellular level of superoxide via the respiratory chain (Cassanova, O'Brien, Stahl, McClure, & Poyton, 2005; Castello et al., 2006; Liu et al., 2000). The importance of Yhb1 as both an antioxidative and antinitrosative stress agent is supported by the cytosolic and mitochondrial matrix localization of Yhb1. In respiratory-deficient cells, Yhb1 is mostly present within the cytosol, since there is no mitochondrial-dependent superoxide production; in contrast, during hypoxic conditions, Yhb1 is mostly distributed within the mitochondria because NO is produced primarily via mitochondrial activity (Cassanova et al., 2005).

4.2 *S*-Nitrosoglutathione Reductase

RSNO has been extensively studied, since NO-derived molecules essentially function as signalling molecules for numerous physiological events (Foster, McMahon, & Stamler, 2003). However, an excessive amount of RSNO leads to severe nitrosative stress, which in turn causes cell damage. Within the cellular system, GSNO is often used as a marker of intracellular RSNO levels, since glutathione (GSH), which is the substrate for GSNO synthesis, is the most abundant thiol-containing compound in eukaryotic cells. GSNO is formed via the nitrosation reaction of GSH with nitrosonium cation or other nitrosation agents, such as N_2O_3, NO_x, and ONOO–, as well as metal–NO (Kharitonov, Sundquist, & Sharma, 1995; Singh, Wishnok, Keshive, Deen, & Tannenbaum, 1996).

GSH plays an important role in maintaining intracellular redox homeostasis and prevents the development of oxidative stress by acting as a radical scavenger and a lipid peroxidation inhibitor. The intracellular ratio between GSH and its oxidized forms, including glutathione disulphide (GSSG) and GSNO, is used to monitor the occurrence of oxidative stress (Schafer & Buettner, 2001). Reduced GSH should be maintained at a higher level than its oxidized forms for cellular health. Indeed, in normal cells, reduced GSH constitutes up to 98% of cellular GSH (Owen & Butterfield, 2010). Thus, the ratio of reduced GSH to its oxidized forms should be strictly regulated. An imbalance between GSH and GSNO may be indicative of either oxidative or nitrosative stress conditions. GSNO is reduced to GSSG and ammonia via GSNOR (Liu et al., 2001). Sequential reduction is necessary to convert GSSG to two GSH molecules via glutathione-disulphide reductase activity, thereby increasing intracellular reduced GSH levels (López-Mirabal & Winther, 2008) and overcoming redox imbalance.

Interestingly, the known activity of GSNOR is attributed to GSH-dependent formaldehyde dehydrogenase (GS-FDH) (Liu et al., 2001). This enzyme is important for converting toxic formaldehyde to nontoxic formylglutathione, in addition to its function as an NO detoxification enzyme (Staab et al., 2008). GSNOR is a zinc-containing enzyme and functions in a dimeric form. GSNOR belongs to GS-FDH class III, and members of this family contain alcohol dehydrogenase-like NADH-binding and dimerization domains (Tillmann et al., 2015; Xu, Guerra, Lee, & Vierling, 2013). The GSNOR monomer consists of catalytic and coenzyme domains and contains two zinc atoms. One zinc atom (the catalytic zinc) may activate the substrate GSNO during catalysis, since this zinc ion coordinates the hydroxyl and carbonyl groups of the substrate to activate the hydride transfer reaction. The second zinc atom (the structural zinc) is believed to have only a structural function. In addition, the five residues Thr49, Asp58, Glu60, Arg117, and Lys287 of *Solanum lycopersicum* GSNOR are involved in binding to *S*-(hydroxymethyl) glutathione (HMGSH), a GSH adduct (Kubienová et al., 2013).

Our latest findings confirmed that, of the three GS-FDH encoding genes within the *S. pombe* genome, Fmd2 encoded by *fmd2*$^{+}$ is responsible for GSNOR (Astuti et al., 2016). We have also observed that the activity of Fmd2 is highly related to mitochondrial-dependent NO synthesis. Interestingly, the amino acid sequence of Fmd2 showed low homology to other GS-FDHs. Thus, it is likely that Fmd2 has regulatory properties distinct from those of previously known GS-FDHs. Four cysteine residues that bind with the structural zinc atom are conserved between Fmd2 and other GS-FDHs; however, of the four conserved cysteine residues engaged in binding the catalytic zinc atom, one is replaced by aspartate in Fmd2. Interestingly, not all of the five residues reported to bind HMGSH are conserved in Fmd2 (Astuti et al., 2016). In contrast, the amino acid residues important for the enzymatic properties of GSNOR are conserved in Sfa1, an *S. cerevisiae* GSNOR homologue (Li et al., 2011). In addition to Sfa1, Adh6 (alcohol dehydrogenase 6) has been reported to possess denitrosylase activity specifically for the substrate of nitrosylated coenzyme A (SNO-CoA). Adh6 is phylogenetically conserved and is analogous to the deacetylases, which regulate CoA-mediated protein acetylation (Anand et al., 2014). In *C. albicans*, the Fdh3 protein exhibits GSNOR activity. Fdh3 is 65% identical to GS-FDH type III protein and ChiChi alcohol dehydrogenase in human (Tillmann et al., 2015).

Loss of GNOR activity has been reported to increase the susceptibility of cells to nitrosative stress (Li et al., 2011; Sahoo et al., 2006; Tillmann et al., 2015). As a cysteine-rich cytosolic enzyme, GSNOR is also a target protein of nitrosation reactions. Indeed, amino acid substitution of the cysteine residues of GSNOR, which are important for binding both the catalytic and structural zinc atoms, prevents the potential inhibition of GSNOR following nitroso treatment. The nitrosation of GSNOR might impair its catalytic activity, leading to the accumulation of RSNO and subsequent nitrosative stress conditions (Guerra, Ballard, Truebridge, & Vierling, 2016). In *C. albicans*, the oxidative stress response is unchanged following loss of Fdh3 activity, but the nitrosative stress response, which compromises virulence capacity, is altered (Tillmann et al., 2015). Disjunction of a signal transduction pathway has been suggested to induce an oxidative or nitrosative stress response in this yeast species.

5. PHYSIOLOGICAL ROLE OF NO IN YEAST (DUAL EFFECTS)

As a signalling molecule, NO exhibits complex regulatory activities in various organisms. Both NO and its derivatives, RSNO (Gaston et al., 2006; Palmer et al., 2007; Smith & Marletta, 2012), NO_2^- (Bryan et al., 2005; Gladwin, 2005), NO_3 (Alboresi et al., 2005), and PN (Liaudet, Vassalli, & Pacher, 2009; Pacher et al., 2007), have been reported to act as signalling molecules for numerous physiological and pathological events in both prokaryotes and eukaryotes. Interestingly, NO exhibits dual effects depending on its level in cells, as well as on environmental conditions (Fig. 5). Generally, lower levels of NO are essential for signal transduction events, but at higher levels or in the presence of superoxide, NO can damage cellular systems due to the development of nitrosative stress (Chiang et al., 2000; Domitrovic et al., 2003). In addition, posttranslational modification of

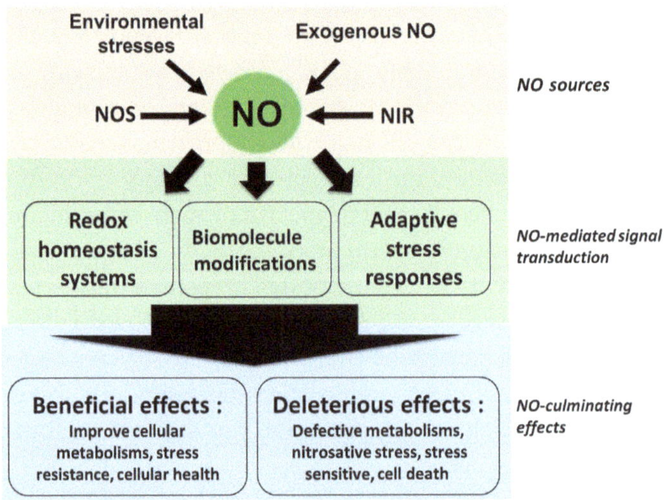

Fig. 5 Fine-tuned NO signalling in yeast. Sequential pathways of NO synthesis leading to detoxification mechanisms are essential to maintain intracellular redox balance. However, the regulatory mechanisms, both at the transcriptional and translational levels of these pathways, remain unclear. Yeast may employ distinct NO synthesis mechanisms according to the physiological conditions. Both nitrite reductase (NIR) and nitric oxide synthase (NOS) can generate NO. Generally, NOS serves as the main NO source in cells with low mitochondrial respiratory activity, whereas NIR acts as the main NO producer in highly respiring cells. In *S. cerevisiae*, the Pro-Arg pathway is the upstream route for NO synthesis by NOS, but the initial pathway for NO_2-dependent NO synthesis via NIR is unknown. NO and its *S*-nitrosothiol (RSNO) derivatives are detoxified via NO dioxygenase (NOD) and *S*-nitrosoglutathione reductase (GSNOR), since the accumulation of NO and RSNO generates nitrosative stress conditions. NOD converts NO to nitrate, whereas GSNOR reduces *S*-nitrosoglutathione (GSNO) to glutathione disulphide (GSSG), an oxidative glutathione, which is further converted to fully reduced glutathione (GSH) by glutathione reductase (GR). In highly respiring cells, mitochondria may generate O_2^- (superoxide) that easily reacts with NO to form peroxynitrite (ONOO–), a nitrosation agent of GSH, thereby yielding GSNO. The presence of O_2^- triggers H_2O_2 generation via superoxide dismutase (SOD), and this H_2O_2 may oxidize GSH to GSSG via glutathione peroxidase (GPx) activity. The lack of a certain pathway within these NO signalling systems potentially causes redox imbalance, resulting in oxidative stress that affects cell health.

biomolecules via nitrosation, nitrosylation, or nitration either inhibits or induces certain pathways, thereby supporting cell growth or triggering apoptosis (Almeida et al., 2007; Silva et al., 2011).

5.1 Beneficial Effects of NO Signalling

In mammalian cells, NO modulates cellular responses mostly via the activation of sGC, leading to the production of cyclic guanidine monophosphate (cGMP). This NO-sGC-cGMP pathway appears to play an essential role in smooth muscle relaxation (Lincoln, Wu, Sellak, Dey, & Choi, 2006), platelet inhibition (Walter & Gambaryan, 2009), antiapoptotic effects (Noguchi, Takada, Nakayama, & Ishikawa, 2008), and antiinflammatory effects (Ahluwalia et al., 2004). Similarly, nitrated cGMP is involved in the activity of guard cells in *Arabidopsis thaliana* (Joudoi et al., 2013). Despite previous reports that sGC activity converts guanidine triphosphate into cGMP, and that cGMP was detected in *S. cerevisiae* cells (Eckstein & Schlobohm, 1997; Kuo, Kanadia, & Mcnabb, 1998), little is currently known regarding NO-sGC-cGMP signalling in unicellular eukaryotes. In *S. pombe*, a study using potential sGC inhibitors suggested that NO regulates ascus and spore formation partly via the cGMP pathway (Kig & Temizkan, 2009).

In yeast cells, NO has been reported to exert cytoprotective actions under a variety of environmental stress conditions, particularly oxidative stress, by regulating certain transcription factors, including *S. cerevisiae* Mac1 (Nasuno et al., 2014) and *S. pombe* Pap1 (Kang et al., 2011; Kim, Jung, & Lim, 2008), Sty1 (Astuti et al., 2016), and Rst2 (Kato, Zhou, & Ma, 2013), which modulate the respective mechanisms. Activation of Mac1 by NO produced under a high temperature stress condition is important for activation of the copper-dependent superoxide dismutase Sod1, which is a crucial antioxidative enzyme (Nasuno et al., 2014) (Fig. 6A). Several reports have suggested that the defensive mechanisms against nitrosative and oxidative stresses partly overlap in yeast. When yeast cells are exposed to mild oxidative stress following treatment with a small amount of NO donor (Lushchak & Lushchak, 2008), NO appears to function in the cellular adaptation to oxidative stress. Consistent with this, our latest findings demonstrate that NO may play a role in the adaptive response to oxidative stress by activating peroxide scavenger functions, as well as by limiting the availability of ferrous iron to hamper Fenton reactions in *S. pombe* (Astuti et al., 2016) (Fig. 6B). Intriguingly, different genes downstream of

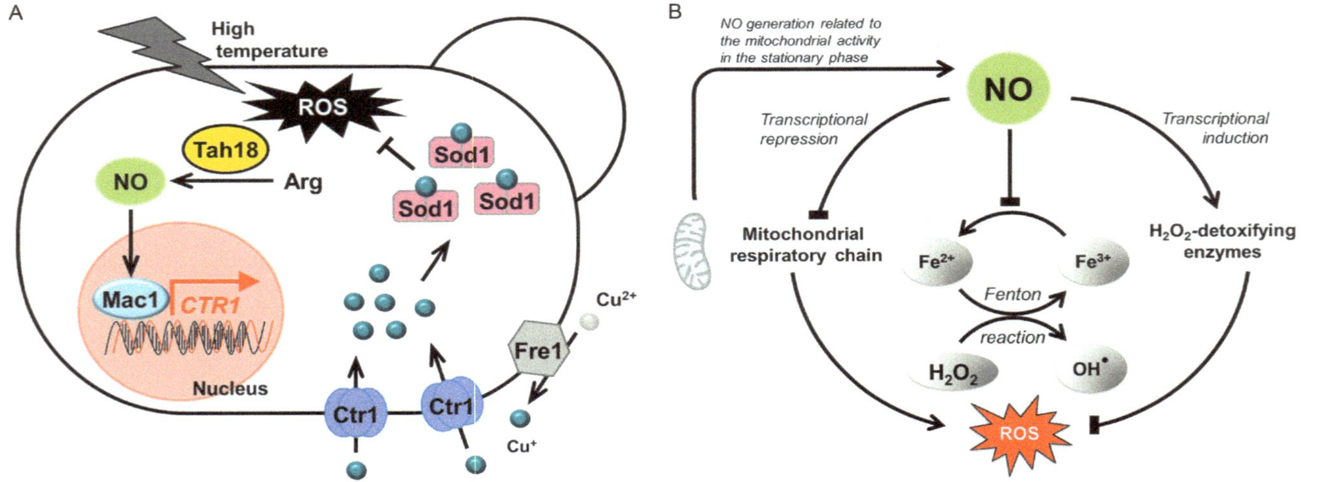

Fig. 6 Proposed models for NO-mediated oxidative stress responses in yeast. (A) When *S. cerevisiae* cells are exposed to high-temperature stress, NO is produced from Arg by an NOS-like activity. NO-mediated Mac1 is activated by posttranslational modification, then upregulates the expression of *CTR1*, which encodes a copper transporter. Protein names: *Ccs1*, copper chaperone; *Ctr1*, copper ion transporter; *Fre1*, cupric reductase; *Mac1*, transcription factor responsible for copper metabolism; *Sod1*, Cu,Zn-superoxide dismutase. (B) In *S. pombe*, NO is mainly generated by the activities of an unidentified NOS and the MRC complex III during log phase growth and is detoxified by Yhb1 (NOD) and Fmd2 (GSNOR) in stationary phase. NO triggers at least three antioxidant mechanisms: (i) inhibition of the conversion of Fe^{3+} into Fe^{2+}, (ii) upregulation of the genes encoding H_2O_2-detoxifying enzymes, and (iii) downregulation of the MRC genes.

Pap1 are regulated by NO donor-induced nitrosative stress. For example, genes regulating meiosis, the cell cycle, the spliceosome, and oxidative phosphorylation are affected most under these particular stress conditions (Biswas & Ghosh, 2015).

NO also plays a pivotal role in the resistance of cells against heat shock, hydrostatic pressure (Domitrovic et al., 2003), and high temperature in *S. cerevisiae* (Liu, Yuan, Li, & Lu, 2015; Nishimura et al., 2013, 2010). Our studies showed that an increased NO level resulting from the *PUT1*- and *MPR1*-mediated Pro-Arg metabolic pathways conferred tolerance to high temperature stress in *S. cerevisiae* (Fig. 2) (Nishimura et al., 2013, 2010). A more recent study revealed that loss of Cka2, an α' catalytic subunit of casein kinase 2 (CK2), represses the activity of NOS, which further decreases H_2O_2-induced apoptosis as well as the viability of cells under high temperature stress (Liu et al., 2015). In a study of possible biotechnological applications of NO, Sasano et al. (2012) constructed a self-cloning diploid baker's yeast strain exhibiting enhanced Pro and NO synthesis. The intracellular NO level of this engineered strain increased in response to air-drying stress and was tolerant not only to oxidative stress but also to air-drying and freeze–thaw stresses due to the reduced intracellular ROS level. This strain was shown to retain higher leavening activity in bread dough after air-drying and freeze–thaw stress than the wild-type strain. This result suggests that NO is synthesized in baker's yeast in response to oxidative stresses that induce ROS generation, and that increased NO levels play an important role in baking-associated stress tolerance. The enhancement of proline and NO synthesis levels could be a promising approach for breeding novel baker's yeast strains (Sasano et al., 2012).

The regulatory properties of NO, especially in response to oxidative stress, may provide a conceptual perspective into certain diseases in higher eukaryotic systems, since oxidative stresses are the main cause of both pathological events and the loss of cell function. Indeed, the induction of mitochondrial-dependent NO synthesis by calorie restriction in *S. cerevisiae* leads to a metabolic shift that improves metabolism and cellular response to age-related accumulation of oxidative stress molecules (Li et al., 2011), suggesting that NO may help to combat the effects of aging as well as age-related diseases. Moreover, a comprehensive study in a yeast model revealed the pathology of Batten disease, which results from reduced NO production due to the loss of Cln3, an important protein for the maintenance of intracellular Arg levels and pH homeostasis (Osório et al., 2007).

5.2 Deleterious Effects of NO Signalling

In general, NO has toxic effects at higher concentrations. For instance, treatment of *S. cerevisiae* cells with 1 mM sodium nitroprusside (SNP), an NO donor, showed a cytoprotective effect upon heat shock treatment, whereas 2 mM SNP severely reduced cell growth (Domitrovic et al., 2003). The cytotoxic effects caused by excessive levels of NO occur through various mechanisms, such as an increase in metal toxicity (Chiang et al., 2000) and the promotion of glyceraldehyde-3-phosphate dehydrogenase (GAPDH)-targeted proteolysis by metacaspase through *S*-nitrosation of GAPDH (Almeida et al., 2007; Silva et al., 2011). In *S. pombe*, the NO donors ONOO– and SNP have been reported to decrease the ratio of GSH to GSSG, suggesting that ONOO– and SNP-treated cells suffer from oxidative stress (Sahoo et al., 2006). Consistent with this, the lack of Pap1, an oxidative stress response regulator, leads to increased sensitivity to nitrosative stress (Kang et al., 2011; Majumdar, Biswas, Sarkar, Maiti, & Ghosh, 2012; Sahoo et al., 2006). Moreover, treatment with the NO donor (*Z*)-1-[2-(2-aminoethyl)-*N*-(2-ammonioethyl)amino]diazen-1-ium-1,2-diolate (DETA NONOate) delays mitosis in *S. pombe* cells by activating the G2/M damage checkpoint. This response is highly correlated with the inactivation of Cdc25, a Cdc2-activated protein kinase, due to *S*-nitrosylation of this kinase (Majumdar et al., 2012). NO has also been reported to induce sexual development in the fungus *Aspergillus nidulans* (Baidya, Cary, Grayburn, & Calvo, 2011). Our recent data using exogenous NO suggest a feedback regulatory effect of NO towards mitochondrial activity at the transcriptional level in *S. pombe* (Astuti et al., 2016). MRCs, including complexes I, III, and IV, were reported to be targets of NO modulation, resulting in detrimental alteration of mitochondrial activity (Brown, 2007).

On the other hand, endogenous NO also exerts deleterious effects, even in unicellular microorganisms such as yeast. NO generated by treating cells with H_2O_2 or cadmium induces cell death in *S. cerevisiae* (Almeida et al., 2007; Liu et al., 2015; Wu, Chen, Gao, Yin, & Huang, 2016; Yoshikawa et al., 2016). Almeida et al. (2007) reported that *S*-nitrosation of GAPDH is crucial for H_2O_2-induced apoptosis. Our latest data indicate that the Tah18–Dre2 complex regulates cell death, acting as a molecular switch via Tah18-dependent NOS-like activity in response to environmental changes (Yoshikawa et al., 2016) (Fig. 3). In addition, the deficiency in apoptosis in *S. cerevisiae* led to a loss of cell regrowth after long-term culture due to the accumulation of damaged cells. Therefore, apoptosis functions to remove damaged cells from the cell population, thereby helping maintain the health of the remaining cells (Herker et al., 2004). The existence of endogenous NO-mediated cell death indicates the significance of cell suicide even in unicellular organisms.

6. CONCLUSIONS

The presence of NO signalling in yeast cells strengthens the concept of evolutionarily conserved NO-dependent physiological events. Although neither yeast NOS nor NIR encoding genes have been identified, many studies support the occurrence of both NO synthetic mechanisms in a variety of yeast species. The lack of gene orthologs for NOS in yeast genomes makes the study of NO in yeast quite challenging. In addition, the contribution of the MRC to the production of NO via NIR activity remains to be determined in detail, yet it is likely that yeast cells employ this mechanism as a strategy for maintaining oxygen homeostasis, thus preventing the development of fully anoxic conditions. Fine-tuned coordination of NO synthesis and detoxification has been demonstrated in yeast. NOD and GSNOR play main roles in NO detoxification. Deficiency in NO detoxification activities results in the accumulation of NO and its derivatives, potentially leading to cell damage. Interestingly, yeast cells have developed overlapping pathways for responding to oxidative and nitrosative stress conditions, suggesting that the regulatory mechanisms of NO and ROS-induced stress are at least partially identical. The dual functions of NO found in higher eukaryotic organisms also occur in yeast. NO contributes to physiological events in yeast in a manner dependent on its intracellular level, as well as on the environmental conditions.

7. FUTURE PERSPECTIVES

Studies on NO in yeast have provided essential information on NO signalling in mammals and plants, since certain yeast genera are applicable as models for higher eukaryotic systems. Indeed, NO signalling is involved in the pathology of certain natural processes such as aging, and of degenerative diseases such as cancer and neurodegenerative syndromes. Many studies have demonstrated that yeast is a relevant microorganism for use as a model of degenerative diseases (Khurana & Lindquist, 2010), mitochondria-related diseases (Lasserre et al., 2015), and aging phenomena (Li et al., 2011), and may ultimately help reveal the underlying mechanisms of these NO-related diseases.

The fact that NO is involved in the stress response of industrially beneficial yeasts suggests the potential development of commercially important yeast cells by NO-targeted engineering at both the physiological and genetic level. The modulation of NO biosynthesis to enhance the fermentation ability of baker's yeast has been reported previously (Sasano et al., 2012). Thus, advanced engineering strategies for the construction of industrial yeast strains are required to obtain cells that can cope with various environmental stresses during fermentation while simultaneously exhibiting improved fermentation rates. Ultimately, therefore, it will be important to clarify the mechanisms by which NO regulates the genes or proteins engaged in the stress response or metabolic pathways in order to realize successful engineering strategies.

ACKNOWLEDGEMENTS

This study was funded by Grants for a Grant-in-Aid for Young Scientists (B) (15K21165) from the Japan Society for the Promotion of Science to R.N., a Grant-in-Aid for Scientific Research (A) (25252065) from the Japan Society for the Promotion of Science to H.T., a Grant-in-Aid for Scientific Research on Innovative Area (ROS Signal) (23117711) and (Oxygen Biology) (14430289) from the Ministry of Education, Science, Culture, Sports, and Technology of Japan to H.T., and Institute for Fermentation, Osaka to H.T.

REFERENCES

Adak, S., Aulak, K. S., & Stuehr, D. J. (2002). Direct evidence for nitric oxide production by a nitric-oxide synthase-like protein from *Bacillus subtilis*. *The Journal of Biological Chemistry*, *277*, 16167–16171.

Agapie, T., Suseno, S., Woodward, J. J., Stoll, S., Britt, R. D., & Marletta, M. A. (2009). NO formation by a catalytically self-sufficient bacterial nitric oxide synthase from *Sorangium cellulosum*. *Proceedings of the National Academy of Sciences of the United States of America*, *106*, 16221–16226.

Ahluwalia, A., Foster, P., Scotland, R. S., Mc Lean, P. G., Mathur, A., Perretti, M., et al. (2004). Antiinflammatory activity of soluble guanylate cyclase: cGMP-dependent down-regulation of P-selectin expression and leukocyte recruitment. *Proceedings of the National Academy of Sciences of the United States of America*, *101*, 1386–1391.

Alboresi, A., Gestin, C., Leydecker, M. T., Bedu, M., Meyer, C., & Truong, H. N. (2005). Nitrate, a signal relieving seed dormancy in *Arabidopsis*. *Plant, Cell & Environment*, *28*, 500–512.

Almeida, B., Buttner, S., Olheimer, S., Silva, A., Mesquita, A., Sampaio-Marques, B., et al. (2007). NO-mediated apoptosis in yeast. *Journal of Cell Science*, *120*, 3279–3288.

Anand, P., Hausladen, A., Wang, Y. J., Zhang, G. F., Stomberski, C., Brunengraber, H., et al. (2014). Identification of S-nitroso-CoA reductases that regulate protein S-nitrosylation. *Proceedings of the National Academy of Sciences*, *111*, 18572–18577.

Andrew, P. J., & Mayer, B. (1999). Enzymatic function of nitric oxide synthases. *Cardiovascular Research, 43*, 521–531.

Arnelle, D. R., & Stamler, J. S. (1995). NO^+, NO•, and NO^- donation by S-nitrosothiols: Implications for regulation of physiological functions by S-nitrosylation and acceleration of disulfide formation. *Archives of Biochemistry and Biophysics, 318*, 279–285.

Astuti, R. I., Watanabe, D., & Takagi, H. (2016). Nitric oxide signaling and its role in oxidative stress response in *Schizosaccharomyces pombe*. *Nitric Oxide, 52*, 29–40.

Baidya, S., Cary, J. W., Grayburn, W. S., & Calvo, A. M. (2011). Role of nitric oxide and flavohemoglobin homolog genes in *Aspergillus nidulans* sexual development and mycotoxin production. *Applied and Environmental Microbiology, 77*, 5524–5528.

Ball, K. A., Nelson, A. W., Foster, D. G., & Poyton, R. O. (2012). Nitric oxide produced by cytochrome c oxidase helps stabilize HIF-1α in hypoxic mammalian cells. *Biochemical and Biophysical Research Communications, 420*, 727–732.

Basu, S., Azarova, N. A., Font, M. D., King, S. B., Hogg, N., Gladwin, M. T., et al. (2008). Nitrite reductase activity of cytochrome c. *The Journal of Biological Chemistry, 283*, 32590–32597.

Benamar, A., Rolletschek, H., Borisjuk, L., Avelange-Macherel, M., Curien, G., Mostefai, H. A., et al. (2008). Nitrite–nitric oxide control of mitochondrial respiration at the frontier of anoxia. *Biochimica et Biophysica Acta, 1777*, 1268–1275.

Bergeron, M., & Olivier, M. (2006). Trypanosoma cruzi-mediated IFN-gamma-inducible nitric oxide output in macrophages is regulated by iNOS mRNA stability. *Journal of Immunology, 177*, 6271–6280.

Bhattacharjee, A., Majumdar, U., Maity, D., Sarkar, T. S., Goswami, A. M., Sahoo, R., et al. (2009). In vivo protein tyrosine nitration in *S. cerevisiae*: Identification of tyrosine-nitrated proteins in mitochondria. *Biochemical and Biophysical Research Communications, 388*, 612–617.

Bird, L. E., Ren, J., Zhang, J., Foxwell, N., Hawkins, A. R., Charles, I. G., et al. (2009). Crystal structure of SANOS, a bacterial nitric oxide synthase oxygenase protein from *Staphylococcus aureus*. *Structure, 10*, 1687–1696.

Biswas, P., & Ghosh, S. (2015). Global transcriptomic profiling of *Schizosaccharomyces pombe* in response to nitrosative stress. *Gene, 558*, 241–253.

Bonamore, A., & Boffi, A. (2008). Flavohemoglobin: Structure and reactivity. *IUBMB Life, 60*, 19–28. http://iubmb.onlinelibrary.wiley.com/hub/journal/10.1002/(ISSN)1521-6551/.

Bredt, D. S., & Snyder, S. H. (1990). Isolation of nitric oxide synthetase, a calmodulin-requiring enzyme. *Proceedings of the National Academy of Sciences of the United States of America, 87*, 682–685.

Brown, G. C. (2007). Nitric oxide and mitochondria. *Frontiers in Bioscience: A Journal and Virtual Library, 12*, 1024–1033.

Brown, G. C., & Borutaite, V. (2004). Inhibition of mitochondrial respiratory complex I by nitric oxide, peroxynitrite and *S*-nitrosothiols. *Biochimica et Biophysica Acta, 1658*, 44–49.

Bryan, N. S., Fernandez, B. O., Bauer, S. M., Garcia-Saura, M. F., Milsom, A. B., Rassaf, T., et al. (2005). Nitrite is a signaling molecule and regulator of gene expression in mammalian tissues. *Nature Chemical Biology, 1*, 290–297.

Calvo, I. A., Ayté, J., & Hidalgo, E. (2013). Reversible thiol oxidation in the H_2O_2-dependent activation of the transcription factor Pap1. *Journal of Cell Science, 126*, 2279–2284.

Campbell, M. G., Smith, B. C., Potter, C. S., Carragher, B., & Marletta, M. A. (2014). Molecular architecture of mammalian nitric oxide synthases. *Proceedings of the National Academy of Sciences of the United States of America, 111*, 3614–3623.

Cassanova, N., O'Brien, K. M., Stahl, B. T., McClure, T., & Poyton, R. O. (2005). Yeast flavohemoglobin, a nitric oxide oxidoreductase, is located in both the cytosol and the mitochondrial matrix: Effects of respiration, anoxia, and the mitochondrial genome on its intracellular level and distribution. *The Journal of Biological Chemistry, 280*, 7645–7653.

Castello, P. R., David, P. S., Mcclure, T., Crook, Z., & Poyton, R. O. (2006). Mitochondrial cytochrome oxidase produces nitric oxide under hypoxic conditions: Implications for oxygen sensing and hypoxic signaling in eukaryotes. *Cell Metabolism*, *3*, 277–287.

Castello, P. R., Woo, D. K., Ball, K., Wojcik, J., Liu, L., & Poyton, R. O. (2008). Oxygen-regulated isoforms of cytochrome c oxidase have differential effects on its nitric oxide production and on hypoxic signaling. *Proceedings of the National Academy of Sciences of the United States of America*, *105*, 8203–8208.

Chardwiriyapreecha, S., Mukaiyama, H., Sekito, T., Iwaki, T., Takegawa, K., & Kakinuma, Y. (2010). Avt5p is required for vacuolar uptake of amino acids in the fission yeast *Schizosaccharomyces pombe*. *FEBS Letters*, *584*, 2339–2345.

Chen, Y., Panda, K., & Stuehr, D. J. (2002). Control of nitric oxide synthase dimer assembly by a heme-NO-dependent mechanism. *Biochemistry*, *41*, 4618–4625.

Chen, Y., & Rosazza, J. P. (1995). Purification and characterization of nitric oxide synthase (NOSNoc) from a *Nocardia species*. *Journal of Bacteriology*, *177*, 5122–5128.

Chiang, K. T., Shinyashiki, M., Switzer, C. H., Valentine, J. S., Gralla, E. B., Thiele, D. J., et al. (2000). Effects of nitric oxide on the copper-responsive transcription factor Ace1 in *Saccharomyces cerevisiae*: Cytotoxic and cytoprotective actions of nitric oxide. *Archives of Biochemistry and Biophysics*, *377*, 296–303.

Chiranand, W., Mcleod, I., Zhou, H., Lynn, J. J., Vega, L. A., Myers, H., et al. (2008). CTA4 transcription factor mediates induction of nitrosative stress response in *Candida albicans*. *Eukaryotic Cell*, 7, 268–278.

Coste, A. T., Ramsdale, M., Ischer, F., & Sanglard, D. (2008). Divergent functions of three *Candida albicans* zinc-cluster transcription factors (*CTA4*, *ASG1* and *CTF1*) complementing pleiotropic drug resistance in *Saccharomyces cerevisiae*. *Microbiology*, *154*, 1491–1501.

Crawford, M. J., Sherman, D. R., & Goldberg, D. E. (1995). Regulation of *Saccharomyces cerevisiae* flavohemoglobin gene expression. *The Journal of Biological Chemistry*, *270*, 6991–6996.

Dijkers, P. F., & O'Farrell, P. H. (2009). Dissection of a hypoxia-induced, nitric oxide-mediated signaling cascade. *Molecular Biology of the Cell*, *20*, 4083–4090.

Domitrovic, T., Palhano, F. L., Barja-Fidalgo, C., DeFreitas, M., Orlando, M. T., & Fernandes, P. M. (2003). Role of nitric oxide in the response of *Saccharomyces cerevisiae* cells to heat shock and high hydrostatic pressure. *FEMS Yeast Research*, *3*, 341–346.

Dym, O., & Eisenberg, D. (2001). Sequence-structure analysis of FAD-containing proteins. *Protein Science*, *10*, 1712–1728.

Eckstein, H., & Schlobohm, H. (1997). A particulate guanylate cyclase (EC 4.6. 1.2) from growing yeast cells *Saccharomyces cerevisiae*. *Zeitschrift für Naturforschung Section C*, *52*, 373–379.

Esplugues, J. V. (2002). NO as a signalling molecule in the nervous system. *British Journal of Pharmacology*, *135*, 1079–1095.

Ferrer-Sueta, G., & Radi, R. (2009). Chemical biology of peroxynitrite: Kinetics, diffusion, and radicals. *ACS Chemical Biology*, *4*, 161–177.

Ford, P. C., Fernandez, B. O., & Lim, M. D. (2005). Mechanisms of reductive nitrosylation in iron and copper models relevant to biological systems. *Chemical Reviews*, *105*, 2439–2456.

Forman, H. J., Fukuto, J. M., & Torres, M. (2004). Redox signaling: Thiol chemistry defines which reactive oxygen and nitrogen species can act as second messengers. *American Journal of Physiology. Cell Physiology*, *287*, C246–C256.

Foster, M. W., McMahon, T. J., & Stamler, J. S. (2003). *S*-Nitrosylation in health and disease. *Trends in Molecular Medicine*, *9*, 160–168.

Fukuto, J., Carrington, S., Tantillo, D., Harrison, J., Ignarro, L., Freeman, B., et al. (2012). Small molecule signaling agents: The integrated chemistry and biochemistry of nitrogen oxides, oxides of carbon, dioxygen, hydrogen sulfide, and their derived species. *Chemical Research in Toxicology*, *25*, 769–793.

Furchgott, R. F., & Zawadzki, J. V. (1980). The obligatory role of endothelial cells in the relaxation of arterial smooth muscle by acetylcholine. *Nature*, *288*, 373–376.

Gardner, P. R. (2005). Nitric oxide dioxygenase function and mechanism of flavohemoglobin, hemoglobin, myoglobin and their associated reductases. *Journal of Inorganic Biochemistry*, *99*, 247–266.

Gardner, P. R. (2012). Hemoglobin: A nitric-oxide dioxygenase. *Scientifica*, *2012*, 683729 34 pp. https://doi.org/10.6064/2012/683729.

Gardner, P. R., Gardner, A. M., Martin, L. A., Dou, Y., Li, T., Olson, J. S., et al. (2000). Nitric-oxide dioxygenase activity and function of flavohemoglobins sensitivity to nitric oxide and carbon monoxide inhibition. *Journal of Biological Chemistry*, *275*, 31581–31587.

Gardner, P. R., Gardner, A. M., Martin, L. A., & Salzman, A. L. (1998). Nitric oxide dioxygenase: An enzymic function for flavohemoglobin. *Proceedings of the National Academy of Sciences of the United States of America*, *95*, 10378–10383.

Gaston, B. M., Carver, J., Doctor, A., & Palmer, L. A. (2003). *S*-Nitrosylation signaling in cell biology. *Molecular Interventions*, *3*, 253–263.

Gaston, B., Singel, D., Doctor, A., & Stamler, J. S. (2006). S-nitrosothiol signaling in respiratory biology. *American Journal of Respiratory and Critical Care Medicine*, *173*, 1186–1193.

Gladwin, M. T. (2005). Nitrite as an intrinsic signaling molecule. *Nature Chemical Biology*, *1*, 245–246.

Gladwin, M. T., & Kim-Saphiro, D. B. (2008). The functional nitrite reductase activity of the heme-globins. *Blood*, *112*, 2636–2647.

Guarente, L. (2005). NO link between calorie restriction and mitochondria. *Nature Chemical Biology*, *1*, 355–356.

Guerra, D., Ballard, K., Truebridge, I., & Vierling, E. (2016). *S*-Nitrosation of conserved cysteines modulates activity and stability of *S*-nitrosoglutathione reductase (GSNOR). *Biochemistry*, *55*, 2452–2464.

Gunaydin, H., & Houk, K. N. (2009). Mechanisms of peroxynitrite mediated nitration of tyrosine. *Chemical Research in Toxicology*, *22*, 894–898.

Gusarov, I., Starodubtseva, M., Wang, Z. Q., McQuade, L., Lippard, S. J., Stuehr, D. J., et al. (2008). Bacterial nitric-oxide synthases operate without a dedicated redox partner. *The Journal of Biological Chemistry*, *283*, 13140–13147.

Heck, D. E. (2001). *NO, RSNO, ONOO-, NO+, *NOO, NOx—Dynamic regulation of oxidant scavenging, nitric oxide stores, and cyclic GMP-independent cell signaling. *Antioxidants & Redox Signaling*, *3*, 249–260.

Heinrich, T. A., Silva, R. S., Miranda, K. M., Switzer, C. H., Wink, D. A., & Fukuto, J. M. (2013). Biological nitric oxide signalling: Chemistry and terminology. *British Journal of Pharmacology*, *169*, 1417–1429.

Helmick, R. A., Fletcher, A. E., Gardner, A. M., Gessner, C. R., Hvitved, A. N., Gustin, M. C., et al. (2005). Imidazole antibiotics inhibit the nitric oxide dioxygenase function of microbial flavohemoglobin. *Antimicrobial Agents and Chemotherapy*, *49*, 1837–1843.

Hendgen-Cotta, U. B., Kelm, M., & Rassaf, T. (2010). A highlight of myoglobin diversity: The nitrite reductase activity during myocardial ischemia–reperfusion. *Nitric Oxide*, *22*, 75–82.

Herker, E., Jungwirth, H., Lehmann, K. A., Maldener, C., Fröhlich, K., Wissing, S., et al. (2004). Chronological aging leads to apoptosis in yeast. *Journal of Cell Biology*, *164*, 501–507.

Hess, D. T., Matsumoto, A., Nudelman, R., & Stamler, J. S. (2001). *S*-Nitrosylation: Spectrum and specificity. *Nature Cell Biology*, *3*, E46–E49.

Holden, J. K., Li, H., Jing, Q., Kang, S., Richo, J., Silverman, R. B., et al. (2013). Structural and biological studies on bacterial nitric oxide synthase inhibitors. *Proceedings of the National Academy of Sciences of the United States of America, 110*, 18127–18131.

Hong, I. S., Kim, Y. K., Choi, W. S., Seo, D. W., Yoon, J. W., Han, J. W., et al. (2003). Purification and characterization of nitric oxide synthase from *Staphylococcus aureus*. *FEMS Microbiology Letters, 222*, 177–182.

Horan, S., Bourges, I., & Meunier, B. (2006). Transcriptional response to nitrosative stress in *Saccharomyces cerevisiae*. *Yeast, 23*, 519–535.

Horchani, F., Prevot, M., Boscari, A., Evangelisti, E., Meilhoc, E., Bruand, C., et al. (2011). Both plant and bacterial nitrate reductases contribute to nitric oxide production in *Medicago truncatula* nitrogen-fixing nodules. *Plant Physiology, 155*, 1023–1036.

Hromatka, B. S., Noble, S. M., & Johnson, A. D. (2005). Transcriptional response of *Candida albicans* to nitric oxide and the role of the *YHB1* gene in nitrosative stress and virulence. *Molecular Biology of the Cell, 16*, 4814–4826.

Huang, Z., Shiva, S., Kim-Shapiro, D. B., Patel, R. P., Ringwood, L. A., Irby, C. E., et al. (2005). Enzymatic function of hemoglobin as a nitrite reductase that produces NO under allosteric control. *Journal of Clinical Investigation, 115*, 2099–2107.

Hughes, M. N. (2008). Chemistry of nitric oxide and related species. *Methods in Enzymology, 436*, 3–19.

Hunt, A. P., & Lehnert, N. (2015). Heme-nitrosyls: Electronic structure implications for function in biology. *Accounts of Chemical Research, 48*, 2117–2125.

Ignarro, L. J., Buga, G. M., Wood, K. S., Byrns, R. E., & Chaudhuri, G. (1987). Endothelium-derived relaxing factor produced and released from artery and vein is nitric oxide. *Proceedings of the National Academy of Sciences of the United States of America, 84*, 9265–9269.

Jia, J., Arif, A., Terenzi, F., Willard, B., Plow, E. F., Hazen, S. L., et al. (2014). Target-selective protein *S*-nitrosylation by sequence motif recognition. *Cell, 159*, 623–634.

Joudoi, T., Shichiri, Y., Kamizono, N., Akaike, T., Sawa, T., Yoshitake, J., et al. (2013). Nitrated cyclic GMP modulates guard cell signaling in Arabidopsis. *Plant Cell, 25*, 558–571.

Kanadia, R. N., Kuo, W. N., Mcnabb, M., & Botchway, A. (1998). Constitutive nitric oxide synthase in *Saccharomyces cerevisiae*. *Biochemistry and Molecular Biology International, 45*, 1081–1087.

Kang, M. H., Jung, H. J., Hyun, D. H., Park, E. H., & Lim, C. J. (2011). Protective roles and Pap1-dependent regulation of the *Schizosaccharomyces pombe spy1* gene under nitrosative and nutritional stresses. *Molecular Biology Reports, 38*, 1129–1136.

Kato, T., Zhou, X., & Ma, Y. (2013). Possible involvement of nitric oxide and reactive oxygen species in glucose deprivation-induced activation of transcription factor *rst2*. *PLoS One, 8*, e78012.

Kers, J. A., Wach, M. J., Krasnoff, S. B., Widom, J., Cameron, K. D., Bukhalid, R. A., et al. (2004). Nitration of a peptide phytotoxin by bacterial nitric oxide synthase. *Nature, 429*, 79–82.

Kharitonov, V. G., Sundquist, A. R., & Sharma, V. S. (1995). Kinetics of nitrosation of thiols by nitric oxide in the presence of oxygen. *The Journal of Biological Chemistry, 270*, 28158–28164.

Khurana, V., & Lindquist, S. (2010). Modelling neurodegeneration in *Saccharomyces cerevisiae*: Why cook with baker's yeast? *Nature Reviews. Neuroscience, 11*, 436–449.

Kig, C., & Temizkan, G. (2009). Nitric oxide as signaling molecule in the fission yeast *Schizosaccharomyces pombe*. *Protoplasma, 238*, 59–66.

Kim, H. J., Jung, H. Y., & Lim, C. J. (2008). The *pap1*$^{+}$ gene of fission yeast is transcriptionally regulated by nitrosative and nutritional stress. *FEMS Microbiology Letters, 280*, 176–181.

Kovacs, I., & Lindermayr, C. (2013). Nitric oxide-based protein modification: Formation and site-specificity of protein *S*-nitrosylation. *Frontiers in Plant Science, 4*, 137.

Kubienová, L., Kopečný, D., Tylichová, M., Briozzo, P., Skopalová, J., Šebela, M., et al. (2013). Structural and functional characterization of a plant *S*-nitrosoglutathione reductase from *Solanum lycopersicum*. *Biochemie*, *95*, 889–902.

Kuo, W. N., Jn-Baptiste, J. B., Kanadia, R. N., McNabb, L. D., Zhai, L., Weeks, K., et al. (1996). Immunoreactivities of m-calpain, calpastatin, nitric oxide synthase, myelin basic protein and dynamin II in baker's yeast, wheat germ and lobster tail muscle. *Cytobios*, *87*, 251–263.

Kuo, W. N., Kanadia, R. N., & Mcnabb, M. (1998). Soluble guanylate cyclase in *Saccharomyces cerevisiae*. *IUBMB Life*, *45*, 125–131.

Larsen, F. J., Weitzberg, E., Lundberg, J. O., & Ekblom, B. (2007). Effects of dietary nitrate on oxygen cost during exercise. *Acta Physiologica*, *191*, 59–66.

Lasserre, J. P., Dautant, A., Aiyar, R. S., Kucharczyk, R., Glatigny, A., Tribouillard-Tanvier, D., et al. (2015). Yeast as a system for modeling mitochondrial disease mechanisms and discovering therapies. *Disease Models & Mechanisms*, *8*, 509–526.

Lewinska, A., & Bartosz, G. (2006). Yeast flavohemoglobin protects against nitrosative stress and controls ferric reductase activity. *Redox Report: Communications in Free Radical Research*, *11*, 231–239.

Li, B., Skinner, C., Castello, P. R., Kato, M., Easlon, E., Xie, L., et al. (2011). Identification of potential calorie restriction-mimicking yeast mutants with increased mitochondrial respiratory chain and nitric oxide levels. *Journal of Aging Research*, *201*, 1–16. https://doi.org/10.4061/2011/673185.

Liaudet, L., Vassalli, G., & Pacher, P. (2009). Role of peroxynitrite in the redox regulation of cell signal transduction pathways. *Frontiers in Bioscience*, *14*, 4809–4914.

Lincoln, T. M., Wu, X., Sellak, H., Dey, N., & Choi, C. S. (2006). Regulation of vascular smooth muscle cell phenotype by cyclic GMP and cyclic GMP-dependent protein kinase. *Front Bioscience*, *11*, 356–367.

Liu, L., Hausladen, A., Zeng, M., Que, L., Heitman, J., & Stamler, J. S. (2001). A metabolic enzyme for S-nitrosothiol conserved from bacteria to human. *Nature*, *410*, 490–494.

Liu, V. W., & Huang, P. L. (2008). Cardiovascular roles of nitric oxide: A review of insights from nitric oxide synthase gene disrupted mice. *Cardiovascular Research*, *77*, 19–29.

Liu, W. C., Yuan, H. M., Li, Y. H., & Lu, Y. T. (2015). *CKA2* functions in H_2O_2-induced apoptosis and high-temperature stress tolerance by regulating NO accumulation in yeast. *FEMS Yeast Research 15*. fov051. https://doi.org/10.1093/femsyr/fov051.

Liu, L., Zeng, M., Hausladen, A., Heitman, J., & Stamler, J. S. (2000). Protection from nitrosative stress by yeast flavohemoglobin. *Proceedings of the National Academy of Sciences of the United States of America*, *97*, 4672–4676.

López-Lluch, G., Hunt, N., Jones, B., Zhu, M., Jamieson, H., Hilmer, S., et al. (2006). Calorie restriction induces mitochondrial biogenesis and bioenergetic efficiency. *Proceedings of the National Academy of Sciences of the United States of America*, *103*, 1768–1773.

López-Mirabal, H. R., & Winther, J. R. (2008). Redox characteristics of the eukaryotic cytosol. *Biochimica et Biophysica Acta*, *1783*, 629–640.

Lushchak, O. V., & Lushchak, V. I. (2008). Sodium nitroprusside induces mild oxidative stress in *Saccharomyces cerevisiae*. *Redox Report: Communications in Free Radical Research*, *13*, 144–152.

Majumdar, U., Biswas, P., Sarkar, T. S., Maiti, D., & Ghosh, S. (2012). Regulation of cell cycle and stress responses under nitrosative stress in *Schizosaccharomyces pombe*. *Free Radical Biology & Medicine*, *52*, 2186–2200.

McMurry, J. L., Chrestensen, C. A., Scott, I. M., Lee, E. W., Rahn, A. M., Johansen, A. M., et al. (2011). Rate, affinity and calcium dependence of nitric oxide synthase isoform binding to the primary physiological regulator calmodulin. *FEBS Journal*, *278*, 4943–4954.

Morita, Y., Nakamori, S., & Takagi, H. (2003). L-Proline accumulation and freeze tolerance in *Saccharomyces cerevisiae* are caused by a mutation in the *PRO1* gene encoding γ-glutamyl kinase. *Applied and Environmental Microbiology*, *69*, 212–219.

Murad, F., Mittal, C. K., Arnold, W. P., Katsuki, S., & Kimura, H. (1978). Guanylate cyclase: Activation by azide, nitro compounds, nitric oxide, and hydroxyl radical and inhibition by hemoglobin and myoglobin. *Advances in Cyclic Nucleotide Research*, *9*, 145–158.

Nasuno, R., Aitoku, M., Manago, Y., Nishimura, A., Sasano, Y., & Takagi, H. (2014). Nitric oxide-mediated antioxidative mechanism in yeast through the activation of the transcription factor Mac1. *PLoS One 9*, e113788. https://doi.org/10.1371/journal.pone.0113788.

Netz, D. J., Stümpfig, M., Doré, C., Mühlenhoff, U., Pierik, A. J., & Lill, R. (2010). Tah18 transfers electrons to Dre2 in cytosolic iron-sulfur protein biogenesis. *Nature Chemical Biology*, *6*, 758–765.

Nishimura, A., Kawahara, N., & Takagi, H. (2013). The flavoprotein Tah18-dependent NO synthesis confers high-temperature stress tolerance on yeast cells. *Biochemical and Biophysical Research Communications*, *430*, 137–143.

Nishimura, A., Kotani, T., Sasano, Y., & Takagi, H. (2010). An antioxidative mechanism mediated by the yeast N-acetyltransferase Mpr1: Oxidative stress-induced arginine synthesis and its physiological role. *FEMS Yeast Research*, *10*, 687–698.

Noguchi, A., Takada, M., Nakayama, K., & Ishikawa, T. (2008). cGMP-independent anti-apoptotic effect of nitric oxide on thapsigargin-induced apoptosis in the pancreatic beta-cell line INS-1. *Life Sciences*, *83*, 865–870.

Nohl, H., Staniek, K., Sobhian, B., Bahrami, S., Redl, H., & Kozlov, A. V. (2000). Mitochondria recycle nitrite back to the bioregulator nitric monoxide. *Acta Biochimica Polonica*, *47*, 913–921.

Osório, N. S., Carvalho, A., Almeida, A. J., Padilla-Lopez, S., Leão, C., Laranjinha, J., et al. (2007). Nitric oxide signaling is disrupted in the yeast model for Batten disease. *Molecular Biology of the Cell*, *18*, 2755–2767.

Owen, J. B., & Butterfield, D. A. (2010). Measurement of oxidized/reduced glutathione ratio. *Methods in Molecular Biology*, *648*, 269–277.

Pacher, P., Beckman, J. S., & Liaudet, L. (2007). Nitric oxide and peroxynitrite in health and disease. *Physiological Reviews*, *87*, 315–424.

Packer, M. A., Porteous, C. M., & Murphy, M. P. (1996). Superoxide production by mitochondria in the presence of nitric oxide forms peroxynitrite. *IUBMB Life*, *40*, 527–534.

Palmer, L. A., Doctor, A., Chhabra, P., Sheram, M. L., Laubach, V. E., Karlinsey, M. Z., et al. (2007). *S*-Nitrosothiols signal hypoxia-mimetic vascular pathology. *Journal of Clinical Investigation*, *117*, 2592–2601.

Pant, K., Bilwes, A. M., Adak, S., Stuehr, D. J., & Crane, B. R. (2002). Structure of a nitric oxide synthase heme protein from *Bacillus subtilis*. *Biochemistry*, *41*, 11071–11079.

Radi, R. (2004). Nitric oxide, oxidants, and protein tyrosine nitration. *Proceedings of the National Academy of Sciences of the United States of America*, *101*, 4003–4008.

Radi, R. (2013). Protein tyrosine nitration: Biochemical mechanisms and structural basis of functional effects. *Accounts of Chemical Research*, *46*, 550–559.

Ridnour, L. A., Thomas, D. D., Mancardi, D., Espey, M. G., Miranda, K. M., Paolocci, N., et al. (2004). The chemistry of nitrosative stress induced by nitric oxide and reactive nitrogen oxide species. Putting perspective on stressful biological situations. *Biological Chemistry*, *385*, 1–10.

Rocco, N. M., Carmen, J. C., & Klein, B. S. (2011). *Blastomyces dermatitidis* yeast cells inhibit nitric oxide production by alveolar macrophage inducible nitric oxide synthase. *Infection and Immunity*, *79*, 2385–2395.

Roma, G. W., Crowley, L. J., Davis, C. A., & Barber, M. J. (2005). Mutagenesis of glycine 179 modulates both catalytic efficiency and reduced pyridine nucleotide specificity in cytochrome b5 reductase. *Biochemistry*, *44*, 13467–13476.

Sahoo, R., Dutta, T., Das, A., Ray, S. S., Sengupta, R., & Ghosh, S. (2006). Effect of nitrosative stress on *Schizosaccharomyces pombe*: Inactivation of glutathione reductase by peroxynitrite. *Free Radical Biology & Medicine, 40*, 625–631.

Sarver, A., & DeRisi, J. (2005). Fzf1p regulates an inducible response to nitrosative stress in *Saccharomyces cerevisiae. Molecular Biology of the Cell, 16*, 4781–4791.

Sasano, Y., Haitani, Y., Hashida, K., Ohtsu, I., Shima, J., & Takagi, H. (2012). Enhancement of the proline and nitric oxide synthetic pathway improves fermentation ability under multiple baking-associated stress conditions in industrial baker's yeast. *Microbial Cell Factories, 11*, 1. https://doi.org/10.1186/1475-2859-11-40.

Schafer, F. Q., & Buettner, G. R. (2001). Redox environment of the cell as viewed through the redox state of the glutathione disulfide/glutathione couple. *Free Radical Biology & Medicine, 30*, 1191–1212.

Shiva, S. (2013). Nitrite: A physiological store of nitric oxide and modulator of mitochondrial function. *Redox Biology, 1*, 40–44.

Silva, A., Almeida, B., Sampaio-Marques, B., Reis, M. I., Ohlmeier, S., Rodrigues, F., et al. (2011). Glyceraldehyde-3-phosphate dehydrogenase (GAPDH) is a specific substrate of yeast metacaspase. *Biochimica et Biophysica Acta, 1813*, 2044–2049.

Singh, S. P., Wishnok, J. S., Keshive, M., Deen, W. M., & Tannenbaum, S. R. (1996). The chemistry of the *S*-nitrosoglutathione/glutathione system. *Proceedings of the National Academy of Sciences of the United States of America, 93*, 14428–14433.

Smith, B. C., & Marletta, M. A. (2012). Mechanisms of *S*-nitrosothiol formation and selectivity in nitric oxide signaling. *Current Opinion in Chemical Biology, 16*, 498–506.

Sudhamsu, J., & Crane, B. R. (2009). Bacterial nitric oxide synthases: What are they good for? *Trends in Microbiology, 17*, 212–218.

Staab, C. A., Ålander, J., Brandt, M., Lengqvist, J., Morgenstern, R., Grafström, R. C., et al. (2008). Reduction of *S*-nitrosoglutathione by alcohol dehydrogenase 3 is facilitated by substrate alcohols *via* direct cofactor recycling and leads to GSH-controlled formation of glutathione transferase inhibitors. *Biochemical Journal, 413*, 493–504.

Stuehr, D. J. (1999). Mammalian nitric oxide synthases. *Biochimica et Biophysica Acta, 411*, 217–230.

Takagi, H. (2008). Proline as a stress protectant in yeast: Physiological functions, metabolic regulations and biotechnological applications. *Applied Microbiology and Biotechnology, 81*, 211–223.

Takagi, H., Taguchi, J., & Kaino, T. (2016). Proline accumulation protects *Saccharomyces cerevisiae* cells in the stationary phase from ethanol stress by reducing reactive oxygen species levels. *Yeast, 33*, 355–363.

Tejero, J., & Stuehr, D. (2013). Tetrahydrobiopterin in nitric oxide synthase. *IUBMB Life, 65*, 358–365.

Tillmann, A. T., Strijbis, K., Cameron, G., Radmaneshfar, E., Thiel, M., Munro, C. A., et al. (2015). Contribution of Fdh3 and Glr1 to glutathione redox state, stress adaptation and virulence in *Candida albicans. PLoS One10*, e0126940. https://doi.org/10.1371/journal.pone.0126940.

Tischner, R., Planchet, E., & Kaiser, W. M. (2004). Mitochondrial electron transport as a source for nitric oxide in the unicellular green alga *Chlorella sorokiniana. FEBS Letters, 576*, 151–155.

Totzeck, M., Hendgen-Cotta, U. B., Luedike, P., Berenbrink, M., Klare, J. P., Steinhoff, H. J., et al. (2012). Nitrite regulates hypoxic vasodilation via myoglobin-dependent nitric oxide generation. *Circulation, 126*, 325–334.

Ullmann, B. D., Myers, H., Chiranand, W., Lazzell, A. L., Zhao, Q., Vega, L. A., et al. (2004). Inducible defense mechanism against nitric oxide in *Candida albicans. Eukaryotic Cell, 3*, 715–723.

Versari, D., Daghini, E., Virdis, A., Ghiadoni, L., & Taddei, S. (2009). Endothelial dysfunction as a target for prevention of cardiovascular disease. *Diabetes Care, 32*, S314–S321.

Walter, U., & Gambaryan, S. (2009). cGMP and cGMP-dependent protein kinase in platelets and blood cells. *Handbook of Experimental Pharmacology, 191*, 533–548.

Wang, Y., & Golledge, J. (2013). Neuronal nitric oxide synthase and sympathetic nerve activity in neurovascular and metabolic systems. *Current Neurovascular Research, 10*, 81–89.

Wang, Z. Q., Lawson, R. J., Buddha, M. R., Wei, C. C., Crane, B. R., Munro, A. W., et al. (2007). Bacterial flavodoxins support nitric oxide production by *Bacillus subtilis* nitric-oxide synthase. *The Journal of Biological Chemistry, 282*, 2196–2202.

Webb, A., Bond, R., McLean, P., Uppal, R., Benjamin, N., & Ahluwalia, A. (2004). Reduction of nitrite to nitric oxide during ischemia protects against myocardial ischemia–reperfusion damage. *Proceedings of the National Academy of Sciences of the United States of America, 101*, 13683–13688.

Wu, L., Chen, Y., Gao, H., Yin, J., & Huang, L. (2016). Cadmium-induced cell killing in *Saccharomyces cerevisiae* involves increases in intracellular NO levels. *FEMS Microbiology Letters, 363*, 1–5. https://doi.org/10.1093/femsle/fnw032.

Xu, S., Guerra, D., Lee, U., & Vierling, E. (2013). *S*-Nitrosoglutathione reductases are low-copy number, cysteine-rich proteins in plants that control multiple developmental and defense responses in Arabidopsis. *Frontiers in Plant Science 4*, 430. https://doi.org/10.3389/fpls.2013.00430.

Yoshikawa, Y., Nasuno, R., Kawahara, N., Nishimura, A., Watanabe, D., & Takagi, H. (2016). Regulatory mechanism of the flavoprotein Tah18-dependent nitric oxide synthesis and cell death in yeast. *Nitric Oxide, 57*, 85–91.

Zamora, R., Vodovotz, Y., & Billiar, T. R. (2000). Inducible nitric oxide synthase and inflammatory diseases. *Molecular Medicine, 6*, 347–373.

FURTHER READING

Abello, N., Kerstjens, H. A., Postma, D. S., & Bischoff, R. (2009). Protein tyrosine nitration: Selectivity, physicochemical and biological consequences, denitration, and proteomics methods for the identification of tyrosine-nitrated proteins. *Journal of Proteome Research, 8*, 3222–3238.

Bosworth, C. A., Toledo, J. C., Zmijewski, J. W., Li, Q., & Lancaster, J. R. (2009). Dinitrosyliron complexes and the mechanism(s) of cellular protein nitrosothiol formation from nitric oxide. *Proceedings of the National Academy of Sciences of the United States of America, 106*, 4671–4676.

Brown, G. C. (2001). Regulation of mitochondrial respiration by nitric oxide inhibition of cytochrome c oxidase. *Biochimica et Biophysica Acta, 1504*, 46–57.

Crow, J. P., Ye, Y. Z., Strong, M., Kirk, M., Barnes, S., & Beckman, J. S. (1997). Superoxide dismutase catalyzes nitration of tyrosines by peroxynitrite in the rod and head domains of neurofilament-L. *Journal of Neurochemistry, 69*, 1945–1953.

Doulias, P., Greene, J., Greco, T., Tenopoulou, M., Seeholzer, S., Dunbrack, R., et al. (2010). Structural profiling of endogenous *S*-nitrosocysteine residues reveals unique features that accommodate diverse mechanisms for protein S-nitrosylation. *Proceedings of the National Academy of Sciences of the United States of America, 107*, 16958–16963.

Gow, A. J., Farkouh, C. R., Munson, D. A., Posencheg, M. A., & Ischiropoulos, H. (2004). Biological significance of nitric oxide-mediated protein modifications. *American Journal of Physiology. Lung Cellular and Molecular Physiology, 287*, L262–L268.

Greco, T., Hodara, R., Parastatidis, I., Heijnen, H., Dennehy, M., Liebler, D., et al. (2006). Identification of *S*-nitrosylation motifs by site-specific mapping of the *S*-nitrosocysteine proteome in human vascular smooth muscle cells. *Proceedings of the National Academy of Sciences of the United States of America, 103*, 7420–7425.

Keszler, A., Zhang, Y., & Hogg, N. (2010). Reaction between nitric oxide, glutathione, and oxygen in the presence and absence of protein: How are *S*-nitrosothiols formed? *Free Radical Biology & Medicine, 48*, 55–64.

Martínez-Ruiz, A., & Lamas, S. (2004). *S*-Nitrosylation: A potential new paradigm in signal transduction. *Cardiovascular Research, 62*, 43–52.

Mengel, A., Chaki, M., Shekariesfahlan, A., & Lindermayr, C. (2013). Effect of nitric oxide on gene transcription—*S*-nitrosylation of nuclear proteins. *Frontiers in Plant Science, 4*, 293. https://doi.org/10.3389/fpls.2013.00293.

Nisoli, E., Clementi, E., Paolucci, C., Cozzi, V., Tonello, C., Sciorati, C., et al. (2003). Mitochondrial biogenesis in mammals: The role of endogenous nitric oxide. *Science, 299*, 896–899.

Romeo, A. A., Capobianco, J. A., & English, A. M. (2002). Heme nitrosylation of deoxyhemoglobin bys-nitrosoglutathione requires copper. *The Journal of Biological Chemistry, 277*, 24135–24141.

Russwurm, M., & Koesling, D. (2004). NO activation of guanylyl cyclase. *The EMBO Journal, 23*, 4443–4450.

Sarti, P., Forte, E., Mastronicola, D., Giuffrè, A., & Arese, M. (2012). Cytochrome c oxidase and nitric oxide in action: Molecular mechanisms and pathophysiological implications. *Biochimica et Biophysica Acta, 1817*, 610–619.

Sawa, T., Akaike, T., & Maeda, H. (2000). Tyrosine nitration by peroxynitrite formed from nitric oxide and superoxide generated by xanthine oxidase. *The Journal of Biological Chemistry, 275*, 32467–32474.

Seth, D., & Stamler, J. S. (2011). The SNO-proteome: Causation and classifications. *Current Opinion in Chemical Biology, 15*, 129–136.

Shinyashiki, M., Chiang, K. T., Switzer, C. H., Gralla, E. B., Valentine, J. S., Thiele, D. J., et al. (2000). The interaction of nitric oxide (NO) with the yeast transcription factor Ace1: A model system for NO-protein thiol interactions with implications to metal metabolism. *Proceedings of the National Academy of Sciences of the United States of America, 97*, 2491–2496.

Stubauer, G., Giuffrè, A., & Sarti, P. (1999). Mechanism of *S*-nitrosothiol formation and degradation mediated by copper ions. *The Journal of Biological Chemistry, 274*, 28128–28133.

Suliman, H. B., Babiker, A., Withers, C. M., Sweeney, T. E., Carraway, M. S., Tatro, L. G., et al. (2011). Nitric oxide synthase-2 regulates mitochondrial Hsp60 chaperone function during bacterial peritonitis in mice. *Free Radical Biology & Medicine, 48*, 736–746.

Thomas, M. L., Wu, X., Sellak, H., Dey, N., & Choi, C. S. (2006). Regulation of vascular smooth muscle cell phenotype by cyclic GMP and cyclic GMP-dependent protein kinase. *Frontiers in Bioscience, 11*, 356–367.

Turko, I. V., & Murad, F. (2002). Protein nitration in cardiovascular disease. *Pharmacological Reviews, 89*, 444–448.

Vanin, A. F., Malenkova, I. V., & Serezhenkov, V. A. (1997). Iron catalyzes both decomposition and synthesis of *S*-nitrosothiols: Optical and electron paramagnetic resonance studies. *Nitric Oxide, 1*, 191–203.

Wade, R. S., & Castro, C. E. (1990). Redox reactivity of iron(III) porphyrins and heme proteins with nitric oxide. Nitrosyl transfer to carbon, oxygen, nitrogen and sulfur. *Chemical Research in Toxicology, 3*, 289–291.

Wright, A. M., Wu, G., & Hayton, T. W. (2010). Structural characterization of a copper nitrosyl complex with a $\{CuNO\}^{10}$ configuration. *Journal of the American Chemical Society, 132*, 14336–14337.

CHAPTER THREE

The Inflammasome: Regulation of Nitric Oxide and Antimicrobial Host Defence

Rajalaksmy A. Ramachandran*, Christopher Lupfer†, Hasan Zaki*,[1]
*UT Southwestern Medical Center, Dallas, TX, United States
†Missouri State University, Springfield, MO, United States
[1]Corresponding author: e-mail address: hasan.zaki@utsouthwestern.edu

Contents

Abstract

Nitric oxide (NO) is a gaseous signalling molecule that plays diverse physiological functions including antimicrobial host defence. During microbial infection, NO is synthesized by inducible NO synthase (iNOS), which is expressed by host immune cells through the recognition of microbial pattern molecules. Therefore, sensing pathogens or their pattern molecules by pattern recognition receptors (PRRs), which are located at the cell surface, endosomal and phagosomal compartment, or in the cytosol, is key in inducing iNOS and eliciting antimicrobial host defence. A group of cytosolic PRRs is involved in inducing NO and other antimicrobial molecules by forming a molecular

Advances in Microbial Physiology, Volume 72
ISSN 0065-2911
https://doi.org/10.1016/bs.ampbs.2018.01.004

complex called the inflammasome. Assembled inflammasomes activate inflammatory caspases, such as caspase-1 and caspase-11, which in turn process proinflammatory cytokines IL-1β and IL-18 into their mature forms and induce pyroptotic cell death. IL-1β and IL-18 play a central role in immunity against microbial infection through activation and recruitment of immune cells, induction of inflammatory molecules, and regulation of antimicrobial mediators including NO. Interestingly, NO can also regulate inflammasome activity in an autocrine and paracrine manner. Here, we discuss molecular mechanisms of inflammasome formation and the inflammasome-mediated regulation of host defence responses during microbial infections.

1. INTRODUCTION

Innate immunity is the first line of defence against microbial infections. Upon encountering the microbes, phagocytic and antigen-presenting cells, such as neutrophils, macrophages, and dendritic cells, utilize their antimicrobial arsenals to kill invading microbes. The antimicrobial host defence response of the innate immune cells involves production of inflammatory and antimicrobial mediators, including nitric oxide (NO), reactive oxygen species (ROS), antimicrobial peptides (AMP), proteolytic enzymes, and cytokines. NO is a gaseous signalling molecule which participates in many biochemical reactions, causing posttranslational modification of proteins and DNA damage (Mayer & Hemmens, 1997; Moncada, Palmer, & Higgs, 1991; Nathan & Hibbs, 1991). Thus, production of NO is an essential effector mechanism for innate host defence against bacteria, viruses, fungi, and parasites (Akaike et al., 1996; Alam et al., 2002; Mishra et al., 2013; Nathan & Hibbs, 1991; Zaki, Akuta, & Akaike, 2005).

The induction of NO is initiated with the activation of several cell signalling pathways, such as NF-κB and mitogen-activated protein kinases (MAPKs), which includes ERK, P38, and JNK. The first step of the activation of these signalling pathways is recognition of the pathogen and their pattern molecules by immune cells through germline-encoded sensors called 'pattern recognition receptors' (PRRs) (Akira, Uematsu, & Takeuchi, 2006; Takeuchi & Akira, 2010). PRRs detect or sense the presence of pathogen-associated molecular patterns (PAMPs) such as microbial cell wall components, flagellin, nucleic acids, and bacterial secretion systems. In addition, PRRs sense many danger-associated molecular patterns (DAMPs) such as uric acid crystals or ATP, which are released by the infected and injured host cells. Based on the subcellular localization, PRRs can be membrane bound or cytosolic. Toll-like receptors (TLRs) and C-type lectin receptors (CLRs) are transmembrane proteins that detect PAMPs and DAMPs extracellularly or at the endosomal compartments. A second class of receptors, including RIG-I-like receptors (RLR), AIM2-like receptors (ALR), and nucleotide-binding domain and leucine-rich repeat-containing (NLR) proteins, are located in the cytoplasm (Kanneganti, Lamkanfi, & Nunez, 2007; Kawai & Akira, 2006; Paludan & Bowie, 2013; Takeuchi & Akira, 2010). Detection of PAMPs and DAMPs by most PRRs leads to the activation of transcription factors such as NF-κB, AP1, and interferon regulatory factors (IRFs) (Akira et al., 2006; Kawai & Akira, 2006). Activation of the transcription factors results in the production of proinflammatory cytokines, AMP, and free radicals such as NO and O_2^- (Akira et al., 2006; Takeuchi & Akira, 2010) (Fig. 1). A subset of NLRs and ALRs respond to cytosolic DAMPs and PAMPs via a distinct, nontranscriptional mechanism. They mediate the assembly of multimeric cytosolic protein complexes known as inflammasomes that activate inflammatory caspases, caspase-1 and caspase-11 (Lamkanfi, 2011; Martinon, Burns, & Tschopp, 2002; Martinon, Mayor, & Tschopp, 2009). Activated caspase-1 cleaves pro-IL-1β and pro-IL-18 resulting in their maturation and secretion of active IL-1β and IL-18 from the cell, and participate in cell death (Martinon et al., 2002).

The inflammasome has recently emerged as a central player of pathogenesis as well as host immune response against diverse diseases including microbial infections. The inflammasome-regulated cytokines IL-1β and IL-18 play critical roles in innate immunity via activation of immune cells, cell signalling pathways, and induction of inflammatory molecules (Dinarello, 1998, 2002, 2007; Franchi, Eigenbrod, Munoz-Planillo, & Nunez, 2009; Sedimbi, Hagglof, & Karlsson, 2013). NO is one of the proinflammatory mediators induced by cell signalling pathways involving the inflammasome. The inflammasome-mediated production of NO involves recruitment of IL-1β and IL-18 to their receptors, IL1R and IL18R, respectively, on the surface of myeloid cells. The recognition of these cytokines by their receptors leads to activation of the downstream cell signalling pathways such as NF-κB, ERK, P38, JNK, NFAT, T-bet, STAT, AKT, etc., leading to the induction of inducible nitric oxide synthase (iNOS) enzyme (Adachi et al., 1998; Dinarello, 2002, 2007; Kalina et al., 2000; Sedimbi et al., 2013; Tsujino et al., 1994) (Fig. 2). IL-18, which was identified as IFN-γ-inducing factors (Gu et al., 1997; Okamura et al., 1995), also induces

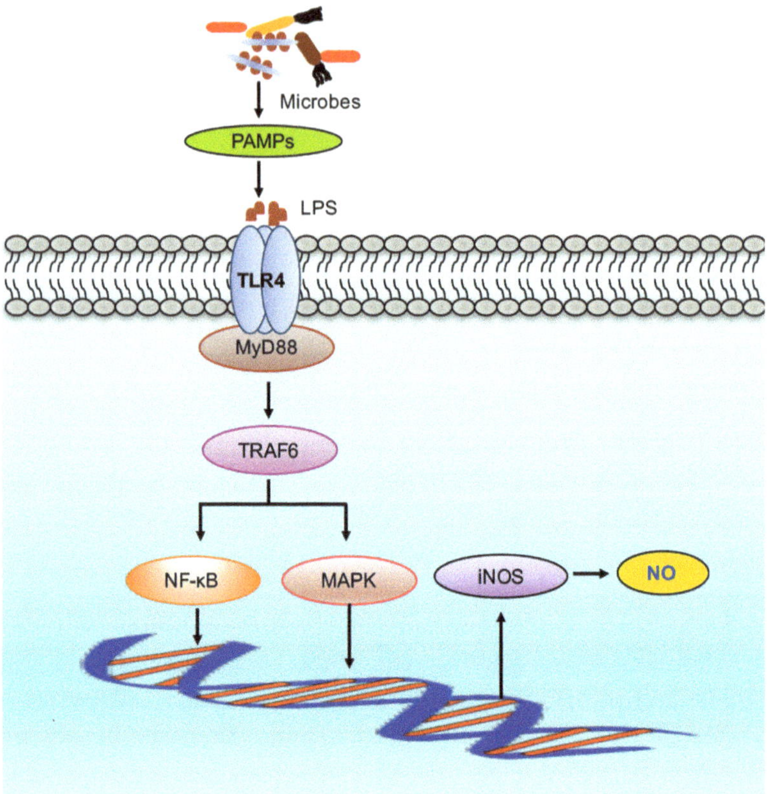

Fig. 1 Induction of iNOS upon recognition of pathogen by TLR. The recognition of pathogens or pathogen-associated molecular patterns (PAMPs) by TLRs results in the activation of MyD88-dependent signalling cascade, which leads to the activation of transcription factors of the NF-κB and MAPK pathways. The NF-κB and MAPK pathways regulate the expression of a myriad of inflammatory genes including iNOS.

iNOS indirectly via IFN-γ production (Maxwell et al., 2006). IL-18 triggers T cell activation and expression of IFN-γ, which is a potent inducer of iNOS through the activation of JAK/STAT1 pathways in the myeloid cells (Fig. 3). While the inflammasome serves as a trigger for NO production, NO also regulates inflammasome activity either positively or negatively (Chlichlia et al., 1998; Dimmeler, Haendeler, Nehls, & Zeiher, 1997; Li, Billiar, Talanian, & Kim, 1997; Li, Bombeck, Yang, Kim, & Billiar, 1999; Mao et al., 2013; Zhou et al., 2000). In this chapter, the latest insights into the mechanisms of inflammasome activation and its relation with NO in the context of antimicrobial host defence will be discussed.

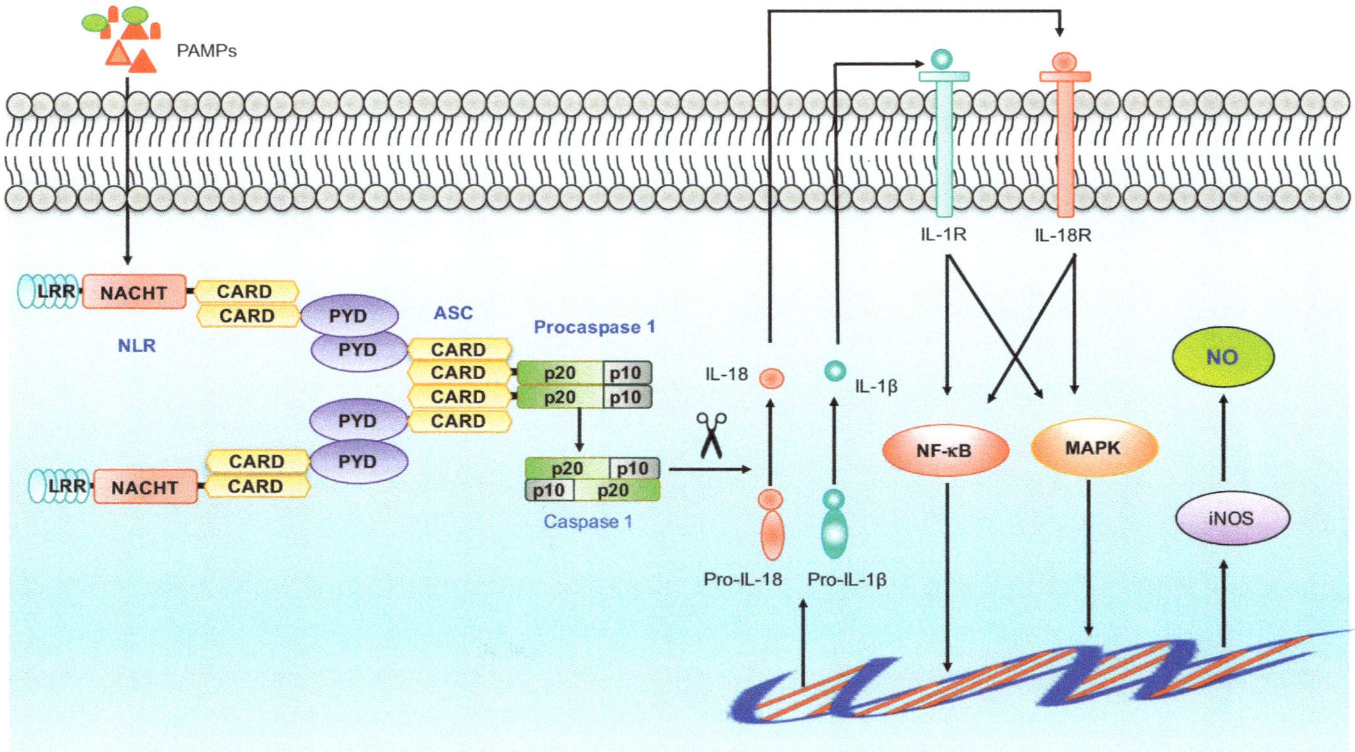

Fig. 2 The inflammasome is a regulator of iNOS. The formation of the inflammasome leads to the production of active IL-1β and IL-18, which are detected by their respective receptors on the cell surface. Activated IL-18 and IL-1β receptors triggers the activation of downstream cell signalling pathways leading to the induction of iNOS and production of NO.

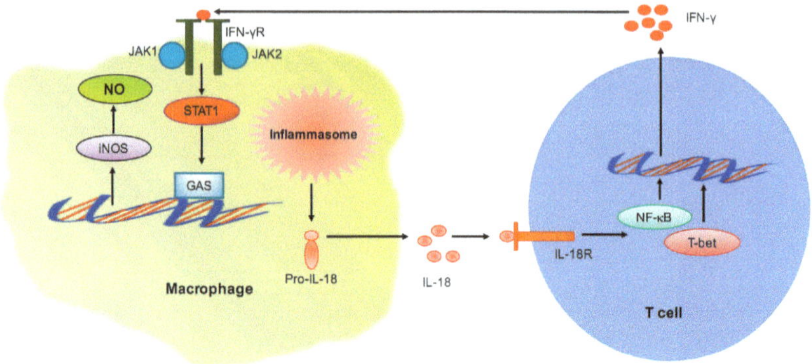

Fig. 3 IL-18 is a critical mediator of NO production. Matured IL-18 produced via inflammasome activation in macrophages binds to the IL-18R of Th1 cells, resulting in the activation of transcription factors NF-κB and T-bet and expression of IFN-γ. When IFN-γ is detected by the IFN-γR on macrophages and dendritic cells, the JAK/STAT1 pathway is activated, which leads to the nuclear translocation of STAT1 to gamma-activated sequence (GAS) elements, resulting in iNOS induction and NO production.

2. NO IS AN EFFECTOR MOLECULE IN ANTIMICROBIAL HOST DEFENCE

NO, a free radical produced by almost all types of cells, plays myriads of physiological functions including vascular and neuronal signal transduction, antimicrobial host defence, inflammation, posttranslational modification, cell death, cancer, etc. (Moncada et al., 1991; Murad, 1986; Nathan & Xie, 1994). The synthesis of NO is an enzymatic process involving NOS. There are three isoforms of NOS, namely, neuronal NOS (nNOS or NOS I), inducible NOS (iNOS or NOSII), and endothelial NOS (eNOS or NOSIII). nNOS and eNOS are constitutively expressed in neuronal and endothelial cells, respectively, in a calcium-dependent manner. eNOS-mediated production of NO is critical in maintaining vascular tone and smooth muscle relaxation (Bredt, Hwang, & Snyder, 1990; Murad, 1986). NO in the neuronal cell plays a role in memory, learning process, and signal transduction. On the other hand, iNOS is an inducible enzyme that does not need calcium and expressed in a wide variety of cells in response to infection, inflammation, and stress (Alderton, Cooper, & Knowles, 2001; Chakravortty & Hensel, 2003). All NOS utilize L-arginine as a substrate and molecular oxygen and reduced nicotinamide adenine dinucleotide phosphate (NADPH) as cosubstrates. Flavin adenine dinucleotide (FAD), flavin mononucleotide (FMN), and (6R-)5,6,7,8-tetrahydro-L-biopterin (BH_4) are similarly cofactors of all three isoenzymes (Moncada et al., 1991; Nathan & Xie, 1994). Functional NOS transfers electrons from NADPH, via the flavins FAD and FMN in the carboxy-terminal reductase domain, to the haem in the amino-terminal oxygenase domain (Alderton et al., 2001; Crane et al., 1998). The oxygenase domain also binds the essential cofactor BH4, molecular oxygen, and L-arginine. At the haem site, O_2 is reduced, which in turn oxidizes L-arginine to L-citrulline, producing NO. There are two steps for NO synthesis. In the first step, NOS hydroxylates L-arginine to *N*-hydroxy-L-arginine. In the second step, NOS oxidizes *N*-hydroxy-L-arginine to L-citrulline and NO (Mayer & Hemmens, 1997). All isoforms of NOS bind to calmodulin, and this facilitates the flow of electrons from NADPH in the reductase domain to the haem in the oxygenase domain. NO formed by NOS can then act on different enzymes and proteins; most notably, soluble guanylyl cyclase and the generation of cyclic guanosine monophosphate (GMP) as a secondary messenger (Murad, 1986).

Macrophages are the major cell type for iNOS induction, but other immune cells as well as epithelial cells and glial cells also express iNOS (Kleinert, Schwarz, & Forstermann, 2003). In the presence of microbes, the TLRs, NLRs, and other PRRs are activated by PAMPs such as LPS, MDP, flagellin, type III secretion systems, DNA, RNA, etc., leading to the activation of the NF-κB and MAPK pathways. The transcription factors NF-κB and AP-1 induce iNOS expression and NO synthesis (Xie, Whisnant, & Nathan, 1993). iNOS can also be induced by many cytokines such as IL-1β, IL-12, IL-18, IFN-γ, and TNFα (Fig. 2) (Drapier, Wietzerbin, & Hibbs, 1988; Kamijo et al., 1994; Martin, Nathan, & Xie, 1994). NO produced by iNOS primarily targets microbes in the phagolysosome compartment or in the cytosol. NO is cell permeable and, therefore, NO secreted by the activated macrophages can also kill extracellular pathogens. Multiple mechanisms are involved in NO-mediated microbicidal activity, including inhibition of enzymes of the mitochondrial electron transport chain and metabolic pathway, posttranslational modification of proteins involved in DNA replication, transcription and translation, and induction of DNA damage (Cooper, 2002; Grisham, Jourd'Heuil, & Wink, 1999; Mayer & Hemmens, 1997; Moncada et al., 1991; Nathan & Hibbs, 1991). All of these biochemical processes involving NO exert a cytotoxic effect on the pathogens. However, NO per se is not very reactive and depends on its reactive species for the biochemical reactions and modifications. The generation of RNS requires NO to react with ROS such as O_2^- and H_2O_2, leading to the formation of peroxynitrite ($ONOO^-$), nitrogen dioxide (NO_2), and other RNS (Beckman & Koppenol, 1996). RNS are biochemically highly reactive causing oxidation and nitration of proteins, lipids, and nucleic acids. The classic examples of reactivity of RNS is nitration and S-nitrosylation of proteins and S-guanylation of nucleic acids. Nitration of the amino acid tyrosine by $ONOO^-$ forming 3-nitrotyrosine has been observed in many pathophysiological contexts including microbial infections as detected by immunohistochemical analysis (Eiserich et al., 1998; Radi, 2004; Schopfer, Baker, & Freeman, 2003). Similarly, S-nitrosylation of protein occurs when NO derivatives react with cysteine amino acid (Akaike, 2000). Nucleic acids are also modified by nitration reaction when guanine or its derivatives are nitrated to form 8-nitroguanine, 8-nitroguanosine, and 8-nitro-cGMP (Akaike et al., 2003; Akuta, Zaki, Yoshitake, Okamoto, & Akaike, 2006; Sawa et al., 2003, 2007). All of these NO-mediated modifications of biological molecules play profound roles in disease pathogenesis as well as host defence response. The biochemical reactions involving NO not only targets host molecules but also pathogens. Such modifications are the basis for the antiviral or antimicrobial effects of NO (Akuta et al., 2006; James, 1995; Liew, 1995; Nathan & Hibbs, 1991; Zaki et al., 2005). Other indirect effects on pathogens include phagosome maturation and immune modulation of inflammatory cells to protect against infectious agents.

3. NOD-LIKE RECEPTORS

The first step of iNOS induction during microbial infection is recognition of PAMPs by PRRs of the host immune cells leading to the activation of inflammatory signalling pathways. The most well-characterized PRRs are TLRs which sense the pathogen and PAMPs at the cell surface or endosomal compartment leading to the activation of NF-κB and MAPK. Recently, NLRs, a family of cytosolic PRRs, have emerged as critical regulators of innate immune responses. The human NLR family has 22 proteins and at least 33 NLR genes have been identified in mice. NLRs are primarily expressed in lymphocytes and myeloid cells. However, NLRs can also be found in some nonimmune cells such as epithelial, endothelial, and mesothelial cells. NLRs share a common tripartite structure consisting of a central NOD (NACHT: NAIP, CIITA, HET-E, and TP-2), N-terminal effector domain, and C-terminal leucine-rich repeats (LRRs) (Ting et al., 2008). The NOD domain consists of seven different conserved motifs including the ATP/GTPase-specific P-loop, the Mg^{2+}-binding site, and five more-specific motifs. The central NOD domain is involved in the dNTPase activity and self-oligomerization of NLRs (Koonin & Aravind, 2000). The C-terminal LRR domain performs the ligand-sensing function, and the N-terminal domain is involved in the protein–protein interactions resulting in NLR-specific effector functions. Based on N-terminal domain structure, NLRs are broadly classified into four subfamilies: NLRA contains an acidic transactivation domain, NLRB has the baculoviral inhibitory repeat-like domain, NLRC possesses a caspase activation and recruitment domain, and NLRP has a pyrin domain (PYD) (Fig. 4) (Saxena & Yeretssian, 2014). In contrast to TLRs that are primarily involved in activation of NF-κB and MAPK, NLRs vary in their functions. Two NLR members, NOD1 and NOD2, activate the NF-κB and MAPK pathways in response to the bacterial cell wall component peptidoglycan. At least three other NLR proteins, namely, NLRP1, NLRP3, and NLRC4 are involved in the regulation of cell death and cytokine secretion by forming

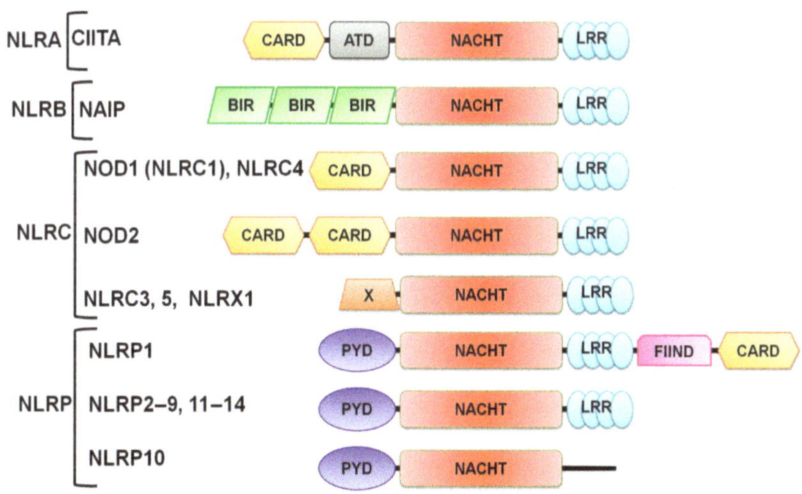

Fig. 4 Human NLR genes. Human NLR family of genes consist of 22 members which share a common three-domain structure with an N-terminal signalling domain, central nucleotide binding and oligomerization domain, and a C-terminal ligand-sensing domain. NLRs are subdivided into subgroups based on the N-terminal domain. NLRA has an acid transactivation domain (ATD), NLRB contains baculovirus IAP repeat (BIR) domain, and NLRC has caspase recruitment domain.

a multiprotein complex called the inflammasome. Some other NLRs, such as NLRP12 and NLRX1, were shown to negatively regulate cell signalling pathways involved in inflammation.

4. THE INFLAMMASOME

The term inflammasome was first coined in 2002 to describe a molecular machinery in the physiological system involved in the activation of inflammatory caspases (Martinon et al., 2002). Since then the inflammasome biology has been rapidly evolving to explain many pathophysiological processes. The inflammasome is composed of an inflammasome sensor, an adaptor protein called ASC, and caspase-1 (Martinon et al., 2002). Sensing PAMPs or DAMPs by inflammasome sensors leads to its oligomerization and recruitment of adapter ASC to its PYD. ASC is a bipartite protein consisting of a PYD and a caspase recruitment domain (CARD). ASC interacts with the inflammasome sensor via homotypic interaction of the PYD domains leaving the CARD domain to interact with the CARD domain of caspase-1. Caspase-1 is synthesized as an inactive form—procaspase-1, which must be catalytically processed into its active form. The recruitment of caspase-1 in the inflammasome complex allows its dimerization leading to its autoprocessing and formation of active caspase-1. Caspase-1 possesses multiple biological functions. The most important function of caspase-1 is catalytically processing precursor forms of cytokines IL-1β and IL-18 into their active and mature forms (Kanneganti et al., 2007; Lamkanfi & Dixit, 2014). Caspase-1 is also involved in a form of cell death called pyroptosis.

The inflammasome is formed in response to different pathophysiological stimuli. The first step of inflammasome activation is recognition of PAMPs and DAMPs by PRRs. There is distinct specificity in recognizing different PAMPs and DAMPs. Until now, five PRRs have been characterized as inflammasome sensors, including three NLR members—NLRP1, NLRP3, and NLRC4, and two non-NLR family proteins—absent in melanoma 2 (AIM2) and pyrin. Additional NLRs such as NLRP2, NLRP6, NLRP7, NLRP12 and ALR protein IFI16 are also suggested to assemble inflammasomes, but the precise mechanisms of inflammasome activation by these sensor proteins are not well defined. In addition to these canonical inflammasome pathways, recently noncanonical pathways of inflammasome activation have been described in which caspase-11, instead of caspase-1, is activated and participate in cell death (Kayagaki et al., 2011). Caspase-1 and -11 interact with other proteins via large prodomains. Interaction with the death domain superfamily motifs brings zymogens into the activating protein complex (Lamkanfi & Dixit, 2012; Riedl & Salvesen, 2007). These interaction domains typically consist of seven antiparallel helices. Based on the relative alignment of these domains, they are classified as CARD, PYD, death domain (DD), or death effector domain (DED). Exact structural rearrangements during procaspases-1 and -11 activation are not well characterized. Canonical inflammosome activation is a three-step process involving several features similar to caspase-9 activation. First, canonical inflammasomes recruit and promote the activation of a CARD-containing initiator caspase. Second, the ATP attachment results in a wheel structure with 7–8 symmetry (Duncan et al., 2007; Faustin et al., 2007; Riedl & Salvesen, 2007). Third is the recruitment of procaspase to the protein complex which results in the proteolytic cleavage to release active caspase. The mechanism of activation of the common inflammasome pathways are discussed later.

5. THE NLRP1 INFLAMMASOME

NLRP1 was the first NLR family member shown to form an inflammasome complex and caspase-1 activation (Martinon et al., 2002). Human NLRP1 consists an N-terminal PYD, followed by nucleotide-binding NOD/NACHT domain, five tandem LRR domains, a function-to-find domain (FIIND), and a C-terminal CARD (Chu et al., 2001). Using purified recombinant protein, NLRP1 was shown to form the inflammasome and caspase-1 activation in response to muramyl dipeptide (MDP), a component of peptidoglycan (Faustin et al., 2007). However, such an MDP-mediated activation of NLRP1 inflammasome has not been observed in mouse macrophages (Kovarova et al., 2012). The murine orthologues of human NLRP1 are *Nlrp1a*, *Nlrp1b*, and *Nlrp1c* (Boyden & Dietrich, 2006). Nlrp1b was shown to activate the inflammasomes during *Bacillus anthracis* infection (Boyden & Dietrich, 2006; Sastalla et al., 2013). Anthrax toxin is a 3-component exotoxin consisting of cell-binding protective antigen (PA) and effector proteins edema factor (EF) and lethal factor (LF). PA forms a channel that translocate the zinc metalloproteinase LF into the host cytosol, where it inactivates immune signalling by cleaving MAPK. LF activates the inflammasome by its metalloprotease activity (Fink, Bergsbaken, & Cookson, 2008) (Fig. 5). Recent animal studies show that NLRP1b can be cleaved at the N-terminal end by LF, which results in inflammasome formation and caspase-1 activation (Chavarría-Smith,

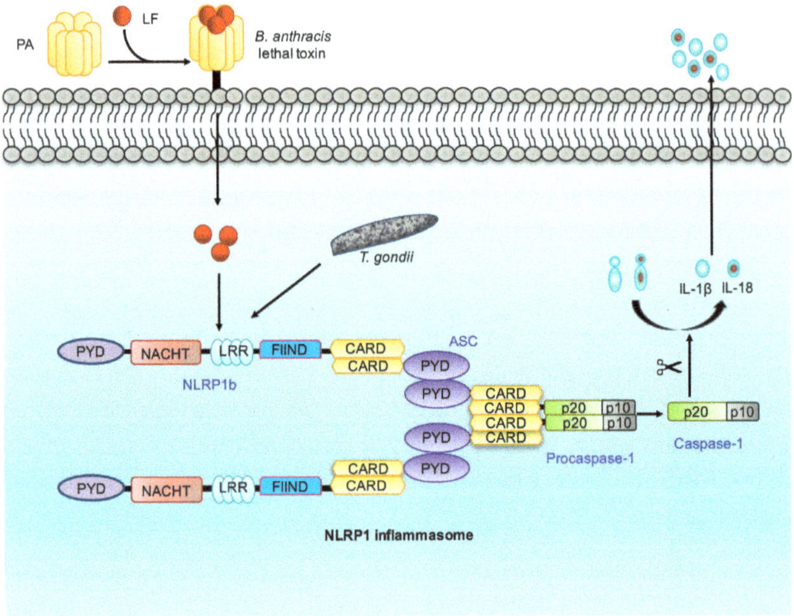

Fig. 5 The NLRP1 inflammasome. NLRP1 recognizes *Bacillus anthracis* lethal factor (LF) and *Toxoplasma gondii*. Sensing of LF or *T. gondii* by NLRP1 leads to the formation of NLRP1 inflammasome together with caspase-1 and ASC. Active caspase-1 formed via autoproteolytic processing cleaves the pro-IL-1β and pro-IL-18 into mature IL-1β and IL-18.

2013; Hellmich et al., 2012). NLRP1b can recruit caspase-1 via CARD–CARD domain interactions, but interaction between NLRP1b and the adaptor protein ASC (which contains both PYD and CADR domains) through PYD–PYD interactions stabilizes the inflammasome complex. Although ASC-mediated recruitment and dimerization of caspase-1 is a critical step for its activation, studies show that *B. anthracis* lethal toxin (LeTx) can mediate pyroptosis and IL-1β secretion even without ASC (Opdenbosch et al., 2014). Recently, Chavirra et al. showed that specific proteolytic cleavage of the N-terminal linker region was sufficient to activate human NLRP1, similar to mouse NLRP1b (Chavarria-Smith, Mitchell, Ho, Daugherty, & Vance, 2016) (Fig. 5). Apart from *B. anthracis*, NLRP1 serves as an inflammasome sensor for *Toxoplasma gondii*. However, in contrast to LeTx-mediated activation by *B. anthracis*, there was no NLRP1 proteolytic processing observed during *T. gondii* infection (Ewald, Chavarria-Smith, & Boothroyd, 2014). Thus, the exact mechanism of *T. gondii* recognition by NLRP1 is presently unclear.

NLRP1 is known to mediate sterile inflammation or inflammation in the absence of any detectable pathogen. NLRP1 polymorphisms are associated with autoimmune conditions such as arthritis and multiple sclerosis (MS). MS patients with a missense variant in the NLRP1 gene (G587S) showed hyperactivation of inflammatory pathways (Maver et al., 2017). Mice with the Q593P point mutation in the FIIND domain of *Nlrp1a* have constitutively active Nlrp1a. This leads to a systemic neutrophilic inflammatory disease mediated by excessive IL-1β production and pyroptosis of haematopoietic progenitor cells (Masters et al., 2012). The mice also develop cytopenia after lymphocytic choriomeningitis virus (LCMV) infection or chemotherapy. Further biochemical studies are required to understand the mechanisms behind NLRP1/NLRP1a activation mutations. A potential mechanism for sterile inflammation is the recent link between NLRP1 and endoplasmic reticulum (ER) stress, as transcription factor ATF4 induces NLRP1 expression during ER stress (D'Osualdo et al., 2015).

6. THE NLRP3 INFLAMMASOME

NLRP3 is clinically one of the most well-studied inflammasomes. NLRP3 responds to a diverse set of stimuli: microbial pathogens including viruses, bacteria, fungi, protozoa; crystalline and particulate structures such as uric acid crystals, silica, asbestos, and alum; extracellular ATP; bacterial RNA–DNA hybrids; and virulence factors such as pore-forming toxins (Broz & Dixit, 2016; Lamkanfi, 2011; Lupfer, Malik, & Kanneganti, 2015). Interaction of the N-terminal PYD of NLRP3 with the PYD of ASC leads to NLRP3 inflammasome assembly. The NLRP3 inflammasome requires two signals for its activation (Fig. 6). First is a priming signal that is required for the upregulation for NLRP3 and IL-1β and a second signal results in NLRP3 multimerization and inflammasome assembly. NLRP3 priming is initiated by PAMPs through TLRs, CLRs, RLRs, or other NLRs, which leads to NF-κB activation and induction of pro-IL-1β and NLRP3, and regulation of posttranslational modifications that facilitate NLRP3 activation (Lamkanfi & Dixit, 2014). NLRP3-activating agents then provide the second signal for inflammasome assembly (Lamkanfi, 2011). Various PAMPs and DAMPs of bacterial and viral pathogens activate NLRP3 (Franchi, Munoz-Planillo, & Nunez, 2012). In a few instances, priming can be bypassed. For example, acute LPS stimulation itself can activate the NLRP3 inflammasome independent of NLRP3 induction by NF-κB (Schroder et al., 2012) and in some conditions TLR pathway

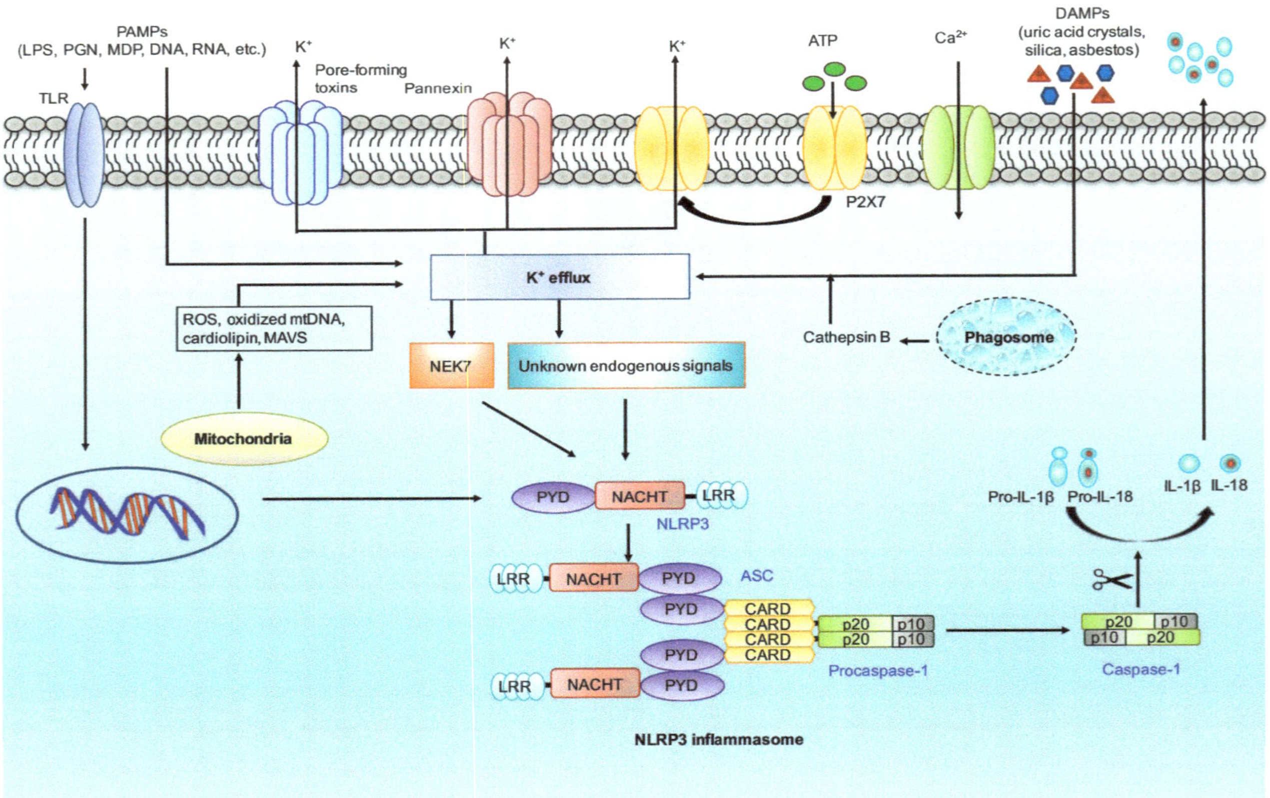

Fig. 6 The NLRP3 inflammasome. NLRP3 is activated by external stimuli such as uric acid crystals, silica, asbestos, extracellular ATP, pathogen-associated molecular patterns (e.g., LPS, PGN, RNA, DNA), and pore-forming toxins. Reactive oxygen species (ROS), mitochondrial oxidized DNA, cardiolipin, and mitochondrial antiviral-signalling protein (MAVS) released from the mitochondria and cathepsin released from the damaged phagosome also contribute to NLRP3 activation. A common physiological event in response to diverse stimuli for NLRP3 activation is potassium (K^+) efflux, which leads to induction of unknown signals. One of such signals downstream of K^+ efflux is the activation of NIMA-related kinase NEK7, which binds to NLRP3 leading to its oligomerization and formation of the inflammasome complex. Autoproteolytic processing releases active caspase-1, resulting in the maturation and release of IL-1β and IL-18.

adaptors MyD88, IRAK1, and IRAK4 directly activate NLRP3 inflammasome assembly independent of priming (Fernandes-Alnemri et al., 2013; Juliana et al., 2012; Lin et al., 2014). Apart from these signals, NLRP3 inflammasome activation can be regulated by deubiquitination (Juliana et al., 2012; Py, Kim, Vakifahmetoglu-Norberg, & Yuan, 2013).

Although priming of NLRP3 is fairly straightforward, the precise mechanism of NLRP3 inflammasome activation is not clear. Many models have been put forward to explain the underlying mechanism of NLRP3 activation. NLRP3 can be activated by the production of mitochondrial reactive oxygen, release of oxidized mitochondrial DNA and cardiolipin, cytosolic release of lysosomal cathepsins, changes in intracellular calcium concentration, and K^+ efflux from the cell (Fig. 6) (Ainscough, Gerberick, Kimber, & Dearman, 2015; Carlos et al., 2017; Iyer et al., 2013; Misawa et al., 2013; Muñoz-Planillo et al., 2013; Murakami et al., 2012; O'Neill, 2013; Orlowski et al., 2017; Saiga et al., 2012; Zhou, Yazdi, Menu, & Tschopp, 2011). Inflammasome activation results in NLRP3 and adaptor ASC colocalization with mitochondria, allowing the mitochondrial protein MAVS to regulate NLRP3 inflammasome activation (Subramanian, Natarajan, Clatworthy, Wang, & Germain, 2013). Some studies link various adaptor proteins such as thioredoxin (TRX)-interacting protein (TXNIP) (Zhou et al., 2011), guanylate-binding protein (GBP) 5 (Shenoy et al., 2012), linear ubiquitin assembly complex (LUBAC) component HOIL-1L (Rodgers et al., 2014), DExD/H box helicase family protein DHX33 (Mitoma et al., 2013), dsRNA-dependent protein kinase (PKR) (Chen et al., 2012), calcium-sensing receptors (Lee et al., 2012), mitochondrial protein mitofusin 2 (Ichinohe, Yamazaki, Koshiba, & Yanagi, 2013), serine–threonine kinase RIP1 and RIP3 (Wang et al., 2014), and NIMA-related kinase family protein NEK7 regulating the DNA damage and mitotic cell division (He, Zeng, Yang, Motro, & Nunez, 2016; Schmid-Burgk et al., 2016) all as positive regulators of NLRP3 inflammasome activation. Finally, intracellular K^+ concentration serves as a ckeckpoint to regulate the threshold of inflammasome assembly (Arlehamn, Petrilli, Gross, Tschopp, & Evans, 2010; Cain, Langlais, Sun, Brown, & Cohen, 2001). Truly, it is astounding how many signals and interactions appear to mediate NLRP3 inflammasome activation. The question that remains is whether there is a single defining mechanism that activates NLRP3, or if a multitude of adaptors and interacting partners facilitate sensing of the plethora of activating signals.

Negative regulation of NLRP3 is essential to maintain immune homeostasis. With so many activating signals, unintentional or unnecessary inflammasome activation would seem inevitable. Earlier studies showed that tripartite-motif protein 30 (TRIM30) negatively regulates NLRP3 inflammasome activation by modulating ROS production (Hu et al., 2010). NO inhibits NLRP3 assembly and will be discussed in greater detail later in the review (Mao et al., 2013). Carbon monoxide (CO) inhibits NLRP3-mediated IL-1β production by regulating mitochondria (Jung et al., 2015). The adaptor protein caspase recruitment domain-containing protein 8 (CARD8) negatively regulates NLRP3 activation (Ito, Hara, & Kubota, 2014). NF-κB inhibitor A20 inhibits NLRP3 activation and reduces pathology associated with arthritis (Vande Walle et al., 2014). Leucine-rich repeat Fli-I-interacting protein 2 (LRRFIP2) inhibits NLRP3 inflammasome activation by recruiting the caspase-1 inhibitor Flightless-I (Jin, Yu, et al., 2013). Nuclear receptor family protein SHP interacts with NLRP3 and negatively regulates inflammasome activation by regulating mitochondria (Yang et al., 2015). Aryl hydrocarbon receptor binds the xenobiotic response element in the NLRP3 promotor and inhibits transcription of NLRP3, thus limiting priming (Huai et al., 2014). IKKα kinase inhibits NLRP3 inflammasome activation by interacting with the adaptor protein ASC (Martin et al., 2014). Further, the NLRP3 inflammasome can be regulated by autophagy and microRNAs (Chuang et al., 2013; Deretic et al., 2015; Harris et al., 2011; Johnnidis et al., 2008; Lupfer et al., 2013; Pedraza-Alva, Perez-Martinez, Valdez-Hernandez, Meza-Sosa, & Ando-Kuri, 2015; Saitoh et al., 2008; Shi et al., 2012; van der Burgh et al., 2014; Zhou et al., 2015). In all, the reported mechanisms for inhibiting NLRP3 are just as numerous as those for activating this inflammasome.

7. THE NAIP/NLRC4 INFLAMMASOME

NLRC4 assembles inflammasomes in response to cytosolic flagellin from pathogenic bacteria (Franchi et al., 2006; Mariathasan et al., 2004; Miao et al., 2006) or proteins from bacterial type III and type IV secretion systems (Lightfield et al., 2011; Miao et al., 2006; Zhao et al., 2011). Activation of NLRC4 requires interaction with specific NAIPs (Kofoed & Vance, 2011; Rayamajhi, Zak, Chavarria-Smith, Vance, & Miao, 2013; Zhao et al., 2011). NAIP proteins have three sequential baculovirus inhibitor of apoptosis (BIR) domains in their N-terminus, a central NOD domain, and a C-terminal LRR domain (Herman et al., 2009; Hu, Zhou, et al., 2015). NAIP has seven paralogs in mice (NAIP1–NAIP7) but only one single gene in humans (NAIP1) (Yang, Zhao, Shi, & Shao, 2013). Specific NAIPs bind to distinct PAMPs and physically associate with NLRC4 leading to its activation. NLRC4 has an N-terminal CARD motif, which interacts with procaspase-1 via ASC (Broz, von Moltke, Jones, Vance, & Monack, 2010). Mouse NAIP5 and NAIP6 sense bacterial flagellin (Kofoed & Vance, 2011; Zhao et al., 2011), whereas NAIP1 and NAIP2 detect needle and rod components of the bacterial type III secretion system (T3SS) (Kofoed & Vance, 2011; Rayamajhi et al., 2013; Yang et al., 2013; Zhao et al., 2011). In humans, the single NAIP binds to bacterial T3SS (Yang et al., 2013; Zhao et al., 2011). Once activated, NAIPs bind to NLRC4, resulting in oligomeric complex formation with ASC and caspase-1, which leads to activation of caspase-1 and maturation of IL-18 and IL-1β (Fig. 7). Moreover, cathepsin B induced by the NAIP/NLRC4 inflammasome mediates pyroptosis (Lage et al., 2013). Exact molecular mechanisms behind NAIP/NLRC4 inflammasome formation are not very clear, but recent studies show a conserved N-terminal region from flagellin is capable of inducing inflammasome formation (Halff et al., 2012). Unlike other PRRs, ligand sensing by the NAIPs is not conferred by the LRR domain but by the NOD domain-associated α-helical domain (Tenthorey, Kofoed, Daugherty, Malik, & Vance, 2014). Posttranslational modifications, including phosphorylation, also contribute to NLRC4 activation (Qu et al., 2012). In general, interaction with NAIP and its ligands and the association of NLRC4 with NAIP induces conformational changes in these proteins, resulting in self-oligomerization and activation (Halff et al., 2012; Qu et al., 2012). Structural models for the NAIP/NLRC4 inflammasome suggest that these complexes contain multiple NLRC4 molecules for each NAIP and NLRC4 recruits procaspase-1 directly or via the ASC adaptor. Procaspase-1 and ASC interactions then result in autoproteolysis to release active caspase-1, whereas ASC-independent complexes directly activate procaspase-1 without autoproteolysis.

8. THE AIM2 INFLAMMASOME

Absent in melanoma 2 or AIM2 is a member of the interferon-inducible gene HIN-200 family. AIM2 has been characterized as a cytosolic DNA sensor (Burckstummer et al., 2009; Fernandes-Alnemri, Yu, Datta, Wu, & Alnemri, 2009; Hornung et al., 2009). AIM2 contains a

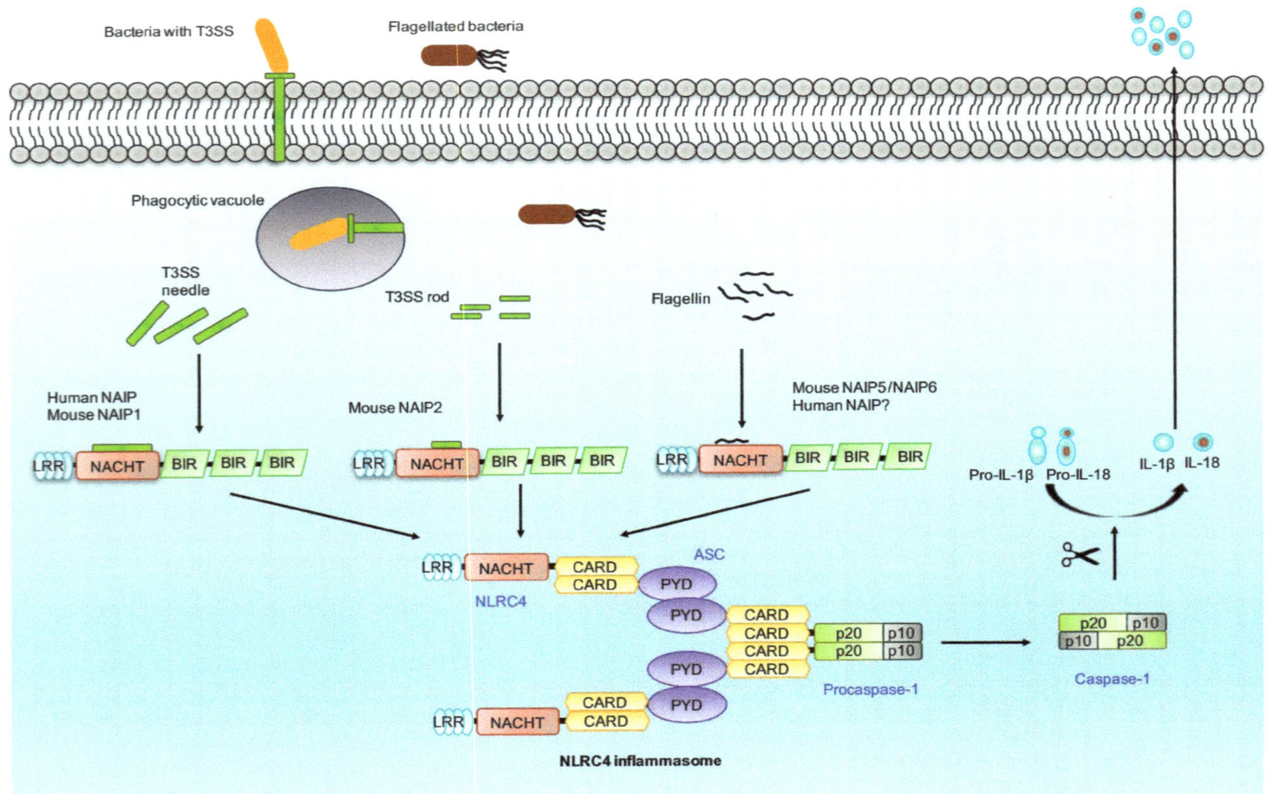

Fig. 7 The NLRC4 inflammosome. NLR apoptosis inhibitor proteins (NAIPs) act as direct receptors for bacterial flagellin and type III secretion systems (T3SS). Upon sensing needle and rod of T3SS and flagellin, NAIPs become activated and are recruited to NLRC4. Activated NLRC4 interacts with ASC and caspase-1 to form the NLRC4 inflammasome. The activated caspase-1 cleaves pro-IL-1β and pro-IL-18 to produce active IL-1β and IL-18.

C-terminal HIN-200 DNA-binding domain and an N-terminal PYD which interacts with the ASC adaptor protein (Jin et al., 2012). Sensing cytosolic DNA at its C-terminal leads to binding of ASC at its N-terminal PYD domain. The CARD domain of ASC then binds with the CARD of procaspase-1, and a functional inflammasome is formed (Fig. 8) (Man & Kanneganti, 2015). Many pathogens including *Listeria monocytogenes*, *Francisella tularensis*, *Streptococcus pneumoniae*, *Porphyromonas gingivalis*, and murine cytomegalovirus activate the inflammasome via AIM2 (Fang et al., 2014; Fernandes-Alnemri et al., 2010; Jones et al., 2010; Kim et al., 2010; Park et al., 2014; Rathinam et al., 2010). AIM2 recognizes DNA in a sequence-independent manner and requires approximately an 80 bp long DNA fragment to trigger activation (Morrone et al., 2015). dsDNA binds with the HIN-200 domain and PYD interacts with ASC to form a filamentous supramolecular inflammasome complex (Jin et al., 2012). Crystallograph studies demonstrate the PYD of AIM2 can self-oligomerize to activate the inflammasome (Hou & Niu, 2015;

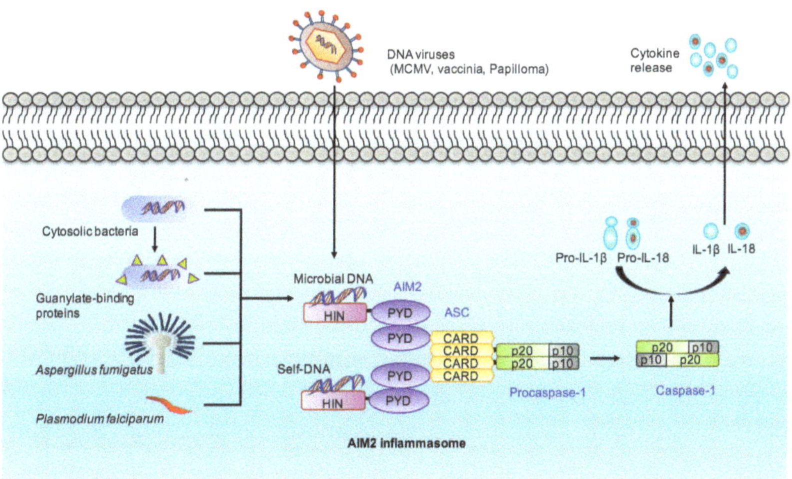

Fig. 8 The AIM2 inflammasome. AIM2 detects the cellular or microbial dsDNA in the cytosol and results in inflammosome activation. DNA viruses such as murine cytomegalovirus, vaccinia, and papilloma virus can activate AIM2. Many intracellular bacterial pathogens such as *Francisella tularensis* subsp. *Novicida*, *Listeria monocytogenes*, *Streptococcus pneumoniae*, *Mycobacterium tuberculosis*, *Porphyromonas gingivalis*, *Staphylococcus aureus*, *Brucella abortus*, and *Chlamydia muridarum* activate the inflammasome via AIM2. Recognition of bacterial and viral DNA as well as self-DNA leads to the recruitment of ASC and caspase-1 to AIM2, resulting in the formation of the AIM2 inflammasome.

Lu, Kabaleeswaran, Fu, Magupalli, & Wu, 2014). Earlier studies suggest that at homeostasis, PYD and HIN-200 domains of AIM2 form an intramolecular complex causing an autoinhibitory state (Jin, Perry, Smith, Jiang, & Xiao, 2013). Recent studies show that AIM2 is not autoinhibited. Instead, the size of the DNA serves as the 'molecular ruler' to regulate AIM2 oligomerization. DNA shorter than 80 bp does not provide a sufficiently large binding platform for AIM2 oligomerization and inflammasome formation (Morrone et al., 2015). Phosphorylation of the adaptor ASC also controls AIM2 inflammasome activation (Hara et al., 2013). Finally, type I IFN signalling through the IRF1 transcription factor triggers expression of IFN-inducible GBP, which are important for AIM2 inflammasome activation during bacterial infections (Fang et al., 2014; Henry, Brotcke, Weiss, Thompson, & Monack, 2007; Man & Kanneganti, 2015). GBPs appear to be required for lysing pathogens or exposing pathogen DNA and making it accessible to AIM2 for detection. (Finethy et al., 2015; Man, Place, Kuriakose, & Kanneganti, 2017).

It is not surprising that AIM2 is important for inflammasome activation for such a wide variety of pathogens due to its ligand, DNA. What is surprising is the involvement of AIM2 in the immune response to RNA virus infection. However, one report suggests that AIM2 recognizes host cell DNA released during viral infection, where extracellular host DNA may serve as a DAMP for the activation of AIM2. In agreement with this, mutations or expression changes in AIM2 are associated with autoimmune or autoinflammatory conditions. Increased AIM2 expression is associated with psoriasis (Dombrowski et al., 2011) and atopic dermatitis (de Koning et al., 2012). AIM2 overexpression and epigenetic modifications have been observed among systemic lupus erythematosus (SLE) patients (Javierre et al., 2010; Kimkong, Avihingsanon, & Hirankarn, 2009). Alternatively, mice deficient in *Aim2* are protected from many of the symptoms associated with lupus, such as nephritis (Kimkong et al., 2009). AIM2 hyperactivation contributes to neuronal death and acute brain injury (Adamczak et al., 2014; Bartesaghi et al., 2015). Finally, AIM2 can trigger autoinflammation by sensing self-DNA leading to polyarthritis (Jakobs, Perner, & Hornung, 2015). These reports show ample evidence that AIM2 responds to self-DNA, but it remains to be seen how this pathway becomes dysregulated in autoimmunity and autoinflammation.

AIM2 was initially recognized for its antitumor functions and was found to be mutated or deleted in melanoma (DeYoung et al., 1997). Clinical studies correlate reduced AIM2 expression with colorectal cancer and prostate cancer (Dihlmann et al., 2014; Ponomareva et al., 2013; Schulmann et al., 2005; Woerner et al., 2007). Whereas overexpression of AIM2 is associated with nasopharyngeal carcinoma, oral squamous cell carcinoma, and lung adenocarcinoma (Chen et al., 2012; Kondo et al., 2012; Kong et al., 2015). Recently, AIM2 was shown to play a critical role in suppression of inflammation and tumorigenesis in the colon (Hu, Peng, et al., 2015; Man, Karki, & Kanneganti, 2016; Wilson et al., 2015). Interestingly, inflammasome-independent function of AIM2 in the regulation of tumorigenesis was attributed to its role in inhibition of AKT via DNA-PK (Wilson et al., 2015). However, future studies should define the precise role of inflammasome-dependent and -independent functions of AIM2 in tumorigenesis.

9. PYRIN INFLAMMASOME

Pyrin (also known as marenostrin/TRIM200) is encoded by the *Mefv* gene and has a N-terminal PYD, 2 B-box zing finger domains, a coiled-coil domain, and a C-terminal B30.2 domain (Broz & Dixit, 2016). During infection, pyrin detects the posttranslational modification of cellular proteins carried out by bacterial toxins. Physiological functions of pyrin include the innate sensing of *Clostridium difficile* toxin B, *Clostridium botulinum* toxin C3, *Vibrio parahaemolyticus* effector protein VopS, and *Histophilus somni* IbpA (Xu et al., 2014). These bacterial toxins target the Rho subfamily of proteins involved in actin cytoskeleton rearrangement and modify the switch-I region to inactivate RHOA. This modification results in pyrin activation and inflammasome assembly. *Burkholderia cenocepacia* evades the host immune response by inactivating Rho GTPase resulting in diminished pyrin activation (Xu et al., 2014). Pyrin is also involved in IL-18 driven autoinflammation and is regulated by the actin-depolymerizing cofactor Wdr1 in mice (Kim et al., 2015). Thus, cytoskeletal signalling during pathogenic conditions may regulate inflammasome activation via pyrin.

Pyrin also belongs to the tripartite motif (TRIM) protein family, which mediates autophagy (Kimura et al., 2017, 2015). Pyrin suppresses inflammasome activation leading to precision autophagy and mutation in the *Mefv* gene can lead to impaired autophagy resulting in autoimmune pathogenesis. *Mefv* gene mutation is associated with the autoinflammatory condition familial Mediterranean fever (FMF). In this disease, a B30.2 mutation spontaneously induces an interaction with the adaptor ASC leading to inflammasome assembly, caspase-1 activation and IL-1β release (Chae et al., 2006; Kovarova et al., 2012). A separate study demonstrated that insertion of the mutant B30.2 domain into pyrin could induce inflammation via inflammasome activation (Chae et al., 2011). Another *Mefv* mutation resulting in the loss of the 14-3-3-binding motif at phosphorylated Ser32 is associated with an autoinflammatory condition characterized by childhood-onset of recurrent neurotrophic dermatosis and fever (Masters et al., 2016).

10. OTHER INFLAMMASOME COMPLEXES

There are a variety of other proposed inflammasomes such as NLRP6, NLRP12, and IFI16. Recent studies demonstrated that NLRP6 is a critical regulator of intestinal homeostasis. In the absence of NLRP6, ASC or caspase-1, mice developed dysbiotic microbiota and were hypersusceptible to chemically induced colitis and tumorigenesis (Elinav et al., 2011). Microbial metabolites produced in the gut microenvironment, such as taurine, spermine, and histamine, play positive or negative regulatory roles against NLRP6-mediated IL-18 production and subsequent inflammation (Levy et al., 2015). In addition, NLRP6 has been shown to suppress NF-κB and MAPK signalling (Anand et al., 2012), regulate goblet cell mucus secretion (Wlodarska et al., 2014), and participate in intestinal antiviral immunity (Wang et al., 2015). Thus, NLRP6 may have different roles in different pathological conditions based on the stimulus, but the specific stimuli that regulate these responses have not been identified yet.

Similar to NLRP6, NLRP12 is another example of NLR family member that plays both inflammasome-dependent and -independent functions. An early study demonstrated that NLRP12 is involved in the activation of NF-κB and caspase-1 (Wang et al., 2002). In contrast to this finding, other in vitro studies demonstrated that NLRP12 negatively regulates inflammatory response via suppression of the canonical and noncanonical NF-κB pathways (Arthur, Lich, Aziz, Kotb, & Ting, 2007; Lich et al., 2007; Williams et al., 2005). Consistently, mice deficient in Nlrp12 are susceptible to colitis, colorectal tumorigenesis, and autoimmune encephalitis (Allen et al., 2012; Lukens et al., 2015; Zaki et al., 2011). The protective function of inflammation and tumorigenesis was attributed to NLRP12-mediated negative regulation of macrophages, T cells, and nonmyeloid cells (Allen et al., 2012; Lukens et al., 2015; Zaki et al., 2011). Vladimir et al. first demonstrated that NLRP12 participate in inflammasome activation in response to *Yersinia pestis* infection (Vladimer et al., 2012). It is not clear what specific PAMPs of *Yersinia* activates the NLRP12 inflammasome. A role for NLRP12 in inflammasome activation and regulation of NF-κB has been tested in some other bacterial studies. While NLRP12 was shown to inhibit NF-κB-mediated inflammatory response during *Salmonella typhimurium* and *Brucella abortus*, such an effect was not observed in response to *Klebsiella* and *Mycobacterium* (Allen et al., 2013; Silveira et al., 2017; Zaki, Man, Vogel, Lamkanfi, & Kanneganti, 2014). An association of missense mutations in *NLRP12* has been observed in autoinflammatory diseases such as periodic fever syndrome and atopic dermatitis (Arthur et al., 2010; Borghini et al., 2011; Jeru et al., 2008; Macaluso et al., 2007). However, it is not clear whether inflammasome-dependent or -independent function of NLRP12 is involved in the pathogenesis of these autoinflammatory diseases.

The human PYHIN family protein IFI16 and its mouse orthologue IFI204 act as DNA sensors involved in the regulation of IFN production during viral and bacterial infections via the adaptor protein STING (Storek, Gertsvolf, Ohlson, & Monack, 2015; Unterholzner et al., 2010). IFI16 also interacts with the adaptor protein ASC and procaspase-1 to form an inflammasome during Kaposi's sarcoma virus infection (Kerur et al., 2011) and HIV infection (Doitsh et al., 2014). IFI16 detects the partially reverse transcribed HIV genome in the cytosol and evokes caspase-1-mediated pyroptosis in $CD4^+$ T cells, contributing to immunosuppression and AIDS progression (Doitsh et al., 2014; Monroe et al., 2014).

11. NONCANONICAL INFLAMMASOME ACTIVATION

The inflammasome activation pathways as discussed earlier are referred as canonical in which caspase-1 is recruited to the inflammasome sensors via ASC. However, with the discovery that *caspase-1*$^{-/-}$ mice are also deficient in caspase-11 (Kayagaki et al., 2011), researchers started looking into the roles of caspase-11 in the process of inflammasome activation. Previously, caspase-11 was shown to be essential for LPS-mediated septic shock in mice (Wang et al., 1998). Using bone marrow-derived macrophages from WT and *caspase-11*$^{-/-}$ mice, Kagyaki et al. showed that caspase-11 is dispensable for caspase-1 activation and IL-1β secretion during canonical NLRP3, NLRC4, and AIM2 inflammasome activation. However, cholera toxin B and several Gram-negative bacteria such as *Escherichia coli*, *Citrobacter rodentium*, and *Vibrio cholerae* require caspase-11 to be cleaved for IL-1β processing and secretion (Kayagaki et al., 2011). They used the term noncanonical inflammasome activation to refer the caspase-11-dependent pyroptosis and IL-1β production (Fig. 9). In this pathway, caspase-11 cleaves IL-1β through the activation of caspase-1. Interestingly, caspase-1 activation by caspase-11 is not required

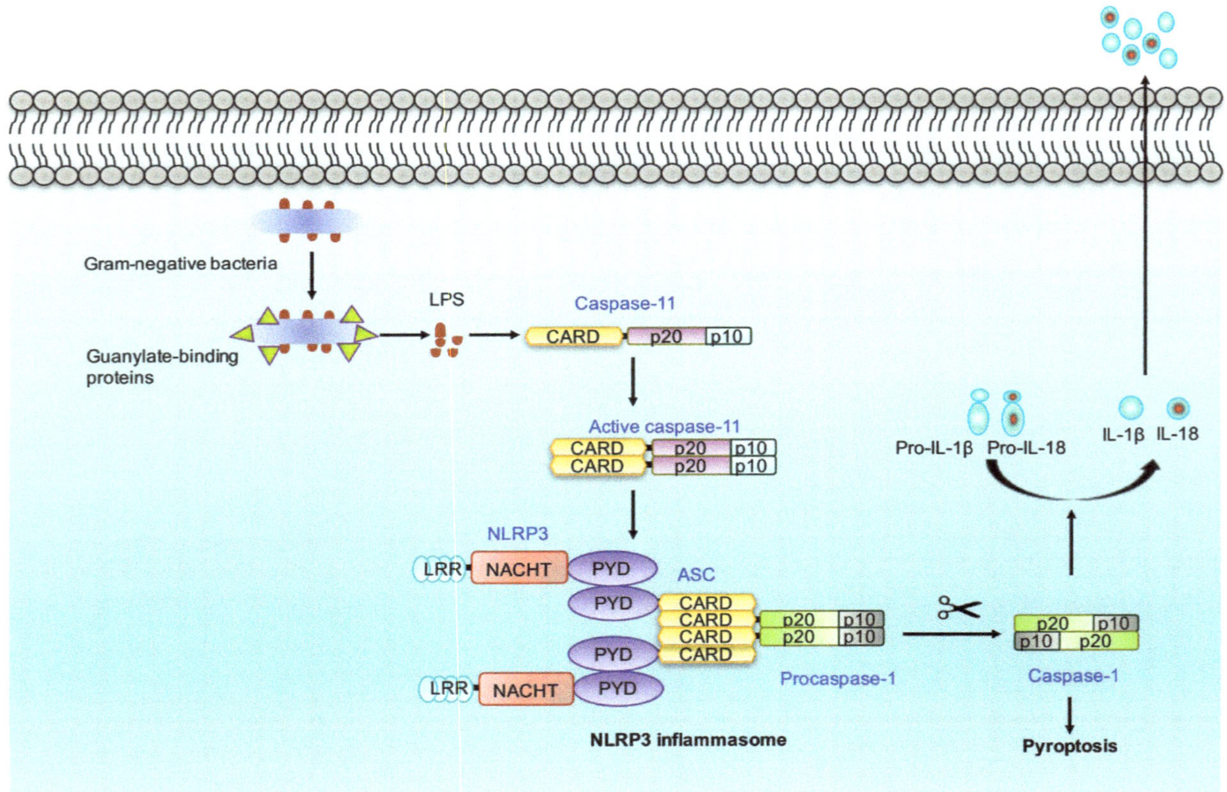

Fig. 9 Noncanonical inflammasome activation. LPS from Gram-negative bacteria activates caspase-11 in mice and caspase-4 or -5 in human which results in the noncanonical NLRP3 activation and oligomerization. Guanylate-binding proteins lyse the bacterial cell and mediate the LPS recognition to activate caspases, resulting in pyroptosis. Activated caspase-11 induces the noncanonical NLRP3 inflammasome formation. Activated caspase-1 cleaves pro-IL-1β and pro-IL-18 into mature IL-1β and IL-18, and induces pyroptosis.

for the noncanonical inflammasome-mediated pyroptotic cell death. Noncanonical inflammasome activation of caspase-11 is essential for the LPS-mediated septic shock (Kayagaki et al., 2011). Caspase-11 expression could be upregulated by the double-stranded RNA mimic poly I:C, and then LPS treatment-mediated lethality even in Tlr4-deficient mice showing that this was not merely an effect of inflammasome priming (Hagar, Powell, Aachoui, Ernst, & Miao, 2013; Kayagaki et al., 2013). The results from these experiments showed that the noncanonical inflammasome component caspase-11 drives LPS-induced endotoximea independent of the TLR4 pathway (Hagar et al., 2013; Kayagaki et al., 2013). In fact, caspase-11 binds directly to LPS in the cytoplasm and does not require other PRRs for ligand recognition (Shi et al., 2014). Caspase-11 is absent in human and the human orthologues of caspase-11 are caspase-4 and -5. Similar to caspase-11, caspase-4 and -5 were shown to participate in LPS-mediated activation of caspase-1 and septic shock (Shi et al., 2014). However, how caspase-11 is involved in pyroptossis and IL-1β secretion remained unresolved until recently when it was shown that gasdermin D acts as a downstream target of caspase-11 and is involved in noncanonical inflammasome activation. Using forward genetic screening, *gasderminD*$^{-/-}$ mice were shown to be resistant to LPS-induced septic shock and defective in noncanonical inflammasome activation and IL-1β secretion (Kayagaki et al., 2015). Further studies showed that caspase-11 cleaves gasdermin D and the amino-terminal fragment of gasdermin D is responsible for pyroptosis and Nlrp3-dependent activation of caspase-1 (Kayagaki et al., 2015). However, *gasderminD*$^{-/-}$ macrophages undergo apoptosis following prolonged caspase-1 activation, suggesting that additional caspase-1 substrates are involved in pyroptosis during noncanonical inflammasome activation (Kayagaki et al., 2015).

12. THE INFLAMMASOME IN ANTIMICROBIAL HOST DEFENCE

Being a sensing entity for pathogen and PAMPs, the inflammasome plays a central role in host defence against microbial infection. As discussed earlier, multiple inflammasome pathways exist in our immune system and pathogens activate the inflammasome in a selective manner. The NLRP1b inflammasome was the first characterized inflammasome which was shown to be activated by the *B. anthracis* LeTx (Kovarova et al., 2012; Moayeri et al., 2010). NLRP1b inflammasome activation was shown to be important in host defence mechanism against *B. anthracis*, because defective activation of the NLRP1b inflammasome hampers host defence in mice infected with live *B. anthracis* spores (Moayeri et al., 2010). NLRP1b inflammasome activation mediates LeTx-induced pyroptosis in macrophages (Kovarova et al., 2012; Moayeri et al., 2010).

The biological function of the inflammasome during bacterial infection has been extensively studied using *S. typhimurium*, a Gram-negative bacteria causing enterocolitis in human and salmonellosis, a disease that resembles human typhoid fever, in mice. Initial studies demonstrated that *S. typhimurium* activates the NLRC4 inflammasome through its flagellin and type III secretion system (Franchi et al., 2006; Mariathasan et al., 2004). The Type III secretion system, encoded by *Salmonella* pathogenicity island I (SPI), ensures delivery of flagellin into the cytosol (Sun, Rolan, & Tsolis, 2007). *Salmonella* expresses SPI at logarithmic growth phase and SPI expressing *Salmonella* induces pyroptotic cell death in macrophages very rapidly by the NLRC4 inflammasome (Mariathasan et al., 2004). However, cell death and caspase-1 activation are delayed in macrophages infected with stationary phase grown *S. typhimurium*, which expresses SPII (Broz, Newton, et al., 2010). Interestingly, persistent *S. typhimurium* infection in macrophages induces both NLRP3 and NLRC4 inflammasomes, which play redundant role in host defence (Broz, Newton, et al., 2010). Mice deficient in either Nlrp3 or Nlrc4 are not susceptible to *Salmonella* infection, while double-deficient mice or caspase-1-deficient mice showed increased mortality and bacterial burden (Lara-Tejero et al., 2006; Raupach, Peuschel, Monack, & Zychlinsky, 2006). It was shown that caspase-1-mediated pyroptotic cell death plays an important role in host defence since cell death of infected macrophages allows the *Salmonella* to be released into extracellular milieu that can be targeted by other immune cells and ROS (Miao et al., 2010). However, the precise mechanism of inflammasome-mediated antimicrobial host defence is still obscure. A recent study reported that caspase-11 is involved in *Salmonella*-induced cell death, but this process is not important for host defence against *Salmonella*. Therefore, cell death-independent mechanisms involving IL-1β- and IL-18-mediated innate immunity were proposed to be critical for clearing *Salmonella* in vivo (Broz et al., 2012).

Inflammasome-mediated caspase-1 activation also plays an important role in innate host defence against intracellular bacteria such as *Listeria monocytogenes* and *Legionella pneumophila*. *L. monocytogenes*, a Gram-positive intracellular bacterium, can activate multiple inflammasome pathways including NLRP3, NLRC4, and AIM2 using listeriolysin O (LLO), flagellin, and dsDNA, respectively (Warren, Mao, Rodriguez, Miao, & Aderem, 2008; Wu, Fernandes-Alnemri, & Alnemri, 2010). The importance of inflammasome activation in host defence against *Listeria* was evident in study showing that *caspase-1*$^{-/-}$ mice are susceptible to *Listeria* infection (Tsuji et al., 2004). *Listeria* escapes phagocytic killing by producing LLO, which cause membrane disruption. Mariathasan et al. showed that NLRP3 and ASC are essential for *Listeria*-induced caspase-1 activation and IL-1β production (Mariathasan et al., 2006). *L. pneumophila* is a Gram-negative bacteria that causes Legionnaires disease, a form of pneumoniae in human. When *L. pneumophila* reaches the human lung, it infects alveolar macrophages where the bacterium secrets type IV secretion system Dot/Icm, which recruits endoplasmic reticulum components to the *Legionella*-containing vacuole (LCV). This process inhibits acidification and lysosomal degradation of LCV, allowing replication of the bacteria in the LCV (Zamboni et al., 2006). However, flagellin of *L. pneumophila* activates the inflammasome via NLRC4, which causes pore formation on the infected macrophages, leading to pyroptotic cell death and the release of *L. pneumophila.* It was shown that inflammasome activation plays a critical role in restricting *L. pneumophila* infection both in macrophages and in vivo (Amer et al., 2006; Molofsky et al., 2006; Pereira et al., 2011; Ren, Zamboni, Roy, Dietrich, & Vance, 2006; Zamboni et al., 2006).

Several other bacteria such as *Shigella flexneri*, *Pseudomonas aeruginosa*, and *Burkholderia thailandensis* also activate the inflammasome via NLRC4 which play a role in the host defence against these pathogens. In addition to the typical inflammasome effector mechanisms such as pyroptosis and cytokine production, several other physiological processes are involved in NLRC4-dependent antimicrobial host defence. For example, Naip5/Nlrc4 induces iNOS upon sensing bacterial flagellin. Interestingly, caspase-1, but not IL-18 and IL-1β, was shown to be involved in iNOS induction (Fig. 10). However, the signal transduction pathway in iNOS induction involving Naip5/Nlrc4-caspase-1 is yet to be explored (Buzzo et al., 2010). A separate study showed that cytosolic flagellin activates Naip5/Nlrc4 inflammasome to release prostaglandins and leukotrienes via activation of phospholipase A2, resulting in robust inflammation and vascular fluid loss in vivo. IL-1β and IL-18, but not caspase-1, are dispensable in such physiological processes (von Moltke et al., 2012). NLRC4 was also shown to contribute to phagosomal maturation and autophagosome formation (Fortier, de Chastellier, Balor, & Gros, 2007; Jounai et al., 2011). *Salmonella typhimurium* infection induced HMGB-1 production in an NLRC4-dependent manner (Gross et al., 2012). HMGB-1 is a nucleoprotein that functions as an

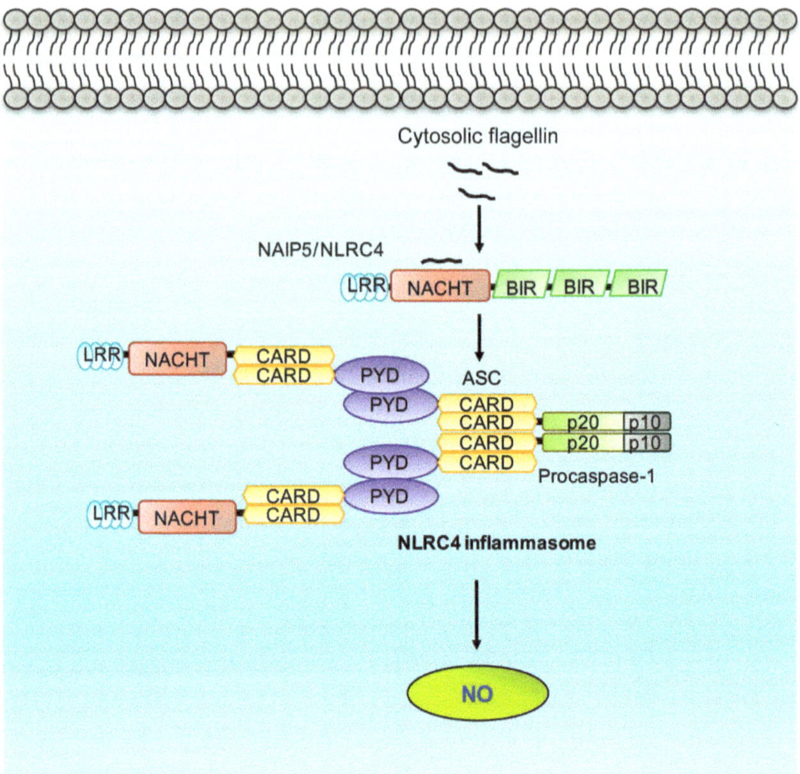

Fig. 10 NAIP5/NLRC4 induces NO production. NAIP5/NLRC4 detects the cytosolic bacterial flagellin resulting in the production of NO, which plays critical roles in antimicrobial host defense.

inflammatory mediator when released from damaged cells. NLRC4 inflammasome activation in dendritic cells by *S. typhimurium*, *Yersinia pseudotuberculosis*, and *P. aeruginosa* inflammasome plays a key role in T cell-mediated immunity. IL-18 produced by activated dendritic cells triggers IFN-γ production by memory $CD8^+$ T cells (Kupz et al., 2012).

The NLRP3 inflammasome has been shown to protect the host from influenza virus infection. Two separate studies showed that mice deficient in Nlrp3 are highly susceptible to influenza virus infection with increased necrotic damage in the alveoli (Allen et al., 2009; Thomas et al., 2009). Although the precise mechanism of NLRP3-mediated protection against influenza is not clear, caspase-1 activation and production of IL-1β and IL-18 have been linked to the host defence function. There was reduced expression of chemokines, neutrophil recruitment, and decreased cell death in *Nlrp3*$^{-/-}$, *ASC*$^{-/-}$, and *caspase-1*$^{-/-}$ mice (Allen et al., 2009; Thomas et al., 2009). A previous study showed the beneficial role for IL-1β in influenza virus clearance (Schmitz, Kurrer, Bachmann, & Kopf, 2005). IL-1β and IL-18 also critical modulator of T cell response which play an important role in host defence against viral infection (Lupfer et al., 2015). However, there are conflicting results showing that ASC, caspase-1, and IL-1R, but not Nlrp3, are essential for antiinfluenza response (Ichinohe, Lee, Ogura, Flavell, & Iwasaki, 2009). The protective function of the inflammasome was attributed to maintaining T cell response (Ichinohe et al., 2009). Furthermore, higher activation of Nlrp3 in *Rip2*$^{-/-}$ mice was shown to be pathogenic but not beneficial (Lupfer et al., 2013). The differential role of NLRP3 in influenza virus infection may lie on the difference in gut microbiota because gut microbial composition regulates immune response against influenza (Chen, Wu, Kuo, & Shih, 2017).

AIM2 is crucial for host defence against pathogenic bacteria such as *Francisella tularensis* subsp. *Novicida*, *Listeria monocytogenes*, *Streptococcus pneumoniae*, *Mycobacterium tuberculosis*, *Porphyromonas gingivalis*, *Staphylococcus aureus*, *Brucella abortus*, and *Chlamydia muridarum* (Fang et al., 2014; Fernandes-Alnemri et al., 2010; Finethy et al., 2015; Hanamsagar, Aldrich, & Kielian, 2014; Jones et al., 2010; Marim et al., 2017; Park et al., 2014; Rathinam et al., 2010; Saiga et al., 2012). Bacterial virulence factors, such as pore-forming complexes or membrane lysing proteins, can contribute towards bacterial invasion into the cytosol and AIM2 activation (Tsuchiya et al., 2010). Aside from bacteria, the AIM2 inflammasome is critical for detecting DNA viruses such as murine cytomegalovirus (MCMV), vaccinia virus, and human papilloma virus (Gerlic et al., 2013; Rathinam et al., 2010; Reinholz et al., 2013). Recent reports further suggest involvement of AIM2-mediated immune responses against RNA viruses such as influenza A virus (Schattgen, Gao, Kurt-Jones, & Fitzgerald, 2016; Zhang et al., 2017), West Nile virus, and Chikungunya virus (Ekchariyawat et al., 2015). Finally, AIM2 is involved in pathogen recognition of the fungus *Aspergillus fumigatus* and the intracellular parasite *Plasmodium falciparum* (Kalantari et al., 2014; Karki et al., 2015).

13. INFLAMMASOMES IN THE REGULATION OF INTESTINAL MICROBIOME AND INFLAMMATORY DISORDERS

The role of inflammasomes has been extensively studied in the context of intestinal inflammatory disorders. Dysregulated immune response to the microbes or PAMPs leads to the development of gastrointestinal inflammatory disorders, for example, *Helicobacter pylori*-induced gastritis, Crohn's disease (CD), ulcerative colitis (UC), irritable bowel syndrome, and colorectal cancer (CRC). Since IL-1β and IL-18 are proinflammatory cytokines, inflammasome activation was considered a trigger for inflammatory disorders. Indeed, higher expression of IL-18 was detected in CD patients' intestinal mucosa (Pizarro et al., 1999). In contrast, several experimental studies in mouse models observed an essential role of the Nlrp3 inflammasome in the protection against colitis (Allen et al., 2009, 2010; Dupaul-Chicoine et al., 2010; Hirota et al., 2011; Zaki, Boyd, et al., 2010; Zaki, Vogel, Body-Malapel, Lamkanfi, & Kanneganti, 2010). Mice deficient in Nlrp3, ASC, and caspase-1 developed greater colitis following administration of dextran sodium sulphate (DSS). In addition to NLRP3, NLRP6 and AIM2 were also shown to participate in inflammasome activation in the intestine, providing protective function against colitis (Chen, Liu, Wang, Bertin, & Nunez, 2011; Elinav et al., 2011; Hu, Peng, et al., 2015; Levy et al., 2015; Man et al., 2016; Wilson et al., 2015). Increased colitis susceptibility of *Nlrp3*$^{-/-}$, *Nlrp6*$^{-/-}$, and *Aim2*$^{-/-}$ mice was associated with reduced levels of IL-18 (Dupaul-Chicoine et al., 2010; Elinav et al., 2011; Hu, Peng, et al., 2015; Zaki, Boyd, et al., 2010). The beneficial role of IL-18 in maintaining mucosal tissue homeostasis was evidenced by the experimental data showing reduced colitis pathogenesis in *caspase-1*$^{-/-}$ mice receiving recombinant IL-18 during DSS-induced colitis (Dupaul-Chicoine et al., 2010; Zaki, Boyd, et al., 2010). Chronic inflammation is a risk factor for CRC, and more than 10% inflammatory bowel disease (IBD) patients develop CRC. Consistently, *caspase-1*$^{-/-}$, *Nlrp3*$^{-/-}$, *Aim2*$^{-/-}$, and *Il18*$^{-/-}$ mice developed greater tumour burden in the colon following treatment with azoxymethane plus DSS (Man et al., 2016; Salcedo et al., 2010; Wilson et al., 2015; Zaki, Vogel, et al., 2010). All these studies strongly support that the inflammasome/IL-18 signalling axis plays a central role in intestinal mucosal homeostasis.

Although the precise mechanism is not clear, the inflammasome/IL-18 may suppress intestinal inflammation and tumorigenesis via regulation of gut microbiota and epithelial cell proliferation. Higher levels of Prevotellaceae and TM7 were detected in *ASC*$^{-/-}$, *caspase-1*$^{-/-}$, and *Nlrp6*$^{-/-}$ mice, suggesting that the inflammasome regulates gut microbiota composition (Elinav et al., 2011). However, the link between increased Prevotellaceae and TM7 with reduced activation of the inflammasome was not confirmed by other studies (Mamantopoulos et al., 2017; Seregin et al., 2017). *Nlrp6*$^{-/-}$ and *Aim2*$^{-/-}$ mice were shown to carry increased levels of *Akkermansia muciniphila* in their gut as compared to wild-type mice

(Man et al., 2015; Seregin et al., 2017). Other studies reported an elevated levels of Enterobacteriaceae in inflammasome-defective mice (Hirota et al., 2011; Hu, Peng, et al., 2015). Consistent to this observation, mice defective in inflammasomes are susceptible to *Citrobacter rodentium* infection (Liu et al., 2012; Wlodarska et al., 2014). The observed discrepancy in microbial composition in mice defective in inflammasome signalling emphasizes that host genetics is not the sole determinant of gut microbial ecology, but other factors such as diet and environment greatly influence the gut microflora. Nevertheless, different line of evidence suggests that the inflammasome is a critical regulator of gut microbiota. In addition to triggering dysbiosis, inflammasome deficiency was shown to increase epithelial barrier permeability, which is linked to defective epithelial cell proliferation during chemical injury (Dupaul-Chicoine et al., 2010; Zaki, Boyd, et al., 2010). Rapid proliferation of epithelial stem cells, located at the base of the intestinal crypt, is a repair mechanism of intestinal injury (Radtke & Clevers, 2005; Rakoff-Nahoum, Paglino, Eslami-Varzaneh, Edberg, & Medzhitov, 2004). Indeed, reduced epithelial cell proliferation in inflammasome-deficient mice leads to translocation of commensal microflora into the lamina propria and further dissemination to other organs like liver and spleen, leading to septicemia and mortality (Zaki, Boyd, et al., 2010).

14. INVOLVEMENT OF NO IN THE EFFECTOR MECHANISM OF INFLAMMASOME-MEDIATED HOST DEFENCE

Increasing evidence suggests an involvement of NO in inflammasome-mediated regulation of host immune response. Reduced expression of iNOS was observed in *Nlrp3*$^{-/-}$ and *caspase-1*$^{-/-}$ mice following DSS-induced colitis (Zaki, Vogel, et al., 2010). Defective NO production in *Nlrp3*$^{-/-}$ and *caspase-1*$^{-/-}$ mice may cause increased bacterial growth in the intestinal lumen during colitis (Zaki, Boyd, et al., 2010). Indeed, the inflammasome is a trigger for antimicrobial host defence response of intestinal epithelial cells. Increased susceptibility of *Aim2*$^{-/-}$ and *caspase-1*$^{-/-}$ mice was associated with reduced expression of Reg3β, Reg3γ, S1008, S1009, and Lcn2 (Hu, Peng, et al., 2015). Defective expression of several AMP was also observed in *Nlrp6*$^{-/-}$ mouse colons during colitis (Levy et al., 2015). These observations were supported by an in vitro study showing that intestinal epithelial cells express AMPs and iNOS upon stimulation with IL-18 (Udden, Waliullah, Harris, & Zaki, 2017). Culture supernatants of IL-18-stimulated cells showed microbicidal activity against *E. coli* (Udden, Waliullah, et al., 2017). Consistently, elevated levels of *E. coli* are present in *caspase-1*$^{-/-}$ and *Il18*$^{-/-}$ mice (Hu, Peng, et al., 2015). The observed role of the inflammasome/IL-18 in controlling *E. coli* growth is clinically significant since increased *E. coli* burden has been observed in IBD and CRC patients' colon (Martin et al., 2004; Martinez-Medina et al., 2014; Martinez-Medina & Garcia-Gil, 2014). Notably, administration of IL-18 remarkably reduced *E. coli* growth, suggesting that activation of the inflammasome could be employed in suppressing colitogenic microbiota (Hu, Peng, et al., 2015). In addition to antimicrobial activity, NO may also contribute to IL-18-mediated cytoprotection. *Caspase-1*$^{-/-}$ mice exhibited increased apoptosis in the colonic mucosa during DSS-induced injury (Dupaul-Chicoine et al., 2010). Ample evidence suggests that NO is a cytoprotective molecule. NO produced by iNOS provide cytoprotective and antimicrobial host defence in vitro and in vivo during *Salmonella* and *Mycobacterium* infection (Alam et al., 2002; Denis, 1991; Mishra et al., 2013; Zaki et al., 2009). The cytoprotective function of NO was shown to be linked with NO-mediated upregulation of cytoprotective molecules, such as HO-1 and other antioxidants, which suppress ROS-mediated cell death (Alam et al., 2002, 2008; Sawa et al., 2007; Umezawa et al., 1997; Zaki et al., 2009). However, the role of inflammasome or its downstream cytokines in the regulation of antioxidants has not been studied in greater detail.

NLRP3 inflammasome-derived IL-1β facilitates host resistance to *Leishmania* infection by triggering iNOS-mediated NO production (Lima-Junior et al., 2013). Similarly, the NAIP5/NLRC4 inflammasome induces iNOS expression. TLR5 senses extracellular flagellin and can induce iNOS (Buzzo et al., 2010). However, intracellular flagellin is sensed via NAIP5 and NLRC4. It was shown that NLRC4 and NAIP5 recognize a separate C-terminal domain of flagellin from *L. pneumophilla* and *S. typhimurium* which is separate from the TLR5 recognition site (Lightfield et al., 2008). Flagellin sensing by NAIP5 and NLRC4 induces inflammasome assembly with subsequent caspase-1 recruitment and activation (Bortoluci & Medzhitov, 2010). iNOS activation by intracellular flagellin was completely abolished in the absence of caspase-1, thus demonstrating that certain inflammasomes can upregulate iNOS expression. Furthermore, iNOS expression was regulated by the inflammasome complex and not by the inflammasome-regulated cytokines, as neutralization of IL-1β or IL-18 had no impact on iNOS induction. This demonstrates a novel IL-1β-independent pathway by which intracellular flagellin induces iNOS production, which is essential for inflammasome-mediated antipathogen effector mechanisms (Buzzo et al., 2010).

The induction of iNOS, AMP, and other inflammatory molecules by IL-18 may involve several cell signalling pathways including NF-κB, STAT1, STAT3, ERK, and AKT in a context-dependent manner (Kalina et al., 2000). In the gut, reduced IL-18 expression is associated with the expression of IFN-γ and STAT1 activation (Zaki, Vogel, et al., 2010). In fact, IL-18 is known as an IFN-γ-inducing factor and IFN-γ activates the STAT1 pathway upon recognition by IFN-γR, which recruits JAK1 and JAK2 leading to the phosphorylation of STAT1 (Fig. 3). IL-18 promotes the production of IFN-γ in activated T cells and NK cells by helping in Th1 cell polarization (Okamura et al., 1995; Takeda et al., 1998). NO produced by macrophages in response to IFN-γ plays a role in host defence against intracellular bacterial infections (Akuta et al., 2006; Nathan & Shiloh, 2000; Zaki et al., 2005). Thus, inflammasomes may exert indirect regulatory role on NO via IL-18-induced IFN-γ during microbial infections.

Inflammasome-mediated induction of NO may not always be beneficial as NO is considered a double-edged sword; while it provides cytoprotection and antimicrobial immunity, excessive NO is often cytotoxic and mutagenic (Akuta et al., 2006). Excessive NO causes DNA damage and mutation during viral infection (Akaike et al., 2003; Yermilov, Rubio, & Ohshima, 1995; Yoshitake et al., 2004). In the contexts of hyperinflammation or inflammasome activation, NO may promote tumorigenesis. For example, several NLR-deficient mice, such as *Nod2*$^{-/-}$ and *Nlrp12*$^{-/-}$, demonstrate increased susceptibility to colorectal tumorigenesis induced by azoxymethane plus DSS because of hyperinflammatory response accompanied by higher induction of iNOS (Udden, Peng, et al., 2017; Zaki et al., 2011). Higher NO production in *Nlrp12*$^{-/-}$ mice is beneficial in the context of bacterial infection as *Nlrp12*$^{-/-}$ mice showed better resistance to *S. typhimurium* infection as compared to wild-type (Zaki et al., 2014). These studies provide ample evidence that NLRs and inflammasomes play regulatory roles in NO production during infection and inflammation.

15. RECIPROCAL REGULATION OF THE INFLAMMASOME BY NO

NO plays a significant role in the regulation of innate immune responses against various intracellular pathogens. One of the mechanisms underlying NO-mediated regulation of immune responses is inhibition of the cysteine proteases. NO prevents IL-1β and IL-18 release from the macrophages by inhibiting caspase-1, a cysteine protease (Kim, Talanian, Li, & Billiar, 1998). It was shown that NO restricts human and murine *Mycobacterium* infection by exerting direct antimicrobial effect on the bacterium or by regulating the antiinflammatory signalling to prevent tissue damage (Chakravortty & Hensel, 2003). During *Mycobacterium* infection, IFN-γ produced by adaptive immune responses inhibits NLRP3 assembly via NO production (Fig. 11). NO produced as a result of adaptive immunity was shown to inhibit the assembly of the NLRP3 inflammasome via N-nitrosylation to prevent neutrophil recruitment and excessive tissue damage (Mishra et al., 2013). Similarly, NO produced by type I IFN was shown to negatively regulates the NLRP3 inflammasome (Hernandez-Cuellar et al., 2012). In addition, NO posttranslationally modifies the ASC–caspase-1 oligomerization events specific to NLRP3 inflammasome via chemical modification of cysteine residues. NO-mediated inflammasome inhibition proved to be an important mechanism by which adaptive immune responses prevent tissue damage during chronic infection. NO suppresses NLRP3 inflammasome-mediated IL-1β secretion and caspase-1 activation in human as well as murine macrophages, providing protection against LPS-induced septic shock (Mao et al., 2013). iNOS depletion

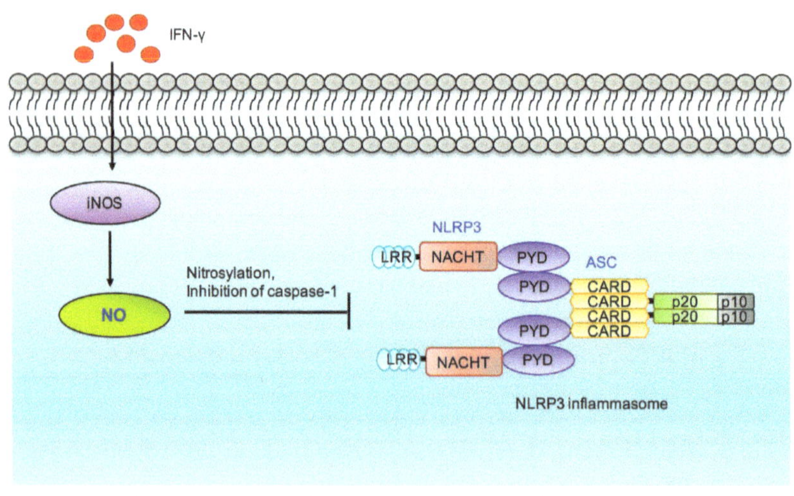

Fig. 11 NO inhibits the NLRP3 inflammasome. IFN-γ produced as a result of adaptive immune response induces NO production. NO negatively regulates the NLRP3 assembly by nitrosylation reaction or inhibiting caspase-1 activation.

increases the accumulation of dysfunctional mitochondria during LPS and ATP stimulation giving rise to increased IL-1β production and caspase-1 activation. iNOS deficiency or pharmacological inhibition of NO production resulted in increased NLRP3 activation and higher mortality during LPS stimulation in murine models, identifying NO as a critical negative regulator of NLRP3 activation to prevent excessive inflammation during bacterial infection. Thus, a potential mechanism through which NO inhibits microbial replication as well as reduces harmful inflammatory responses is regulation of inflammasome activity.

16. CONCLUDING REMARKS

In recent years, overwhelming number of studies on inflammasomes have made a great contribution into our understanding of the host defence response against infections and inflammatory diseases. Great attention has been paid in deciphering the molecular mechanism behind inflammasome assembly and effector mechanisms in response to diverse pathogens and their PAMPs. However, the current studies have not completely resolved the exact mechanisms behind some of the downstream and upstream players involved in inflammasome activation. The downstream signalling events associated with inflammasome signalling during different pathogenic stimuli continue to emerge. Although the inflammasome appears to be a critical regulator of NO, the implication of NO in inflammasome-mediated antimicrobial immune response during diverse types of microbial infection is yet to be explored. Further studies are also needed to understand the mechanism of NO-mediated regulation of inflammasomes and its biological relevance. Ideally, such studies will help to unravel novel therapeutic targets against various infectious and inflammatory diseases.

REFERENCES

Adachi, O., Kawai, T., Takeda, K., Matsumoto, M., Tsutsui, H., Sakagami, M., et al. (1998). Targeted disruption of the MyD88 gene results in loss of IL-1- and IL-18-mediated function. *Immunity*, *9*, 143–150.

Adamczak, S. E., de Rivero Vaccari, J. P., Dale, G., Brand, F. J., 3rd, Nonner, D., Bullock, M. R., et al. (2014). Pyroptotic neuronal cell death mediated by the AIM2 inflammasome. *Journal of Cerebral Blood Flow and Metabolism*, *34*, 621–629.

Ainscough, J. S., Gerberick, G. F., Kimber, I., & Dearman, R. J. (2015). Interleukin-1beta processing is dependent on a calcium-mediated interaction with calmodulin. *The Journal of Biological Chemistry*, *290*, 31151–31161.

Akaike, T. (2000). Mechanisms of biological S-nitrosation and its measurement. *Free Radical Research*, *33*, 461–469.

Akaike, T., Noguchi, Y., Ijiri, S., Setoguchi, K., Suga, M., Zheng, Y. M., et al. (1996). Pathogenesis of influenza virus-induced pneumonia: Involvement of both nitric oxide and oxygen radicals. *Proceedings of the National Academy of Sciences of the United States of America, 93*, 2448–2453.

Akaike, T., Okamoto, S., Sawa, T., Yoshitake, J., Tamura, F., Ichimori, K., et al. (2003). 8-Nitroguanosine formation in viral pneumonia and its implication for pathogenesis. *Proceedings of the National Academy of Sciences of the United States of America, 100*, 685–690.

Akira, S., Uematsu, S., & Takeuchi, O. (2006). Pathogen recognition and innate immunity. *Cell, 124*, 783–801.

Akuta, T., Zaki, M. H., Yoshitake, J., Okamoto, T., & Akaike, T. (2006). Nitrative stress through formation of 8-nitroguanosine: Insights into microbial pathogenesis. *Nitric Oxide, 14*, 101–108.

Alam, M. S., Akaike, T., Okamoto, S., Kubota, T., Yoshitake, J., Sawa, T., et al. (2002). Role of nitric oxide in host defense in murine salmonellosis as a function of its antibacterial and antiapoptotic activities. *Infection and Immunity, 70*, 3130–3142.

Alam, M. S., Zaki, M. H., Sawa, T., Islam, S., Ahmed, K. A., Fujii, S., et al. (2008). Nitric oxide produced in Peyer's patches exhibits antiapoptotic activity contributing to an antimicrobial effect in murine salmonellosis. *Microbiology and Immunology, 52*, 197–208.

Alderton, W. K., Cooper, C. E., & Knowles, R. G. (2001). Nitric oxide synthases: Structure, function and inhibition. *The Biochemical Journal, 357*, 593–615.

Allen, I. C., McElvania-TeKippe, E., Wilson, J. E., Lich, J. D., Arthur, J. C., Sullivan, J. T., et al. (2013). Characterization of NLRP12 during the in vivo host immune response to *Klebsiella pneumoniae* and *Mycobacterium tuberculosis*. *PLoS One, 8*. e60842.

Allen, I. C., Scull, M. A., Moore, C. B., Holl, E. K., McElvania-TeKippe, E., Taxman, D. J., et al. (2009). The NLRP3 inflammasome mediates in vivo innate immunity to influenza A virus through recognition of viral RNA. *Immunity, 30*, 556–565.

Allen, I. C., TeKippe, E. M., Woodford, R. M., Uronis, J. M., Holl, E. K., Rogers, A. B., et al. (2010). The NLRP3 inflammasome functions as a negative regulator of tumorigenesis during colitis-associated cancer. *The Journal of Experimental Medicine, 207*, 1045–1056.

Allen, I. C., Wilson, J. E., Schneider, M., Lich, J. D., Roberts, R. A., Arthur, J. C., et al. (2012). NLRP12 suppresses colon inflammation and tumorigenesis through the negative regulation of noncanonical NF-kappaB signaling. *Immunity, 36*, 742–754.

Amer, A., Franchi, L., Kanneganti, T. D., Body-Malapel, M., Ozoren, N., Brady, G., et al. (2006). Regulation of Legionella phagosome maturation and infection through flagellin and host Ipaf. *The Journal of Biological Chemistry, 281*, 35217–35223.

Anand, P. K., Malireddi, R. K., Lukens, J. R., Vogel, P., Bertin, J., Lamkanfi, M., et al. (2012). NLRP6 negatively regulates innate immunity and host defence against bacterial pathogens. *Nature, 488*, 389–393.

Arlehamn, C. S., Petrilli, V., Gross, O., Tschopp, J., & Evans, T. J. (2010). The role of potassium in inflammasome activation by bacteria. *The Journal of Biological Chemistry, 285*, 10508–10518.

Arthur, J. C., Lich, J. D., Aziz, R. K., Kotb, M., & Ting, J. P. (2007). Heat shock protein 90 associates with monarch-1 and regulates its ability to promote degradation of NF-kappaB-inducing kinase. *Journal of Immunology, 179*, 6291–6296.

Arthur, J. C., Lich, J. D., Ye, Z., Allen, I. C., Gris, D., Wilson, J. E., et al. (2010). Cutting edge: NLRP12 controls dendritic and myeloid cell migration to affect contact hypersensitivity. *Journal of Immunology, 185*, 4515–4519.

Bartesaghi, S., Graziano, V., Galavotti, S., Henriquez, N. V., Betts, J., Saxena, J., et al. (2015). Inhibition of oxidative metabolism leads to p53 genetic inactivation and transformation in neural stem cells. *Proceedings of the National Academy of Sciences of the United States of America, 112*, 1059–1064.

Beckman, J. S., & Koppenol, W. H. (1996). Nitric oxide, superoxide, and peroxynitrite: The good, the bad, and ugly. *The American Journal of Physiology*, *271*, C1424–C1437.
Borghini, S., Tassi, S., Chiesa, S., Caroli, F., Carta, S., Caorsi, R., et al. (2011). Clinical presentation and pathogenesis of cold-induced autoinflammatory disease in a family with recurrence of an NLRP12 mutation. *Arthritis and Rheumatism*, *63*, 830–839.
Bortoluci, K. R., & Medzhitov, R. (2010). Control of infection by pyroptosis and autophagy: Role of TLR and NLR. *Cellular and Molecular Life Sciences*, *67*, 1643–1651.
Boyden, E. D., & Dietrich, W. F. (2006). Nalp1b controls mouse macrophage susceptibility to anthrax lethal toxin. *Nature Genetics*, *38*, 240–244.
Bredt, D. S., Hwang, P. M., & Snyder, S. H. (1990). Localization of nitric oxide synthase indicating a neural role for nitric oxide. *Nature*, *347*, 768–770.
Broz, P., & Dixit, V. M. (2016). Inflammasomes: Mechanism of assembly, regulation and signalling. *Nature Reviews. Immunology*, *16*, 407–420.
Broz, P., Newton, K., Lamkanfi, M., Mariathasan, S., Dixit, V. M., & Monack, D. M. (2010). Redundant roles for inflammasome receptors NLRP3 and NLRC4 in host defense against Salmonella. *The Journal of Experimental Medicine*, *207*, 1745–1755.
Broz, P., Ruby, T., Belhocine, K., Bouley, D. M., Kayagaki, N., Dixit, V. M., et al. (2012). Caspase-11 increases susceptibility to Salmonella infection in the absence of caspase-1. *Nature*, *490*, 288–291.
Broz, P., von Moltke, J., Jones, J. W., Vance, R. E., & Monack, D. M. (2010). Differential requirement for Caspase-1 autoproteolysis in pathogen-induced cell death and cytokine processing. *Cell Host & Microbe*, *8*, 471–483.
Burckstummer, T., Baumann, C., Bluml, S., Dixit, E., Durnberger, G., Jahn, H., et al. (2009). An orthogonal proteomic-genomic screen identifies AIM2 as a cytoplasmic DNA sensor for the inflammasome. *Nature Immunology*, *10*, 266–272.
Buzzo, C. L., Campopiano, J. C., Massis, L. M., Lage, S. L., Cassado, A. A., Leme-Souza, R., et al. (2010). A novel pathway for inducible nitric-oxide synthase activation through inflammasomes. *The Journal of Biological Chemistry*, *285*, 32087–32095.
Cain, K., Langlais, C., Sun, X. M., Brown, D. G., & Cohen, G. M. (2001). Physiological concentrations of K+ inhibit cytochrome c-dependent formation of the apoptosome. *The Journal of Biological Chemistry*, *276*, 41985–41990.
Carlos, D., Costa, F. R., Pereira, C. A., Rocha, F. A., Yaochite, J. N., Oliveira, G. G., et al. (2017). Mitochondrial DNA activates the NLRP3 inflammasome and predisposes to type 1 diabetes in murine model. *Frontiers in Immunology*, *8*, 164.
Chae, J. J., Cho, Y. H., Lee, G. S., Cheng, J., Liu, P. P., Feigenbaum, L., et al. (2011). Gain-of-function Pyrin mutations induce NLRP3 protein-independent interleukin-1beta activation and severe autoinflammation in mice. *Immunity*, *34*, 755–768.
Chae, J. J., Wood, G., Masters, S. L., Richard, K., Park, G., Smith, B. J., et al. (2006). The B30.2 domain of pyrin, the familial Mediterranean fever protein, interacts directly with caspase-1 to modulate IL-1beta production. *Proceedings of the National Academy of Sciences of the United States of America*, *103*, 9982–9987.
Chakravortty, D., & Hensel, M. (2003). Inducible nitric oxide synthase and control of intracellular bacterial pathogens. *Microbes and Infection*, *5*, 621–627.
Chavarría-Smith, J. (2013). Direct proteolytic cleavage of NLRP1B is necessary and sufficient for inflammasome activation by anthrax lethal factor. *PLoS Pathogens 9*. e1003452. https://doi.org/10.1371/journal.ppat.1003452.
Chavarria-Smith, J., Mitchell, P. S., Ho, A. M., Daugherty, M. D., & Vance, R. E. (2016). Functional and evolutionary analyses identify proteolysis as a general mechanism for NLRP1 inflammasome activation. *PLoS Pathogens*, *12*. e1006052.
Chen, G. Y., Liu, M., Wang, F., Bertin, J., & Nunez, G. (2011). A functional role for nlrp6 in intestinal inflammation and tumorigenesis. *Journal of Immunology*, *186*, 7187–7194.

Chen, L. C., Wang, L. J., Tsang, N. M., Ojcius, D. M., Chen, C. C., Ouyang, C. N., et al. (2012). Tumour inflammasome-derived IL-1beta recruits neutrophils and improves local recurrence-free survival in EBV-induced nasopharyngeal carcinoma. *EMBO Molecular Medicine*, *4*, 1276–1293.

Chen, C. J., Wu, G. H., Kuo, R. L., & Shih, S. R. (2017). Role of the intestinal microbiota in the immunomodulation of influenza virus infection. *Microbes and Infection*, *19*, 570–579. https://doi.org/10.1016/j.micinf.2017.09.002.

Chlichlia, K., Peter, M. E., Rocha, M., Scaffidi, C., Bucur, M., Krammer, P. H., et al. (1998). Caspase activation is required for nitric oxide-mediated, CD95(APO-1/Fas)-dependent and independent apoptosis in human neoplastic lymphoid cells. *Blood*, *91*, 4311–4320.

Chu, Z. L., Pio, F., Xie, Z., Welsh, K., Krajewska, M., Krajewski, S., et al. (2001). A novel enhancer of the Apaf1 apoptosome involved in cytochrome c-dependent caspase activation and apoptosis. *The Journal of Biological Chemistry*, *276*, 9239–9245.

Chuang, S. Y., Yang, C. H., Chou, C. C., Chiang, Y. P., Chuang, T. H., & Hsu, L. C. (2013). TLR-induced PAI-2 expression suppresses IL-1beta processing via increasing autophagy and NLRP3 degradation. *Proceedings of the National Academy of Sciences of the United States of America*, *110*, 16079–16084.

Cooper, C. E. (2002). Nitric oxide and cytochrome oxidase: Substrate, inhibitor or effector? *Trends in Biochemical Sciences*, *27*, 33–39.

Crane, B. R., Arvai, A. S., Ghosh, D. K., Wu, C., Getzoff, E. D., Stuehr, D. J., et al. (1998). Structure of nitric oxide synthase oxygenase dimer with pterin and substrate. *Science*, *279*, 2121–2126.

de Koning, H. D., Bergboer, J. G., van den Bogaard, E. H., van Vlijmen-Willems, I. M., Rodijk-Olthuis, D., Simon, A., et al. (2012). Strong induction of AIM2 expression in human epidermis in acute and chronic inflammatory skin conditions. *Experimental Dermatology*, *21*, 961–964.

Denis, M. (1991). Interferon-gamma-treated murine macrophages inhibit growth of tubercle bacilli via the generation of reactive nitrogen intermediates. *Cellular Immunology*, *132*, 150–157.

Deretic, V., Kimura, T., Timmins, G., Moseley, P., Chauhan, S., & Mandell, M. (2015). Immunologic manifestations of autophagy. *The Journal of Clinical Investigation*, *125*, 75–84.

DeYoung, K. L., Ray, M. E., YA, S., Anzick, S. L., Johnstone, R. W., Trapani, J. A., et al. (1997). Cloning a novel member of the human interferon-inducible gene family associated with control of tumorigenicity in a model of human melanoma. *Oncogene*, *15*, 453–457.

Dihlmann, S., Tao, S., Echterdiek, F., Herpel, E., Jansen, L., Chang-Claude, J., et al. (2014). Lack of absent in melanoma 2 (AIM2) expression in tumor cells is closely associated with poor survival in colorectal cancer patients. *International Journal of Cancer*, *135*, 2387–2396.

Dimmeler, S., Haendeler, J., Nehls, M., & Zeiher, A. M. (1997). Suppression of apoptosis by nitric oxide via inhibition of interleukin-1beta-converting enzyme (ICE)-like and cysteine protease protein (CPP)-32-like proteases. *The Journal of Experimental Medicine*, *185*, 601–607.

Dinarello, C. A. (1998). Interleukin-1 beta, interleukin-18, and the interleukin-1 beta converting enzyme. *Annals of the New York Academy of Sciences*, *856*, 1–11.

Dinarello, C. A. (2002). The IL-1 family and inflammatory diseases. *Clinical and Experimental Rheumatology*, *20*, S1–S13.

Dinarello, C. A. (2007). Interleukin-18 and the pathogenesis of inflammatory diseases. *Seminars in Nephrology*, *27*, 98–114.

Doitsh, G., Galloway, N. L., Geng, X., Yang, Z., Monroe, K. M., Zepeda, O., et al. (2014). Cell death by pyroptosis drives CD4 T-cell depletion in HIV-1 infection. *Nature*, *505*, 509–514.

Dombrowski, Y., Peric, M., Koglin, S., Kammerbauer, C., Goss, C., Anz, D., et al. (2011). Cytosolic DNA triggers inflammasome activation in keratinocytes in psoriatic lesions. *Science Translational Medicine, 3*. 82ra38.
D'Osualdo, A., Anania, V. G., Yu, K., Lill, J. R., Kaufman, R. J., Matsuzawa, S., et al. (2015). Transcription factor ATF4 induces NLRP1 inflammasome expression during endoplasmic reticulum stress. *PLoS One, 10*. e0130635.
Drapier, J. C., Wietzerbin, J., & Hibbs, J. B., Jr. (1988). Interferon-gamma and tumor necrosis factor induce the L-arginine-dependent cytotoxic effector mechanism in murine macrophages. *European Journal of Immunology, 18*, 1587–1592.
Duncan, J. A., Bergstralh, D. T., Wang, Y., Willingham, S. B., Ye, Z., Zimmermann, A. G., et al. (2007). Cryopyrin/NALP3 binds ATP/dATP, is an ATPase, and requires ATP binding to mediate inflammatory signaling. *Proceedings of the National Academy of Sciences of the United States of America, 104*, 8041–8046.
Dupaul-Chicoine, J., Yeretssian, G., Doiron, K., Bergstrom, K. S., McIntire, C. R., LeBlanc, P. M., et al. (2010). Control of intestinal homeostasis, colitis, and colitis-associated colorectal cancer by the inflammatory caspases. *Immunity, 32*, 367–378.
Eiserich, J. P., Hristova, M., Cross, C. E., Jones, A. D., Freeman, B. A., Halliwell, B., et al. (1998). Formation of nitric oxide-derived inflammatory oxidants by myeloperoxidase in neutrophils. *Nature, 391*, 393–397.
Ekchariyawat, P., Hamel, R., Bernard, E., Wichit, S., Surasombatpattana, P., Talignani, L., et al. (2015). Inflammasome signaling pathways exert antiviral effect against Chikungunya virus in human dermal fibroblasts. *Infection, Genetics and Evolution, 32*, 401–408.
Elinav, E., Strowig, T., Kau, A. L., Henao-Mejia, J., Thaiss, C. A., Booth, C. J., et al. (2011). NLRP6 inflammasome regulates colonic microbial ecology and risk for colitis. *Cell, 145*, 745–757.
Ewald, S. E., Chavarria-Smith, J., & Boothroyd, J. C. (2014). NLRP1 is an inflammasome sensor for *Toxoplasma gondii*. *Infection and Immunity, 82*, 460–468.
Fang, R., Hara, H., Sakai, S., Hernandez-Cuellar, E., Mitsuyama, M., Kawamura, I., et al. (2014). Type I interferon signaling regulates activation of the absent in melanoma 2 inflammasome during *Streptococcus pneumoniae* infection. *Infection and Immunity, 82*, 2310–2317.
Faustin, B., Lartigue, L., Bruey, J. M., Luciano, F., Sergienko, E., Bailly-Maitre, B., et al. (2007). Reconstituted NALP1 inflammasome reveals two-step mechanism of caspase-1 activation. *Molecular Cell, 25*, 713–724.
Fernandes-Alnemri, T., JW, Y., Datta, P., Wu, J., & Alnemri, E. S. (2009). AIM2 activates the inflammasome and cell death in response to cytoplasmic DNA. *Nature, 458*, 509–513.
Fernandes-Alnemri, T., Yu, J. W., Juliana, C., Solorzano, L., Kang, S., Wu, J., et al. (2010). The AIM2 inflammasome is critical for innate immunity to *Francisella tularensis*. *Nature Immunology, 11*, 385–393.
Fernandes-Alnemri, T., Kang, S., Anderson, C., Sagara, J., Fitzgerald, K. A., & Alnemri, E. S. (2013). Cutting edge: TLR signaling licenses IRAK1 for rapid activation of the NLRP3 inflammasome. *Journal of Immunology, 191*, 3995–3999.
Finethy, R., Jorgensen, I., Haldar, A. K., de Zoete, M. R., Strowig, T., Flavell, R. A., et al. (2015). Guanylate binding proteins enable rapid activation of canonical and non-canonical inflammasomes in Chlamydia-infected macrophages. *Infection and Immunity, 83*, 4740–4749.
Fink, S. L., Bergsbaken, T., & Cookson, B. T. (2008). Anthrax lethal toxin and Salmonella elicit the common cell death pathway of caspase-1-dependent pyroptosis via distinct mechanisms. *Proceedings of the National Academy of Sciences of the United States of America, 105*, 4312–4317.

Fortier, A., de Chastellier, C., Balor, S., & Gros, P. (2007). Birc1e/Naip5 rapidly antagonizes modulation of phagosome maturation by *Legionella pneumophila*. *Cellular Microbiology*, *9*, 910–923.

Franchi, L., Amer, A., Body-Malapel, M., Kanneganti, T. D., Ozoren, N., Jagirdar, R., et al. (2006). Cytosolic flagellin requires Ipaf for activation of caspase-1 and interleukin 1beta in salmonella-infected macrophages. *Nature Immunology*, *7*, 576–582.

Franchi, L., Eigenbrod, T., Munoz-Planillo, R., & Nunez, G. (2009). The inflammasome: A caspase-1-activation platform that regulates immune responses and disease pathogenesis. *Nature Immunology*, *10*, 241–247.

Franchi, L., Munoz-Planillo, R., & Nunez, G. (2012). Sensing and reacting to microbes through the inflammasomes. *Nature Immunology*, *13*, 325–332.

Gerlic, M., Faustin, B., Postigo, A., EC, Y., Proell, M., Gombosuren, N., et al. (2013). Vaccinia virus F1L protein promotes virulence by inhibiting inflammasome activation. *Proceedings of the National Academy of Sciences of the United States of America*, *110*, 7808–7813.

Grisham, M. B., Jourd'Heuil, D., & Wink, D. A. (1999). Nitric oxide. I. Physiological chemistry of nitric oxide and its metabolites:implications in inflammation. *The American Journal of Physiology*, *276*, G315–321.

Gross, O., Yazdi, A. S., Thomas, C. J., Masin, M., Heinz, L. X., Guarda, G., et al. (2012). Inflammasome activators induce interleukin-1alpha secretion via distinct pathways with differential requirement for the protease function of caspase-1. *Immunity*, *36*, 388–400.

Gu, Y., Kuida, K., Tsutsui, H., Ku, G., Hsiao, K., Fleming, M. A., et al. (1997). Activation of interferon-gamma inducing factor mediated by interleukin-1beta converting enzyme. *Science*, *275*, 206–209.

Hagar, J. A., Powell, D. A., Aachoui, Y., Ernst, R. K., & Miao, E. A. (2013). Cytoplasmic LPS activates caspase-11: Implications in TLR4-independent endotoxic shock. *Science*, *341*, 1250–1253.

Halff, E. F., Diebolder, C. A., Versteeg, M., Schouten, A., Brondijk, T. H., & Huizinga, E. G. (2012). Formation and structure of a NAIP5-NLRC4 inflammasome induced by direct interactions with conserved N- and C-terminal regions of flagellin. *The Journal of Biological Chemistry*, *287*, 38460–38472.

Hanamsagar, R., Aldrich, A., & Kielian, T. (2014). Critical role for the AIM2 inflammasome during acute CNS bacterial infection. *Journal of Neurochemistry*, *129*, 704–711.

Hara, H., Tsuchiya, K., Kawamura, I., Fang, R., Hernandez-Cuellar, E., Shen, Y., et al. (2013). Phosphorylation of the adaptor ASC acts as a molecular switch that controls the formation of speck-like aggregates and inflammasome activity. *Nature Immunology*, *14*, 1247–1255.

Harris, J., Hartman, M., Roche, C., Zeng, S. G., O'Shea, A., Sharp, F. A., et al. (2011). Autophagy controls IL-1beta secretion by targeting pro-IL-1beta for degradation. *The Journal of Biological Chemistry*, *286*, 9587–9597.

He, Y., Zeng, M. Y., Yang, D., Motro, B., & Nunez, G. (2016). NEK7 is an essential mediator of NLRP3 activation downstream of potassium efflux. *Nature*, *530*, 354–357.

Hellmich, K. A., Levinsohn, J. L., Fattah, R., Newman, Z. L., Maier, N., Sastalla, I., et al. (2012). Anthrax lethal factor cleaves mouse NLRP1B in both toxin-sensitive and toxin-resistant macrophages. *PLoS One*, 7. e49741.

Henry, T., Brotcke, A., Weiss, D. S., Thompson, L. J., & Monack, D. M. (2007). Type I interferon signaling is required for activation of the inflammasome during Francisella infection. *The Journal of Experimental Medicine*, *204*, 987–994.

Herman, M. D., Moche, M., Flodin, S., Welin, M., Tresaugues, L., Johansson, I., et al. (2009). Structures of BIR domains from human NAIP and cIAP2. *Acta Crystallographica. Section F, Structural Biology and Crystallization Communications*, *65*, 1091–1096.

Hernandez-Cuellar, E., Tsuchiya, K., Hara, H., Fang, R., Sakai, S., Kawamura, I., et al. (2012). Cutting edge: Nitric oxide inhibits the NLRP3 inflammasome. *Journal of Immunology*, *189*, 5113–5117.

Hirota, S. A., Ng, J., Lueng, A., Khajah, M., Parhar, K., Li, Y., et al. (2011). NLRP3 inflammasome plays a key role in the regulation of intestinal homeostasis. *Inflammatory Bowel Diseases*, *17*, 1359–1372. https://doi.org/10.1002/ibd.21478.

Hornung, V., Ablasser, A., Charrel-Dennis, M., Bauernfeind, F., Horvath, G., Caffrey, D. R., et al. (2009). AIM2 recognizes cytosolic dsDNA and forms a caspase-1-activating inflammasome with ASC. *Nature*, *458*, 514–518.

Hou, X., & Niu, X. (2015). The NMR solution structure of AIM2 PYD domain from Mus musculus reveals a distinct alpha2-alpha3 helix conformation from its human homologues. *Biochemical and Biophysical Research Communications*, *461*, 396–400.

Hu, Y., Mao, K., Zeng, Y., Chen, S., Tao, Z., Yang, C., et al. (2010). Tripartite-motif protein 30 negatively regulates NLRP3 inflammasome activation by modulating reactive oxygen species production. *Journal of Immunology*, *185*, 7699–7705.

Hu, S., Peng, L., Kwak, Y. T., Tekippe, E. M., Pasare, C., Malter, J. S., et al. (2015). The DNA sensor AIM2 maintains intestinal homeostasis via regulation of epithelial antimicrobial host defense. *Cell Reports*, *13*, 1922–1936.

Hu, Z., Zhou, Q., Zhang, C., Fan, S., Cheng, W., Zhao, Y., et al. (2015). Structural and biochemical basis for induced self-propagation of NLRC4. *Science*, *350*, 399–404.

Huai, W., Zhao, R., Song, H., Zhao, J., Zhang, L., Zhang, L., et al. (2014). Aryl hydrocarbon receptor negatively regulates NLRP3 inflammasome activity by inhibiting NLRP3 transcription. *Nature Communications*, *5*, 4738.

Ichinohe, T., Lee, H. K., Ogura, Y., Flavell, R., & Iwasaki, A. (2009). Inflammasome recognition of influenza virus is essential for adaptive immune responses. *The Journal of Experimental Medicine*, *206*, 79–87.

Ichinohe, T., Yamazaki, T., Koshiba, T., & Yanagi, Y. (2013). Mitochondrial protein mitofusin 2 is required for NLRP3 inflammasome activation after RNA virus infection. *Proceedings of the National Academy of Sciences of the United States of America*, *110*, 17963–17968.

Ito, S., Hara, Y., & Kubota, T. (2014). CARD8 is a negative regulator for NLRP3 inflammasome, but mutant NLRP3 in cryopyrin-associated periodic syndromes escapes the restriction. *Arthritis Research & Therapy*, *16*, R52.

Iyer, S. S., He, Q., Janczy, J. R., Elliott, E. I., Zhong, Z., Olivier, A. K., et al. (2013). Mitochondrial cardiolipin is required for Nlrp3 inflammasome activation. *Immunity*, *39*, 311–323.

Jakobs, C., Perner, S., & Hornung, V. (2015). AIM2 drives joint inflammation in a self-DNA triggered model of chronic polyarthritis. *PLoS One*, *10*. e0131702.

James, S. L. (1995). Role of nitric oxide in parasitic infections. *Microbiological Reviews*, *59*, 533–547.

Javierre, B. M., Fernandez, A. F., Richter, J., Al-Shahrour, F., Martin-Subero, J. I., Rodriguez-Ubreva, J., et al. (2010). Changes in the pattern of DNA methylation associate with twin discordance in systemic lupus erythematosus. *Genome Research*, *20*, 170–179.

Jeru, I., Duquesnoy, P., Fernandes-Alnemri, T., Cochet, E., Yu, J. W., Lackmy-Port-Lis, M., et al. (2008). Mutations in NALP12 cause hereditary periodic fever syndromes. *Proceedings of the National Academy of Sciences of the United States of America*, *105*, 1614–1619.

Jin, T., Perry, A., Jiang, J., Smith, P., Curry, J. A., Unterholzner, L., et al. (2012). Structures of the HIN domain:DNA complexes reveal ligand binding and activation mechanisms of the AIM2 inflammasome and IFI16 receptor. *Immunity*, *36*, 561–571.

Jin, T., Perry, A., Smith, P., Jiang, J., & Xiao, T. S. (2013). Structure of the absent in melanoma 2 (AIM2) pyrin domain provides insights into the mechanisms of AIM2 autoinhibition and inflammasome assembly. *The Journal of Biological Chemistry*, *288*, 13225–13235.

Jin, J., Yu, Q., Han, C., Hu, X., Xu, S., Wang, Q., et al. (2013). LRRFIP2 negatively regulates NLRP3 inflammasome activation in macrophages by promoting Flightless-I-mediated caspase-1 inhibition. *Nature Communications*, *4*, 2075.

Johnnidis, J. B., Harris, M. H., Wheeler, R. T., Stehling-Sun, S., Lam, M. H., Kirak, O., et al. (2008). Regulation of progenitor cell proliferation and granulocyte function by microRNA-223. *Nature*, *451*, 1125–1129.

Jones, J. W., Kayagaki, N., Broz, P., Henry, T., Newton, K., O'Rourke, K., et al. (2010). Absent in melanoma 2 is required for innate immune recognition of *Francisella tularensis*. *Proceedings of the National Academy of Sciences of the United States of America*, *107*, 9771–9776.

Jounai, N., Kobiyama, K., Shiina, M., Ogata, K., Ishii, K. J., & Takeshita, F. (2011). NLRP4 negatively regulates autophagic processes through an association with beclin1. *Journal of Immunology*, *186*, 1646–1655.

Juliana, C., Fernandes-Alnemri, T., Kang, S., Farias, A., Qin, F., & Alnemri, E. S. (2012). Non-transcriptional priming and deubiquitination regulate NLRP3 inflammasome activation. *The Journal of Biological Chemistry*, *287*, 36617–36622.

Jung, S. S., Moon, J. S., JF, X., Ifedigbo, E., Ryter, S. W., Choi, A. M., et al. (2015). Carbon monoxide negatively regulates NLRP3 inflammasome activation in macrophages. *American Journal of Physiology. Lung Cellular and Molecular Physiology*, *308*, L1058–1067.

Kalantari, P., DeOliveira, R. B., Chan, J., Corbett, Y., Rathinam, V., Stutz, A., et al. (2014). Dual engagement of the NLRP3 and AIM2 inflammasomes by plasmodium-derived hemozoin and DNA during malaria. *Cell Reports*, *6*, 196–210.

Kalina, U., Kauschat, D., Koyama, N., Nuernberger, H., Ballas, K., Koschmieder, S., et al. (2000). IL-18 activates STAT3 in the natural killer cell line 92, augments cytotoxic activity, and mediates IFN-gamma production by the stress kinase p38 and by the extracellular regulated kinases p44erk-1 and p42erk-21. *Journal of Immunology*, *165*, 1307–1313.

Kamijo, R., Harada, H., Matsuyama, T., Bosland, M., Gerecitano, J., Shapiro, D., et al. (1994). Requirement for transcription factor IRF-1 in NO synthase induction in macrophages. *Science*, *263*, 1612–1615.

Kanneganti, T. D., Lamkanfi, M., & Nunez, G. (2007). Intracellular NOD-like receptors in host defense and disease. *Immunity*, *27*, 549–559.

Karki, R., Man, S. M., Malireddi, R. K., Gurung, P., Vogel, P., Lamkanfi, M., et al. (2015). Concerted activation of the AIM2 and NLRP3 inflammasomes orchestrates host protection against Aspergillus infection. *Cell Host & Microbe*, *17*, 357–368.

Kawai, T., & Akira, S. (2006). TLR signaling. *Cell Death and Differentiation*, *13*, 816–825.

Kayagaki, N., Stowe, I. B., Lee, B. L., O'Rourke, K., Anderson, K., Warming, S., et al. (2015). Caspase-11 cleaves gasdermin D for non-canonical inflammasome signalling. *Nature*, *526*, 666–671.

Kayagaki, N., Warming, S., Lamkanfi, M., Vande Walle, L., Louie, S., Dong, J., et al. (2011). Non-canonical inflammasome activation targets caspase-11. *Nature*, *479*, 117–121.

Kayagaki, N., Wong, M. T., Stowe, I. B., Ramani, S. R., Gonzalez, L. C., Akashi-Takamura, S., et al. (2013). Noncanonical inflammasome activation by intracellular LPS independent of TLR4. *Science*, *341*, 1246–1249.

Kerur, N., Veettil, M. V., Sharma-Walia, N., Bottero, V., Sadagopan, S., Otageri, P., et al. (2011). IFI16 acts as a nuclear pathogen sensor to induce the inflammasome in response to Kaposi Sarcoma-associated herpesvirus infection. *Cell Host & Microbe*, *9*, 363–375.

Kim, S., Bauernfeind, F., Ablasser, A., Hartmann, G., Fitzgerald, K. A., Latz, E., et al. (2010). *Listeria monocytogenes* is sensed by the NLRP3 and AIM2 inflammasome. *European Journal of Immunology*, *40*, 1545–1551.

Kim, M. L., Chae, J. J., Park, Y. H., De Nardo, D., Stirzaker, R. A., Ko, H. J., et al. (2015). Aberrant actin depolymerization triggers the pyrin inflammasome and autoinflammatory

disease that is dependent on IL-18, not IL-1beta. *The Journal of Experimental Medicine*, *212*, 927–938.
Kim, Y. M., Talanian, R. V., Li, J., & Billiar, T. R. (1998). Nitric oxide prevents IL-1beta and IFN-gamma-inducing factor (IL-18) release from macrophages by inhibiting caspase-1 (IL-1beta-converting enzyme). *Journal of Immunology*, *161*, 4122–4128.
Kimkong, I., Avihingsanon, Y., & Hirankarn, N. (2009). Expression profile of HIN200 in leukocytes and renal biopsy of SLE patients by real-time RT-PCR. *Lupus*, *18*, 1066–1072.
Kimura, T., Jain, A., Choi, S. W., Mandell, M. A., Johansen, T., & Deretic, V. (2017). TRIM-directed selective autophagy regulates immune activation. *Autophagy*, *13*, 989–990.
Kimura, T., Jain, A., Choi, S. W., Mandell, M. A., Schroder, K., Johansen, T., et al. (2015). TRIM-mediated precision autophagy targets cytoplasmic regulators of innate immunity. *The Journal of Cell Biology*, *210*, 973–989.
Kleinert, H., Schwarz, P. M., & Forstermann, U. (2003). Regulation of the expression of inducible nitric oxide synthase. *Biological Chemistry*, *384*, 1343–1364.
Kofoed, E. M., & Vance, R. E. (2011). Innate immune recognition of bacterial ligands by NAIPs determines inflammasome specificity. *Nature*, *477*, 592–595.
Kondo, Y., Nagai, K., Nakahata, S., Saito, Y., Ichikawa, T., Suekane, A., et al. (2012). Overexpression of the DNA sensor proteins, absent in melanoma 2 and interferon-inducible 16, contributes to tumorigenesis of oral squamous cell carcinoma with p53 inactivation. *Cancer Science*, *103*, 782–790.
Kong, H., Wang, Y., Zeng, X., Wang, Z., Wang, H., & Xie, W. (2015). Differential expression of inflammasomes in lung cancer cell lines and tissues. *Tumour Biology*, *36*, 7501–7513.
Koonin, E. V., & Aravind, L. (2000). The NACHT family—A new group of predicted NTPases implicated in apoptosis and MHC transcription activation. *Trends in Biochemical Sciences*, *25*, 223–224.
Kovarova, M., Hesker, P. R., Jania, L., Nguyen, M., Snouwaert, J. N., Xiang, Z., et al. (2012). NLRP1-dependent pyroptosis leads to acute lung injury and morbidity in mice. *Journal of Immunology*, *189*, 2006–2016.
Kupz, A., Guarda, G., Gebhardt, T., Sander, L. E., Short, K. R., Diavatopoulos, D. A., et al. (2012). NLRC4 inflammasomes in dendritic cells regulate noncognate effector function by memory CD8(+) T cells. *Nature Immunology*, *13*, 162–169.
Lage, S. L., Buzzo, C. L., Amaral, E. P., Matteucci, K. C., Massis, L. M., Icimoto, M. Y., et al. (2013). Cytosolic flagellin-induced lysosomal pathway regulates inflammasome-dependent and -independent macrophage responses. *Proceedings of the National Academy of Sciences of the United States of America*, *110*, E3321–3330.
Lamkanfi, M. (2011). Emerging inflammasome effector mechanisms. *Nature Reviews. Immunology*, *11*, 213–220.
Lamkanfi, M., & Dixit, V. M. (2012). Inflammasomes and their roles in health and disease. *Annual Review of Cell and Developmental Biology*, *28*, 137–161.
Lamkanfi, M., & Dixit, V. M. (2014). Mechanisms and functions of inflammasomes. *Cell*, *157*, 1013–1022.
Lara-Tejero, M., Sutterwala, F. S., Ogura, Y., Grant, E. P., Bertin, J., Coyle, A. J., et al. (2006). Role of the caspase-1 inflammasome in *Salmonella typhimurium* pathogenesis. *The Journal of Experimental Medicine*, *203*, 1407–1412.
Lee, G. S., Subramanian, N., Kim, A. I., Aksentijevich, I., Goldbach-Mansky, R., Sacks, D. B., et al. (2012). The calcium-sensing receptor regulates the NLRP3 inflammasome through Ca2+ and cAMP. *Nature*, *492*, 123–127.
Levy, M., Thaiss, C. A., Zeevi, D., Dohnalova, L., Zilberman-Schapira, G., Mahdi, J. A., et al. (2015). Microbiota-modulated metabolites shape the intestinal microenvironment by regulating NLRP6 inflammasome signaling. *Cell*, *163*, 1428–1443.

Li, J., Billiar, T. R., Talanian, R. V., & Kim, Y. M. (1997). Nitric oxide reversibly inhibits seven members of the caspase family via S-nitrosylation. *Biochemical and Biophysical Research Communications*, *240*, 419–424.

Li, J., Bombeck, C. A., Yang, S., Kim, Y. M., & Billiar, T. R. (1999). Nitric oxide suppresses apoptosis via interrupting caspase activation and mitochondrial dysfunction in cultured hepatocytes. *The Journal of Biological Chemistry*, *274*, 17325–17333.

Lich, J. D., Williams, K. L., Moore, C. B., Arthur, J. C., Davis, B. K., Taxman, D. J., et al. (2007). Monarch-1 suppresses non-canonical NF-kappaB activation and p52-dependent chemokine expression in monocytes. *Journal of Immunology*, *178*, 1256–1260.

Liew, F. Y. (1995). Regulation of lymphocyte functions by nitric oxide. *Current Opinion in Immunology*, *7*, 396–399.

Lightfield, K. L., Persson, J., Brubaker, S. W., Witte, C. E., von Moltke, J., Dunipace, E. A., et al. (2008). Critical function for Naip5 in inflammasome activation by a conserved carboxy-terminal domain of flagellin. *Nature Immunology*, *9*, 1171–1178.

Lightfield, K. L., Persson, J., Trinidad, N. J., Brubaker, S. W., Kofoed, E. M., Sauer, J. D., et al. (2011). Differential requirements for NAIP5 in activation of the NLRC4 inflammasome. *Infection and Immunity*, *79*, 1606–1614.

Lima-Junior, D. S., Costa, D. L., Carregaro, V., Cunha, L. D., Silva, A. L., Mineo, T. W., et al. (2013). Inflammasome-derived IL-1beta production induces nitric oxide-mediated resistance to Leishmania. *Nature Medicine*, *19*, 909–915.

Lin, K. M., Hu, W., Troutman, T. D., Jennings, M., Brewer, T., Li, X., et al. (2014). IRAK-1 bypasses priming and directly links TLRs to rapid NLRP3 inflammasome activation. *Proceedings of the National Academy of Sciences of the United States of America*, *111*, 775–780.

Liu, Z., Zaki, M. H., Vogel, P., Gurung, P., Finlay, B. B., Deng, W., et al. (2012). Role of inflammasomes in host defense against *Citrobacter rodentium* infection. *The Journal of Biological Chemistry*, *287*, 16955–16964.

Lu, A., Kabaleeswaran, V., Fu, T., Magupalli, V. G., & Wu, H. (2014). Crystal structure of the F27G AIM2 PYD mutant and similarities of its self-association to DED/DED interactions. *Journal of Molecular Biology*, *426*, 1420–1427.

Lukens, J. R., Gurung, P., Shaw, P. J., Barr, M. J., Zaki, M. H., Brown, S. A., et al. (2015). The NLRP12 sensor negatively regulates autoinflammatory disease by modulating Interleukin-4 production in T cells. *Immunity*, *42*, 654–664.

Lupfer, C., Malik, A., & Kanneganti, T. D. (2015). Inflammasome control of viral infection. *Current Opinion in Virology*, *12*, 38–46.

Lupfer, C., Thomas, P. G., Anand, P. K., Vogel, P., Milasta, S., Martinez, J., et al. (2013). Receptor interacting protein kinase 2-mediated mitophagy regulates inflammasome activation during virus infection. *Nature Immunology*, *14*, 480–488.

Macaluso, F., Nothnagel, M., Parwez, Q., Petrasch-Parwez, E., Bechara, F. G., Epplen, J. T., et al. (2007). Polymorphisms in NACHT-LRR (NLR) genes in atopic dermatitis. *Experimental Dermatology*, *16*, 692–698.

Mamantopoulos, M., Ronchi, F., Van Hauwermeiren, F., Vieira-Silva, S., Yilmaz, B., Martens, L., et al. (2017). Nlrp6- and ASC-dependent inflammasomes do not shape the commensal gut microbiota composition. *Immunity*, *47*. 339-348 e334.

Man, S. M., & Kanneganti, T. D. (2015). Regulation of inflammasome activation. *Immunological Reviews*, *265*, 6–21.

Man, S. M., Karki, R., & Kanneganti, T. D. (2016). AIM2 inflammasome in infection, cancer, and autoimmunity: Role in DNA sensing, inflammation, and innate immunity. *European Journal of Immunology*, *46*, 269–280.

Man, S. M., Place, D. E., Kuriakose, T., & Kanneganti, T. D. (2017). Interferon-inducible guanylate-binding proteins at the interface of cell-autonomous immunity and inflammasome activation. *Journal of Leukocyte Biology*, *101*, 143–150.

Man, S. M., Zhu, Q., Zhu, L., Liu, Z., Karki, R., Malik, A., et al. (2015). Critical role for the DNA sensor AIM2 in stem cell proliferation and cancer. *Cell*, *162*, 45–58.

Mao, K., Chen, S., Chen, M., Ma, Y., Wang, Y., Huang, B., et al. (2013). Nitric oxide suppresses NLRP3 inflammasome activation and protects against LPS-induced septic shock. *Cell Research*, *23*, 201–212.

Mariathasan, S., Newton, K., Monack, D. M., Vucic, D., French, D. M., Lee, W. P., et al. (2004). Differential activation of the inflammasome by caspase-1 adaptors ASC and Ipaf. *Nature*, *430*, 213–218.

Mariathasan, S., Weiss, D. S., Newton, K., McBride, J., O'Rourke, K., Roose-Girma, M., et al. (2006). Cryopyrin activates the inflammasome in response to toxins and ATP. *Nature*, *440*, 228–232.

Marim, F. M., Franco, M. M., Gomes, M. T., Miraglia, M. C., Giambartolomei, G. H., & Oliveira, S. C. (2017). The role of NLRP3 and AIM2 in inflammasome activation during *Brucella abortus* infection. *Seminars in Immunopathology*, *39*, 215–223.

Martin, H. M., Campbell, B. J., Hart, C. A., Mpofu, C., Nayar, M., Singh, R., et al. (2004). Enhanced Escherichia coli adherence and invasion in Crohn's disease and colon cancer. *Gastroenterology*, *127*, 80–93.

Martin, E., Nathan, C., & Xie, Q. W. (1994). Role of interferon regulatory factor 1 in induction of nitric oxide synthase. *The Journal of Experimental Medicine*, *180*, 977–984.

Martin, B. N., Wang, C., Willette-Brown, J., Herjan, T., Gulen, M. F., Zhou, H., et al. (2014). IKKalpha negatively regulates ASC-dependent inflammasome activation. *Nature Communications*, *5*, 4977.

Martinez-Medina, M., Denizot, J., Dreux, N., Robin, F., Billard, E., Bonnet, R., et al. (2014). Western diet induces dysbiosis with increased *E. coli* in CEABAC10 mice, alters host barrier function favouring AIEC colonisation. *Gut*, *63*, 116–124.

Martinez-Medina, M., & Garcia-Gil, L. J. (2014). *Escherichia coli* in chronic inflammatory bowel diseases: An update on adherent invasive *Escherichia coli* pathogenicity. *World Journal of Gastrointestinal Pathophysiology*, *5*, 213–227.

Martinon, F., Burns, K., & Tschopp, J. (2002). The inflammasome: A molecular platform triggering activation of inflammatory caspases and processing of proIL-beta. *Molecular Cell*, *10*, 417–426.

Martinon, F., Mayor, A., & Tschopp, J. (2009). The inflammasomes: Guardians of the body. *Annual Review of Immunology*, *27*, 229–265.

Masters, S. L., Gerlic, M., Metcalf, D., Preston, S., Pellegrini, M., O'Donnell, J. A., et al. (2012). NLRP1 inflammasome activation induces pyroptosis of hematopoietic progenitor cells. *Immunity*, *37*, 1009–1023.

Masters, S. L., Lagou, V., Jeru, I., Baker, P. J., Van Eyck, L., Parry, D. A., et al. (2016). Familial autoinflammation with neutrophilic dermatosis reveals a regulatory mechanism of pyrin activation. *Science Translational Medicine*, *8*, 332ra345.

Maver, A., Lavtar, P., Ristic, S., Stopinsek, S., Simcic, S., Hocevar, K., et al. (2017). Identification of rare genetic variation of NLRP1 gene in familial multiple sclerosis. *Scientific Reports*, *7*, 3715.

Maxwell, J. R., Yadav, R., Rossi, R. J., Ruby, C. E., Weinberg, A. D., Aguila, H. L., et al. (2006). IL-18 bridges innate and adaptive immunity through IFN-gamma and the CD134 pathway. *Journal of Immunology*, *177*, 234–245.

Mayer, B., & Hemmens, B. (1997). Biosynthesis and action of nitric oxide in mammalian cells. *Trends in Biochemical Sciences*, *22*, 477–481.

Miao, E. A., Alpuche-Aranda, C. M., Dors, M., Clark, A. E., Bader, M. W., Miller, S. I., et al. (2006). Cytoplasmic flagellin activates caspase-1 and secretion of interleukin 1beta via Ipaf. *Nature Immunology*, *7*, 569–575.

Miao, E. A., Leaf, I. A., Treuting, P. M., Mao, D. P., Dors, M., Sarkar, A., et al. (2010). Caspase-1-induced pyroptosis is an innate immune effector mechanism against intracellular bacteria. *Nature Immunology*, *11*, 1136–1142.

Misawa, T., Takahama, M., Kozaki, T., Lee, H., Zou, J., Saitoh, T., et al. (2013). Microtubule-driven spatial arrangement of mitochondria promotes activation of the NLRP3 inflammasome. *Nature Immunology, 14*, 454–460.
Mishra, B. B., Rathinam, V. A., Martens, G. W., Martinot, A. J., Kornfeld, H., Fitzgerald, K. A., et al. (2013). Nitric oxide controls the immunopathology of tuberculosis by inhibiting NLRP3 inflammasome-dependent processing of IL-1beta. *Nature Immunology, 14*, 52–60.
Mitoma, H., Hanabuchi, S., Kim, T., Bao, M., Zhang, Z., Sugimoto, N., et al. (2013). The DHX33 RNA helicase senses cytosolic RNA and activates the NLRP3 inflammasome. *Immunity, 39*, 123–135.
Moayeri, M., Crown, D., Newman, Z. L., Okugawa, S., Eckhaus, M., Cataisson, C., et al. (2010). Inflammasome sensor NLRP1B-dependent resistance to anthrax is mediated by caspase-1, IL-1 signaling and neutrophil recruitment. *PLoS Pathogens, 6*, e1001222.
Molofsky, A. B., Byrne, B. G., Whitfield, N. N., Madigan, C. A., Fuse, E. T., Tateda, K., et al. (2006). Cytosolic recognition of flagellin by mouse macrophages restricts *Legionella pneumophila* infection. *The Journal of Experimental Medicine, 203*, 1093–1104.
Moncada, S., Palmer, R. M., & Higgs, E. A. (1991). Nitric oxide: Physiology, pathophysiology, and pharmacology. *Pharmacological Reviews, 43*, 109–142.
Monroe, K. M., Yang, Z., Johnson, J. R., Geng, X., Doitsh, G., Krogan, N. J., et al. (2014). IFI16 DNA sensor is required for death of lymphoid CD4 T cells abortively infected with HIV. *Science, 343*, 428–432.
Morrone, S. R., Matyszewski, M., Yu, X., Delannoy, M., Egelman, E. H., & Sohn, J. (2015). Assembly-driven activation of the AIM2 foreign-dsDNA sensor provides a polymerization template for downstream ASC. *Nature Communications, 6*, 7827.
Muñoz-Planillo, R., Kuffa, P., Martinez-Colon, G., Smith, B. L., Rajendra, T. M., & Nunez, G. (2013). K+ efflux is the common trigger of NLRP3 inflammasome activation by bacterial toxins and particulate matter. *Immunity, 38*, 1142–1153.
Murad, F. (1986). Cyclic guanosine monophosphate as a mediator of vasodilation. *The Journal of Clinical Investigation, 78*, 1–5.
Murakami, T., Ockinger, J., Yu, J., Byles, V., McColl, A., Hofer, A. M., et al. (2012). Critical role for calcium mobilization in activation of the NLRP3 inflammasome. *Proceedings of the National Academy of Sciences of the United States of America, 109*, 11282–11287.
Nathan, C. F., & Hibbs, J. B., Jr. (1991). Role of nitric oxide synthesis in macrophage antimicrobial activity. *Current Opinion in Immunology, 3*, 65–70.
Nathan, C., & Shiloh, M. U. (2000). Reactive oxygen and nitrogen intermediates in the relationship between mammalian hosts and microbial pathogens. *Proceedings of the National Academy of Sciences of the United States of America, 97*, 8841–8848.
Nathan, C., & Xie, Q. W. (1994). Nitric oxide synthases: Roles, tolls, and controls. *Cell, 78*, 915–918.
Okamura, H., Tsutsi, H., Komatsu, T., Yutsudo, M., Hakura, A., Tanimoto, T., et al. (1995). Cloning of a new cytokine that induces IFN-gamma production by T cells. *Nature, 378*, 88–91.
O'Neill, L. A. (2013). Cardiolipin and the Nlrp3 inflammasome. *Cell Metabolism, 18*, 610–612.
Opdenbosch, V., Gurung, P., Vande Walle, L., Fossoul, A., Kanneganti, T. D., & Lamkanfi, M. (2014). Activation of the NLRP1b inflammasome independently of ASC-mediated caspase-1 autoproteolysis and speck formation. *Nature Communications, 5*, 3209.
Orlowski, G. M., Sharma, S., Colbert, J. D., Bogyo, M., Robertson, S. A., Kataoka, H., et al. (2017). Frontline science: Multiple cathepsins promote inflammasome-independent, particle-induced cell death during NLRP3-dependent IL-1beta activation. *Journal of Leukocyte Biology, 102*, 7–17. https://doi.org/10.1189/jlb.3HI0316-152R.

Paludan, S. R., & Bowie, A. G. (2013). Immune sensing of DNA. *Immunity, 38*, 870–880.

Park, E., Na, H. S., Song, Y. R., Shin, S. Y., Kim, Y. M., & Chung, J. (2014). Activation of NLRP3 and AIM2 inflammasomes by *Porphyromonas gingivalis* infection. *Infection and Immunity, 82*, 112–123.

Pedraza-Alva, G., Perez-Martinez, L., Valdez-Hernandez, L., Meza-Sosa, K. F., & Ando-Kuri, M. (2015). Negative regulation of the inflammasome: Keeping inflammation under control. *Immunological Reviews, 265*, 231–257.

Pereira, M. S., Morgantetti, G. F., Massis, L. M., Horta, C. V., Hori, J. I., & Zamboni, D. S. (2011). Activation of NLRC4 by flagellated bacteria triggers caspase-1-dependent and -independent responses to restrict *Legionella pneumophila* replication in macrophages and in vivo. *Journal of Immunology, 187*, 6447–6455.

Pizarro, T. T., Michie, M. H., Bentz, M., Woraratanadharm, J., Smith, M. F., Jr., Foley, E., et al. (1999). IL-18, a novel immunoregulatory cytokine, is up-regulated in Crohn's disease: Expression and localization in intestinal mucosal cells. *Journal of Immunology, 162*, 6829–6835.

Ponomareva, L., Liu, H., Duan, X., Dickerson, E., Shen, H., Panchanathan, R., et al. (2013). AIM2, an IFN-inducible cytosolic DNA sensor, in the development of benign prostate hyperplasia and prostate cancer. *Molecular Cancer Research, 11*, 1193–1202.

Py, B. F., Kim, M. S., Vakifahmetoglu-Norberg, H., & Yuan, J. (2013). Deubiquitination of NLRP3 by BRCC3 critically regulates inflammasome activity. *Molecular Cell, 49*, 331–338.

Qu, Y., Misaghi, S., Izrael-Tomasevic, A., Newton, K., Gilmour, L. L., Lamkanfi, M., et al. (2012). Phosphorylation of NLRC4 is critical for inflammasome activation. *Nature, 490*, 539–542.

Radi, R. (2004). Nitric oxide, oxidants, and protein tyrosine nitration. *Proceedings of the National Academy of Sciences of the United States of America, 101*, 4003–4008.

Radtke, F., & Clevers, H. (2005). Self-renewal and cancer of the gut: Two sides of a coin. *Science, 307*, 1904–1909.

Rakoff-Nahoum, S., Paglino, J., Eslami-Varzaneh, F., Edberg, S., & Medzhitov, R. (2004). Recognition of commensal microflora by toll-like receptors is required for intestinal homeostasis. *Cell, 118*, 229–241.

Rathinam, V. A., Jiang, Z., Waggoner, S. N., Sharma, S., Cole, L. E., Waggoner, L., et al. (2010). The AIM2 inflammasome is essential for host defense against cytosolic bacteria and DNA viruses. *Nature Immunology, 11*, 395–402.

Raupach, B., Peuschel, S. K., Monack, D. M., & Zychlinsky, A. (2006). Caspase-1-mediated activation of interleukin-1beta (IL-1beta) and IL-18 contributes to innate immune defenses against Salmonella enterica serovar Typhimurium infection. *Infection and Immunity, 74*, 4922–4926.

Rayamajhi, M., Zak, D. E., Chavarria-Smith, J., Vance, R. E., & Miao, E. A. (2013). Cutting edge: Mouse NAIP1 detects the type III secretion system needle protein. *Journal of Immunology, 191*, 3986–3989.

Reinholz, M., Kawakami, Y., Salzer, S., Kreuter, A., Dombrowski, Y., Koglin, S., et al. (2013). HPV16 activates the AIM2 inflammasome in keratinocytes. *Archives of Dermatological Research, 305*, 723–732.

Ren, T., Zamboni, D. S., Roy, C. R., Dietrich, W. F., & Vance, R. E. (2006). Flagellin-deficient Legionella mutants evade caspase-1- and Naip5-mediated macrophage immunity. *PLoS Pathogens, 2*, e18.

Riedl, S. J., & Salvesen, G. S. (2007). The apoptosome: Signalling platform of cell death. *Nature Reviews. Molecular Cell Biology, 8*, 405–413.

Rodgers, M. A., Bowman, J. W., Fujita, H., Orazio, N., Shi, M., Liang, Q., et al. (2014). The linear ubiquitin assembly complex (LUBAC) is essential for NLRP3 inflammasome activation. *The Journal of Experimental Medicine, 211*, 1333–1347.

Saiga, H., Kitada, S., Shimada, Y., Kamiyama, N., Okuyama, M., Makino, M., et al. (2012). Critical role of AIM2 in *Mycobacterium tuberculosis* infection. *International Immunology*, *24*, 637–644.
Saitoh, T., Fujita, N., Jang, M. H., Uematsu, S., Yang, B. G., Satoh, T., et al. (2008). Loss of the autophagy protein Atg16L1 enhances endotoxin-induced IL-1beta production. *Nature*, *456*, 264–268.
Salcedo, R., Worschech, A., Cardone, M., Jones, Y., Gyulai, Z., Dai, R. M., et al. (2010). MyD88-mediated signaling prevents development of adenocarcinomas of the colon: Role of interleukin 18. *The Journal of Experimental Medicine*, *207*, 1625–1636.
Sastalla, I., Crown, D., Masters, S. L., McKenzie, A., Leppla, S. H., & Moayeri, M. (2013). Transcriptional analysis of the three Nlrp1 paralogs in mice. *BMC Genomics*, *14*, 188.
Sawa, T., Akaike, T., Ichimori, K., Akuta, T., Kaneko, K., Nakayama, H., et al. (2003). Superoxide generation mediated by 8-nitroguanosine, a highly redox-active nucleic acid derivative. *Biochemical and Biophysical Research Communications*, *311*, 300–306.
Sawa, T., Zaki, M. H., Okamoto, T., Akuta, T., Tokutomi, Y., Kim-Mitsuyama, S., et al. (2007). Protein S-guanylation by the biological signal 8-nitroguanosine 3′,5′-cyclic monophosphate. *Nature Chemical Biology*, *3*, 727–735.
Saxena, M., & Yeretssian, G. (2014). NOD-like receptors: Master regulators of inflammation and cancer. *Frontiers in Immunology*, *5*, 327.
Schattgen, S. A., Gao, G., Kurt-Jones, E. A., & Fitzgerald, K. A. (2016). Cutting edge: DNA in the lung microenvironment during influenza virus infection tempers inflammation by engaging the DNA sensor AIM2. *Journal of Immunology*, *196*, 29–33.
Schmid-Burgk, J. L., Chauhan, D., Schmidt, T., Ebert, T. S., Reinhardt, J., Endl, E., et al. (2016). A genome-wide CRISPR (clustered regularly interspaced short palindromic repeats) screen identifies NEK7 as an essential component of NLRP3 inflammasome activation. *The Journal of Biological Chemistry*, *291*, 103–109.
Schmitz, N., Kurrer, M., Bachmann, M. F., & Kopf, M. (2005). Interleukin-1 is responsible for acute lung immunopathology but increases survival of respiratory influenza virus infection. *Journal of Virology*, *79*, 6441–6448.
Schopfer, F. J., Baker, P. R., & Freeman, B. A. (2003). NO-dependent protein nitration: A cell signaling event or an oxidative inflammatory response? *Trends in Biochemical Sciences*, *28*, 646–654.
Schroder, K., Sagulenko, V., Zamoshnikova, A., Richards, A. A., Cridland, J. A., Irvine, K. M., et al. (2012). Acute lipopolysaccharide priming boosts inflammasome activation independently of inflammasome sensor induction. *Immunobiology*, *217*, 1325–1329.
Schulmann, K., Brasch, F. E., Kunstmann, E., Engel, C., Pagenstecher, C., Vogelsang, H., et al. (2005). HNPCC-associated small bowel cancer: Clinical and molecular characteristics. *Gastroenterology*, *128*, 590–599.
Sedimbi, S. K., Hagglof, T., & Karlsson, M. C. (2013). IL-18 in inflammatory and autoimmune disease. *Cellular and Molecular Life Sciences*, *70*, 4795–4808.
Seregin, S. S., Golovchenko, N., Schaf, B., Chen, J., Pudlo, N. A., Mitchell, J., et al. (2017). NLRP6 protects Il10$^{-/-}$ mice from colitis by limiting colonization of *Akkermansia muciniphila*. *Cell Reports*, *19*, 2174.
Shenoy, A. R., Wellington, D. A., Kumar, P., Kassa, H., Booth, C. J., Cresswell, P., et al. (2012). GBP5 promotes NLRP3 inflammasome assembly and immunity in mammals. *Science*, *336*, 481–485.
Shi, C. S., Shenderov, K., Huang, N. N., Kabat, J., Abu-Asab, M., Fitzgerald, K. A., et al. (2012). Activation of autophagy by inflammatory signals limits IL-1beta production by targeting ubiquitinated inflammasomes for destruction. *Nature Immunology*, *13*, 255–263.
Shi, J., Zhao, Y., Wang, Y., Gao, W., Ding, J., Li, P., et al. (2014). Inflammatory caspases are innate immune receptors for intracellular LPS. *Nature*, *514*, 187–192.

Silveira, T. N., Gomes, M. T., Oliveira, L. S., Campos, P. C., Machado, G. G., & Oliveira, S. C. (2017). NLRP12 negatively regulates proinflammatory cytokine production and host defense against *Brucella abortus*. *European Journal of Immunology*, *47*, 51–59.

Storek, K. M., Gertsvolf, N. A., Ohlson, M. B., & Monack, D. M. (2015). cGAS and Ifi204 cooperate to produce type I IFNs in response to Francisella infection. *Journal of Immunology*, *194*, 3236–3245.

Subramanian, N., Natarajan, K., Clatworthy, M. R., Wang, Z., & Germain, R. N. (2013). The adaptor MAVS promotes NLRP3 mitochondrial localization and inflammasome activation. *Cell*, *153*, 348–361.

Sun, Y. H., Rolan, H. G., & Tsolis, R. M. (2007). Injection of flagellin into the host cell cytosol by Salmonella enterica serotype Typhimurium. *The Journal of Biological Chemistry*, *282*, 33897–33901.

Takeda, K., Tsutsui, H., Yoshimoto, T., Adachi, O., Yoshida, N., Kishimoto, T., et al. (1998). Defective NK cell activity and Th1 response in IL-18-deficient mice. *Immunity*, *8*, 383–390.

Takeuchi, O., & Akira, S. (2010). Pattern recognition receptors and inflammation. *Cell*, *140*, 805–820.

Tenthorey, J. L., Kofoed, E. M., Daugherty, M. D., Malik, H. S., & Vance, R. E. (2014). Molecular basis for specific recognition of bacterial ligands by NAIP/NLRC4 inflammasomes. *Molecular Cell*, *54*, 17–29.

Thomas, P. G., Dash, P., Aldridge, J. R., Jr., Ellebedy, A. H., Reynolds, C., Funk, A. J., et al. (2009). The intracellular sensor NLRP3 mediates key innate and healing responses to influenza A virus via the regulation of caspase-1. *Immunity*, *30*, 566–575.

Ting, J. P., Lovering, R. C., Alnemri, E. S., Bertin, J., Boss, J. M., Davis, B. K., et al. (2008). The NLR gene family: A standard nomenclature. *Immunity*, *28*, 285–287.

Tsuchiya, K., Hara, H., Kawamura, I., Nomura, T., Yamamoto, T., Daim, S., et al. (2010). Involvement of absent in melanoma 2 in inflammasome activation in macrophages infected with *Listeria monocytogenes*. *Journal of Immunology*, *185*, 1186–1195.

Tsuji, N. M., Tsutsui, H., Seki, E., Kuida, K., Okamura, H., Nakanishi, K., et al. (2004). Roles of caspase-1 in Listeria infection in mice. *International Immunology*, *16*, 335–343.

Tsujino, M., Hirata, Y., Imai, T., Kanno, K., Eguchi, S., Ito, H., et al. (1994). Induction of nitric oxide synthase gene by interleukin-1 beta in cultured rat cardiocytes. *Circulation*, *90*, 375–383.

Udden, S. M. N., Peng, L., Gan, J. L., Shelton, J. M., Malter, J. S., Hooper, L. V., et al. (2017). NOD2 suppresses colorectal tumorigenesis via downregulation of the TLR pathways. *Cell Reports*, *19*, 2756–2770.

Udden, S. M., Waliullah, S., Harris, M., & Zaki, H. (2017). The ex vivo colon organ culture and its use in antimicrobial host defense studies. *Journal of Visualized Experiments*, *120*. https://doi.org/10.3791/55347.

Umezawa, K., Akaike, T., Fujii, S., Suga, M., Setoguchi, K., Ozawa, A., et al. (1997). Induction of nitric oxide synthesis and xanthine oxidase and their roles in the antimicrobial mechanism against *Salmonella typhimurium* infection in mice. *Infection and Immunity*, *65*, 2932–2940.

Unterholzner, L., Keating, S. E., Baran, M., Horan, K. A., Jensen, S. B., Sharma, S., et al. (2010). IFI16 is an innate immune sensor for intracellular DNA. *Nature Immunology*, *11*, 997–1004.

van der Burgh, R., Nijhuis, L., Pervolaraki, K., Compeer, E. B., Jongeneel, L. H., van Gijn, M., et al. (2014). Defects in mitochondrial clearance predispose human monocytes to interleukin-1beta hypersecretion. *The Journal of Biological Chemistry*, *289*, 5000–5012.

Vande Walle, L., Van Opdenbosch, N., Jacques, P., Fossoul, A., Verheugen, E., Vogel, P., et al. (2014). Negative regulation of the NLRP3 inflammasome by A20 protects against arthritis. *Nature*, *512*, 69–73.

Vladimer, G. I., Weng, D., Paquette, S. W., Vanaja, S. K., Rathinam, V. A., Aune, M. H., et al. (2012). The NLRP12 inflammasome recognizes *Yersinia pestis*. *Immunity*, *37*, 96–107.

von Moltke, J., Trinidad, N. J., Moayeri, M., Kintzer, A. F., Wang, S. B., van Rooijen, N., et al. (2012). Rapid induction of inflammatory lipid mediators by the inflammasome in vivo. *Nature*, *490*, 107–111.

Wang, X., Jiang, W., Yan, Y., Gong, T., Han, J., Tian, Z., et al. (2014). RNA viruses promote activation of the NLRP3 inflammasome through a RIP1-RIP3-DRP1 signaling pathway. *Nature Immunology*, *15*, 1126–1133.

Wang, L., Manji, G. A., Grenier, J. M., Al-Garawi, A., Merriam, S., Lora, J. M., et al. (2002). PYPAF7, a novel PYRIN-containing Apaf1-like protein that regulates activation of NF-kappa B and caspase-1-dependent cytokine processing. *The Journal of Biological Chemistry*, *277*, 29874–29880.

Wang, S., Miura, M., Jung, Y. K., Zhu, H., Li, E., & Yuan, J. (1998). Murine caspase-11, an ICE-interacting protease, is essential for the activation of ICE. *Cell*, *92*, 501–509.

Wang, P., Zhu, S., Yang, L., Cui, S., Pan, W., Jackson, R., et al. (2015). Nlrp6 regulates intestinal antiviral innate immunity. *Science*, *350*, 826–830.

Warren, S. E., Mao, D. P., Rodriguez, A. E., Miao, E. A., & Aderem, A. (2008). Multiple nod-like receptors activate caspase 1 during *Listeria monocytogenes* infection. *Journal of Immunology*, *180*, 7558–7564.

Williams, K. L., Lich, J. D., Duncan, J. A., Reed, W., Rallabhandi, P., Moore, C., et al. (2005). The CATERPILLER protein monarch-1 is an antagonist of toll-like receptor-, tumor necrosis factor alpha-, and *Mycobacterium tuberculosis*-induced pro-inflammatory signals. *The Journal of Biological Chemistry*, *280*, 39914–39924.

Wilson, J. E., Petrucelli, A. S., Chen, L., Koblansky, A. A., Truax, A. D., Oyama, Y., et al. (2015). Inflammasome-independent role of AIM2 in suppressing colon tumorigenesis via DNA-PK and Akt. *Nature Medicine*, *21*, 906–913.

Wlodarska, M., Thaiss, C. A., Nowarski, R., Henao-Mejia, J., Zhang, J. P., Brown, E. M., et al. (2014). NLRP6 inflammasome orchestrates the colonic host-microbial interface by regulating goblet cell mucus secretion. *Cell*, *156*, 1045–1059.

Woerner, S. M., Kloor, M., Schwitalle, Y., Youmans, H., Doeberitz, M., Gebert, J., et al. (2007). The putative tumor suppressor AIM2 is frequently affected by different genetic alterations in microsatellite unstable colon cancers. *Genes, Chromosomes & Cancer*, *46*, 1080–1089.

Wu, J., Fernandes-Alnemri, T., & Alnemri, E. S. (2010). Involvement of the AIM2, NLRC4, and NLRP3 inflammasomes in caspase-1 activation by *Listeria monocytogenes*. *Journal of Clinical Immunology*, *30*, 693–702.

Xie, Q. W., Whisnant, R., & Nathan, C. (1993). Promoter of the mouse gene encoding calcium-independent nitric oxide synthase confers inducibility by interferon gamma and bacterial lipopolysaccharide. *The Journal of Experimental Medicine*, *177*, 1779–1784.

Xu, H., Yang, J., Gao, W., Li, L., Li, P., Zhang, L., et al. (2014). Innate immune sensing of bacterial modifications of rho GTPases by the Pyrin inflammasome. *Nature*, *513*, 237–241.

Yang, C. S., Kim, J. J., Kim, T. S., Lee, P. Y., Kim, S. Y., Lee, H. M., et al. (2015). Small heterodimer partner interacts with NLRP3 and negatively regulates activation of the NLRP3 inflammasome. *Nature Communications*, *6*, 6115.

Yang, J., Zhao, Y., Shi, J., & Shao, F. (2013). Human NAIP and mouse NAIP1 recognize bacterial type III secretion needle protein for inflammasome activation. *Proceedings of the National Academy of Sciences of the United States of America*, *110*, 14408–14413.

Yermilov, V., Rubio, J., & Ohshima, H. (1995). Formation of 8-nitroguanine in DNA treated with peroxynitrite in vitro and its rapid removal from DNA by depurination. *FEBS Letters*, *376*, 207–210.

Yoshitake, J., Akaike, T., Akuta, T., Tamura, F., Ogura, T., Esumi, H., et al. (2004). Nitric oxide as an endogenous mutagen for Sendai virus without antiviral activity. *Journal of Virology*, *78*, 8709–8719.

Zaki, M. H., Akuta, T., & Akaike, T. (2005). Nitric oxide-induced nitrative stress involved in microbial pathogenesis. *Journal of Pharmacological Sciences*, *98*, 117–129.

Zaki, M. H., Boyd, K. L., Vogel, P., Kastan, M. B., Lamkanfi, M., & Kanneganti, T. D. (2010). The NLRP3 inflammasome protects against loss of epithelial integrity and mortality during experimental colitis. *Immunity*, *32*, 379–391.

Zaki, M. H., Fujii, S., Okamoto, T., Islam, S., Khan, S., Ahmed, K. A., et al. (2009). Cytoprotective function of heme oxygenase 1 induced by a nitrated cyclic nucleotide formed during murine salmonellosis. *Journal of Immunology*, *182*, 3746–3756.

Zaki, M. H., Man, S. M., Vogel, P., Lamkanfi, M., & Kanneganti, T. D. (2014). Salmonella exploits NLRP12-dependent innate immune signaling to suppress host defenses during infection. *Proceedings of the National Academy of Sciences of the United States of America*, *111*, 385–390.

Zaki, M. H., Vogel, P., Body-Malapel, M., Lamkanfi, M., & Kanneganti, T. D. (2010). IL-18 production downstream of the Nlrp3 inflammasome confers protection against colorectal tumor formation. *Journal of Immunology*, *185*, 4912–4920.

Zaki, M. H., Vogel, P., Malireddi, R. K., Body-Malapel, M., Anand, P. K., Bertin, J., et al. (2011). The NOD-like receptor NLRP12 attenuates colon inflammation and tumorigenesis. *Cancer Cell*, *20*, 649–660.

Zamboni, D. S., Kobayashi, K. S., Kohlsdorf, T., Ogura, Y., Long, E. M., Vance, R. E., et al. (2006). The Birc1e cytosolic pattern-recognition receptor contributes to the detection and control of *Legionella pneumophila* infection. *Nature Immunology*, *7*, 318–325.

Zhang, H., Luo, J., Alcorn, J. F., Chen, K., Fan, S., Pilewski, J., et al. (2017). AIM2 inflammasome is critical for influenza-induced lung injury and mortality. *Journal of Immunology*, *198*, 4383–4393.

Zhao, Y., Yang, J., Shi, J., Gong, Y. N., Lu, Q., Xu, H., et al. (2011). The NLRC4 inflammasome receptors for bacterial flagellin and type III secretion apparatus. *Nature*, *477*, 596–600.

Zhou, X., Gordon, S. A., Kim, Y. M., Hoffman, R. A., Chen, Y., Zhang, X. R., et al. (2000). Nitric oxide induces thymocyte apoptosis via a caspase-1-dependent mechanism. *Journal of Immunology*, *165*, 1252–1258.

Zhou, H., Xiao, J., Wu, N., Liu, C., Xu, J., Liu, F., et al. (2015). MicroRNA-223 regulates the differentiation and function of intestinal dendritic cells and macrophages by targeting C/EBPbeta. *Cell Reports*, *13*, 1149–1160.

Zhou, R., Yazdi, A. S., Menu, P., & Tschopp, J. (2011). A role for mitochondria in NLRP3 inflammasome activation. *Nature*, *469*, 221–225.

CHAPTER FOUR

Nitric Oxide, an Old Molecule With Noble Functions in *Pseudomonas aeruginosa* Biology

Masanori Toyofuku[*,1], **Sang-Sun Yoon**[†,1]
[*]Graduate School of Life and Environmental Sciences, University of Tsukuba, Tsukuba, Japan
[†]Brain Korea 21 PLUS Project for Medical Science, Institute for Immunology and Immunological Diseases, Yonsei University College of Medicine, Seoul, Korea
[1]Corresponding authors: e-mail address: toyofuku.masanori.gf@u.tsukuba.ac.jp; sangsun_yoon@yuhs.ac

Contents

Abstract

Pseudomonas aeruginosa, a Gram-negative bacterium, is characterized by its versatility that enables persistent survival under adverse conditions. It can grow on diverse energy sources and readily acquire resistance to antimicrobial agents. As an opportunistic human pathogen, it also causes chronic infections inside the anaerobic mucus airways of cystic fibrosis patients. As a strict respirer, *P. aeruginosa* can grow by anaerobic nitrate (NO_3^-) respiration. Nitric oxide (NO) produced as an intermediate during anaerobic respiration exerts many important effects on the biological characteristics of *P. aeruginosa*.

Advances in Microbial Physiology, Volume 72
ISSN 0065-2911
https://doi.org/10.1016/bs.ampbs.2018.01.005

This review provides information regarding (i) how *P. aeruginosa* grows by anaerobic respiration, (ii) mechanisms by which NO is produced under such growth, and (iii) bacterial adaptation to NO. We also review the clinical relevance of NO in the fitness of *P. aeruginosa* and the use of NO as a potential therapeutic for treating *P. aeruginosa* infection.

1. INTRODUCTION

Nitric oxide (NO) has been studied as a toxic molecule that shows an antimicrobial effect at high concentrations and is employed as a defence mechanism in macrophages to attack bacteria (Fang, 2004; MacMicking, Xie, & Nathan, 1997). Bacteria are not only exposed to NO from outside sources but can also be exposed to endogenous NO that is produced through biological processes such as anaerobic NO_3^- respiration, often called denitrification. Under conditions where the oxygen level is low and N-oxides are available, some bacteria can use N-oxides instead of oxygen for respiration. The evolution of denitrification suggests that bacteria have been exposed to NO since the early origins of life (Chen & Strous, 2013), indicating that bacteria have been coping with this molecule for quite a long time.

Pseudomonas aeruginosa has been extensively studied to understand how it grows by anaerobic respiration and how NO, derived under such growth, influences its physiological processes. Endogenous NO production is controlled by diverse environmental conditions, such as availability of alternative electron acceptors, oxygen potential, and cell density (Toyofuku, O'toole, & Nomura, 2015). Numerous studies carried out in *P. aeruginosa* have provided insights into how NO, a highly reactive molecule, can differentially act either as a signalling cue or stress that deeply affects bacterial physiology.

As an opportunistic pathogen, *P. aeruginosa* can cause persistent infection in patients with an abnormal airway mucus environment. Patients suffering from cystic fibrosis (CF) (Tiddens, Donaldson, Rosenfeld, & Pare, 2010; Yoon & Hassett, 2004), bronchiectasis (Clarke, 1987; Frey & Russi, 1997), and pneumonia (Reynolds, 1987) are especially vulnerable to *P. aeruginosa* infection. Understanding *P. aeruginosa* biology in association with NO is important to delineate *P. aeruginosa* pathogenesis, particularly in the airways of CF patients, and to develop future strategies for treatments. The stagnant mucus layer lining the airway of chronic CF patients was reported to be anaerobic (Worlitzsch et al., 2002; Yoon et al., 2002). The lack of oxygen potential is ascribed to (i) the limited oxygen transport into the mucus layer due to its increased viscosity and (ii) a high rate of oxygen consumption by immune-related and airway epithelial cells. This observation provided an insight into the establishment of *P. aeruginosa* infection under anaerobic conditions. Importantly, NO_3^- and NO_2^- were detected in larger quantities in the sputum or exhaled condensate of patients with pulmonary exacerbation of CF than in normal individuals (Jones et al., 2000; Linnane et al., 1998). This result likely suggests that *P. aeruginosa* proliferates well inside the anaerobic mucus layer by exploiting the compounds produced by the host and can provide an insight into why *P. aeruginosa* has been such a competitive colonizer in patient airways.

In this review, we introduce the mechanisms for endogenous NO production and regulation in *P. aeruginosa* and discuss how these processes impact bacterial behaviour. We further introduce some clinical relevance of the findings. This review of the research on *P. aeruginosa* provides an example of how basic studies can develop and be employed in clinical applications.

2. ANAEROBIC RESPIRATION, A LEAKY PROCESS FOR NO PRODUCTION

2.1 Overview of Denitrification

P. aeruginosa can grow efficiently in environments depleted of oxygen where NO_3^- or NO_2^- is available by switching their respiration from aerobic respiration to denitrification (Schreiber et al., 2007). *P. aeruginosa* is one of the organisms that possesses all sets of denitrification enzymes to reduce NO_3^- to gaseous nitrogen. Denitrification is an important process in the nitrogen cycle of the ecosystem as it releases nitrogen back to the atmosphere. Recent studies show that some bacteria can generate NO with an NO synthase that is evolutionary related to the mammalian enzyme (Adak, Aulak, & Stuehr, 2002; Gusarov & Nudler, 2005; Kinkel et al., 2016), whereas in *P. aeruginosa*, NO is produced through denitrification.

In denitrification, four steps are involved in which nitrate (NO_3^-) is sequentially reduced to nitrite (NO_2^-), nitric oxide (NO), nitrous oxide (N_2O), and finally to nitrogen (N_2). Each step is catalysed by specific enzymes, nitrate reductase (NAR), nitrite reductase (NIR), NO reductase (NOR), and N_2O reductase (NOS). NAR and NOR are located at the inner membrane, whereas NIR and NOS are found in the periplasm. NarK2, an NO_3^- transporter, transports NO_3^- to the cytoplasm and pumps out NO_2^- from the cytoplasm to the periplasm, where it is further reduced by NIR (Sharma, Noriega, & Rowe, 2006). In addition, *P. aeruginosa* possesses a periplasmic NO_3^- reductase NapAB that can perform NO_3^- reduction in the absence of NAR (Van Alst, Sherrill, Iglewski, & Haidaris, 2009).

Results from a series of experiments suggest that denitrification is involved in the pathogenicity of *P. aeruginosa* (Filiatrault, Picardo, Ngai, Passador, & Iglewski, 2006). For example, N_2O gas has been detected from the sputum of *P. aeruginosa*-infected CF patients, and the amount of NO_3^- in the sputum was found to be sufficient to support denitrification in *P. aeruginosa* (Kolpen et al., 2014; Line et al., 2014).

In this review, denitrification is of particular interest as it is the main endogenous source of NO in *P. aeruginosa.* Anaerobic cultures of *P. aeruginosa* have revealed that NO is a key molecule in understanding the impact of anaerobiosis on this bacterium. NO produced through denitrification influences cell morphology, biofilm formation and dispersal, stress responses, and membrane vesicle (MV) production (Barraud et al., 2006; Lee et al., 2012; Toyofuku et al., 2014; Yoon, Lee, Park, & Yoon, 2011). NO can diffuse through the membrane and will not only impact the producer but also influences neighbouring cells (Lee et al., 2012; Yin, Fuangthong, Laratta, & Shapleigh, 2003). Several factors are suggested to control the cellular level of NO in *P. aeruginosa*, further indicating the importance of this molecule. Furthermore, it was recently shown that NOR and the regulatory protein NosR are core proteins in the supercomplex of the denitrification proteins that consists of NAR, NIR, NOR, NOS, and cofactor proteins (Borrero-de Acuna et al., 2015, 2016). This supercomplex also interacts with proteins involved in ATP synthase, protein transport, motility, and many other proteins with diverse functions central to cell physiology. It is tempting to speculate that this protein–protein network allows the bacteria to control and coordinate their functions and maintain the cellular system.

2.2 Chemical and Enzymatic Production of NO

Although *P. aeruginosa* was reported to be able to sustain its anaerobic growth by fermentation using arginine (Luthi, Mercenier, & Haas, 1986; Vander Wauven, Pierard, Kley-Raymann, & Haas, 1984) or pyruvate (Schreiber et al., 2006), *P. aeruginosa* has been long categorized as an obligate respirer. As in most aerobes, molecular oxygen is the most efficient electron acceptor, and *P. aeruginosa* possesses gene clusters that encode multiple terminal oxidases with distinct subunit combinations of cytochromes, which are iron-containing haem proteins (Arai et al., 2014; Hirai, Osamura, Ishii, & Arai, 2016; Kawakami, Kuroki, Ishii, Igarashi, & Arai, 2010). These terminal oxidases receive electrons from cytochrome *c* (Araï et al., 2014), azurin (Parr, Barber, Greenwood, & Brunori, 1977), or ubiquinol (Kawakami et al., 2010) and mediate the reduction of oxygen to water to generate an electrochemical gradient across the inner membrane (Toledo-Cuevas, 1997). The redundant presence of terminal oxidases in *P. aeruginosa* has been attributed to its unparalleled capability to survive under diverse environments that require metabolic versatility (Clarke, 1982; Frimmersdorf, Horatzek, Pelnikevich, Wiehlmann, & Schomburg, 2010).

P. aeruginosa can also generate a pH gradient, which enables ATP biosynthesis using nitrate (NO_3^-) or nitrite (NO_2^-) as alternative electron acceptors (Yoon et al., 2011). Anaerobic respiration has been extensively studied, and its role in supporting *P. aeruginosa* anaerobiosis has been reviewed elsewhere (Arai, 2011; Williams, Zlosnik, & Ryall, 2007; Yoon & Hassett, 2004). Anaerobic growth by NO_3^- reduction is more proliferative than by NO_2^- reduction (Williams, Rowe, Romero, & Eagon, 1978; Yoon et al., 2006), and this difference has been attributed to several factors. First, NO_3^- reduction by NAR is coupled with consumption of two protons in the cytoplasm and concomitant generation of two protons in the periplasm (Jones, Lamont, & Garland, 1980), creating an electrochemical gradient across the inner membrane (Fig. 1). On the other hand, NIR, located in the periplasm (Borrero-de Acuña et al., 2015), consumes one electron during the enzymatic reaction (Rinaldo, Brunori, & Cutruzzola, 2007), without producing any further electron motive force (Fig. 1). Second, slow anaerobic growth using NO_2^- as an alternative electron acceptor is mediated by NO toxicity. NO can be produced from NO_2^- via a pH-dependent chemical reaction (Carlsson, Wiklund, Engstrand, Weitzberg, & Lundberg, 2001; Yoon et al., 2006). In a bacterial culture medium that contains 15 mM NO_2^- with the pH adjusted to 6.5, NO is persistently produced at concentrations as high as ~500 nM for more than 7 h. These chemical reactions, where HNO_2 (nitrous acid) is involved as an intermediate, are presented in detail (Yoon et al., 2006). Identical bacterial killing was observed under two different conditions: 15 mM NO_2^- at pH 6.5 and 1.5 mM NO_2^- at pH 5.5, demonstrating that formation of HNO_2 is strictly determined by the chemical equilibrium between H^+ and NO_2^-, and HNO_2 is a prerequisite for subsequent NO production (Fig. 1) (Yoon et al., 2006). The presence of vitamin C, presumed to decrease the pH of the medium, potentiates the bactericidal effects of acidified NO_2^- against *P. aeruginosa* (Carlsson et al., 2001).

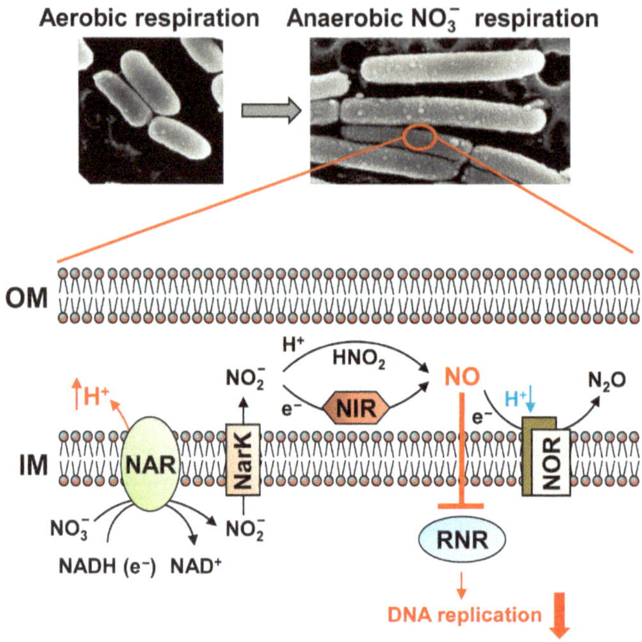

Fig. 1 Denitrification-induced cell elongation in *P. aeruginosa*. Nitrate reductase (NAR) reduces NO_3^-, an alternative electron acceptor, to NO_2^-, which is further reduced to NO by periplasmic nitrite reductase (NIR). In the presence of a proton (H^+), NO_2^- is protonated to HNO_2, which is then decomposed to NO. Endogenously accumulated NO intoxicates ribonucleotide reductase (RNR), which in turn results in decreased synthesis of deoxyribonucleotides.

NO is also enzymatically produced by NIR (Fig. 1). *P. aeruginosa* NIR is a homodimer, and each monomer is composed of cytochrome *c* and d_1. NIR-mediated NO production has been well characterized at the protein level (Rinaldo et al., 2008). The haem group in cytochrome d_1 is the catalytic centre; when reduced, ferrous iron (${d_1}^{2+}$) has high affinity for NO_2^-. With dehydration and oxidization of the iron centre at d_1-haem, NO is produced as $(c^{2+})({d_1}^{3+})$-NO adducts. Then, NO is rapidly displaced with the incoming NO_2^- substrate in the presence of electrons transferred from donors, such as cytochrome *c* (Cutruzzola et al., 2002) or azurin (van de Kamp et al., 1990). This enzymatic activity is pH dependent. At pH 6.2 and 8.0, 11.2 and 1.9 molecules of NO are produced per second from 1 nM of purified NIR, respectively (Rinaldo, Arcovito, Brunori, & Cutruzzola, 2007). This kinetic data can provide a theoretical basis to estimate how much NO is produced when *P. aeruginosa* grows under NO_2^- respiration. Excess amounts of free NO failed to inactivate the enzyme (Rinaldo, Arcovito, et al., 2007). This finding suggests that the enzyme is not subject to product inhibition and can continue to produce NO when NO_2^- is provided. Given that NO has a poisonous effect, this feature implies that NIR activity exerts a very important effect on *P. aeruginosa* anaerobiosis.

2.3 Nitric Oxide Reductase, Another Enzyme That Controls NO Production

NO, produced by NIR, is decomposed to N_2O by NOR, an inner membrane-associated enzyme complex (Borrero-de Acuña et al., 2016). Because NO accumulation in growing cells is also regulated by its degradation rate, understanding the details of this mechanism is important. The NOR complex is composed of two subunits, NorB and NorC, the latter of which contains cytochrome *c* that accepts electrons from outside donors. The NorB subunit is inner membrane bound and accommodates two NO molecules at its catalytic centre (Moenne-Loccoz, 2007). A structural study identified a branched hydrophobic channel, which connects the catalytic centre of the NorB subunit to the surface in the membrane-spanning region, as a channel for the entry of NO (Hino et al., 2010). However, it is not known how fast the NO produced in the periplasmic space by NIR is transferred to the catalytic core of the membrane-associated NorB. This spatial distance can account for diffusional release of NO between these two enzymatic reactions. Consistent with this notion, nanomolar amounts of NO were detected when PAO1, a prototype *P. aeruginosa* laboratory strain, began to grow by anaerobic NO_3^- respiration (Yoon et al., 2007). In a different study analysing gaseous denitrification products, 4% of the initial NO_3^- (~2 mM) was transformed into NO in *P. aeruginosa* (Kalkowski & Conrad, 1991). In the same study, N_2O and N_2 gases were found to occupy 47% and 49% of the consumed NO_3^- at the end of the anaerobic growth period. Together, these results suggest that, while *P. aeruginosa* can control its enzymatic repertoire to suppress NO accumulation, a detectable level of NO is produced during anaerobic respiration. This claim was further supported by mutational studies. When NOR was inactivated, NO accumulation steadily increased as the culture continued and anaerobic growth was ceased (Yoon et al., 2007), while NO accumulation was not detected in a mutant defective in NIR (Yoon et al., 2007, 2011).

NO production during NO_3^- respiration appears to be dependent on the NO_2^- level accumulated in the culture media. In three independent studies, NO_2^- level continuously increased until 8–10 h after inoculation and decreased rapidly thereafter (Kalkowski & Conrad, 1991; Williams et al., 1978; Yoon et al., 2002). On the other hand, the concentration of NO_3^- remaining in the media steadily decreased as the culture proceeded. These findings suggest that (i) NAR exhibits constant activity throughout the growth period, regardless of how much substrate is present, and (ii) NIR is fully activated when the NO_2^- substrate reaches a threshold concentration. Because NO is also produced via chemical decomposition of HNO_2, formation of which is dependent on the amount of NO_2^-, control of NO_2^- accumulation is important to keep NO under nontoxic levels. More experiments are necessary to fully understand the mechanisms by which regulation of multiple denitrification enzymes is orchestrated in a timely manner.

NO reduction by NOR requires two protons, and there is evidence that the protons are provided from the periplasmic space through a hydrogen-bonding network between H_2O and electrically charged residues (Hino et al., 2010). Proton transfer from the periplasmic space leads to a decrease in the proton motive force. The presence of NOR, therefore, is aimed at removing toxic NO rather than supporting anaerobic respiration. Consistent with this notion, a *P. aeruginosa* mutant defective in NOR was less capable of intracellular survival inside murine macrophage cells pretreated with LPS in order to stimulate NO production (Kakishima, Shiratsuchi, Taoka, Nakanishi, & Fukumori, 2007). Likewise, disruption of NOR resulted in reduced virulence towards the silkworm, *Bombyx mori*, an insect species that depends on NO for its innate immunity (Arai & Iiyama, 2013).

3. REGULATION OF DENITRIFICATION

3.1 Signalling Network

As denitrification is the major source of endogenous NO in *P. aeruginosa*, it is important to understand how this metabolic process is regulated. The expression of denitrifying enzymes is induced by low oxygen tension and the presence of N-oxides, including NO (Arai, 2011) (Fig. 2). A hierarchical regulatory network accomplishes this by integrating oxygen and N-oxide as signals to modulate the activation of regulators where low oxygen tension is the initial signal that activates the denitrification pathway (Arai, Kodama, & Igarashi, 1997). In this cascade, the master regulator ANR induces the expression of NarXL, DNR, and NAR. Then, NarXL and

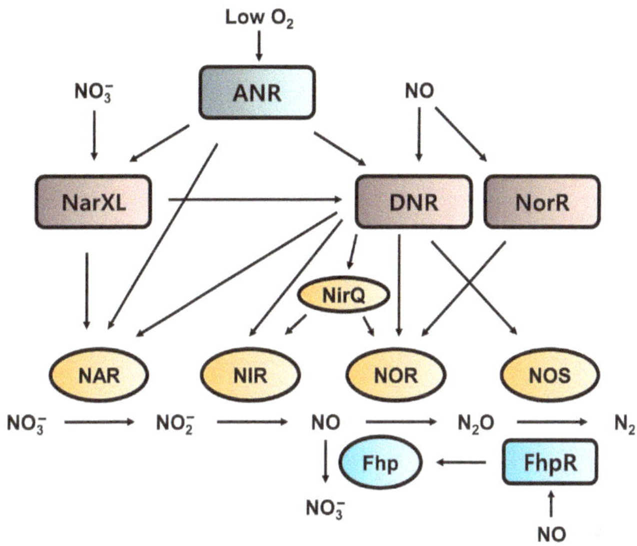

Fig. 2 The regulation of denitrification in *P. aeruginosa*. ANR is the master regulator of denitrification that is activated under low oxygen conditions. Activated ANR induces the expression of NAR, NarXL, and DNR. The two-component regulator, NarXL, senses NO_3^- and stimulates the expression of NAR. DNR induces the expression of NAR, NIR, NOR, and NOS in response to NO. NirQ modulates the activity of NIR and NOR. NO can also be oxidized to NO_3^- by Fhp in the presence of oxygen. FhpR induces *fhp* expression in response to NO.

DNR are activated in response to N-oxides and further induce the upregulation of denitrification genes (Arai et al., 1997; Schreiber et al., 2007). ANR belongs to the CRP/FNR superfamily and contains an Fe—S cluster that is inactivated by oxygen (Zimmermann, Reimmann, Galimand, & Haas, 1991). When oxygen is depleted, active ANR binds to a DNA-binding site known as the ANR box and regulates expression of genes related to fermentation and denitrification, initiating the cell for anaerobic metabolism (Sonnleitner et al., 2011; Trunk et al., 2010). Once the initial gene expression is activated by ANR, DNR, which also belongs to the CRP/FNR superfamily, directly induces the expression of denitrifying genes. In this cascade, ANR activates expression of the *nir*, *nor*, and *nos* genes through DNR; therefore, an *anr* mutant cannot grow under denitrifying conditions (Arai et al., 1997). The major signal detected by DNR is NO (Castiglione, Rinaldo, Giardina, & Cutruzzolà, 2009; Giardina et al., 2008), which presumably allows the cell to fine-tune the denitrification pathway in response to this intermediate. NarX, a sensor kinase, is also involved in the initiation of the denitrification pathway by sensing nitrate. NarX activates the NarL response regulator, which in turn activates the transcription of *narK1*, *nirQ*, and *dnr* (Schreiber et al., 2007). NirQ activates NIR and NOR posttranslationally (Hayashi, Arai, Kodama, & Igarashi, 1998). While initiating denitrification, NarL represses the expression of genes involved in arginine fermentation (Benkert et al., 2008). As the master regulator for anaerobic metabolism, ANR activates both the denitrification and fermentation pathways, causing the regulatory cascade involving NarL to repress the less favourable method of anaerobic energy metabolism (arginine fermentation) by sensing nitrate. In addition to oxygen, NO inactivates ANR by altering the Fe—S cluster (Yoon et al., 2007). The inactivation of ANR also derepresses the expression of two NO-scavenging proteins, homogentisate-1,2-dioxygenase (HmgA) and 4-hydroxyphenylpyruvate dioxygenase (Hpd), which further reduces the NO concentration in the cell. *P. aeruginosa* is also capable of carrying out denitrification under conditions where a sufficient amount of oxygen is present (Chen, Xia, & Ju, 2003). In addition, catabolites of phosphatidylcholine that are abundant in the eukaryotic membrane and host lung surfactant induce *nir* expression in an ANR-dependent manner, even in the presence of oxygen (Jackson et al., 2013). Another NO responding regulator involved in NO detoxification is FhpR. FhpR responds to NO and activates the expression of the flavohaemoglobin Fhp (Arai, Hayashi, Kuroi, Ishii, & Igarashi, 2005; Koskenkorva et al., 2008). Fhp controls the level of NO by converting NO to nitrate under aerobic or microaerobic conditions.

3.2 Roles of Quorum Sensing

As described earlier, ANR, DNR, and NarXL are the major regulators of denitrification activation. In addition to these regulators, quorum sensing (QS) has been found to control denitrification activity and influence NO accumulation (Toyofuku et al., 2007, 2008; Yoon et al., 2002). QS enables bacteria to monitor their cell density in response to signalling molecules and typically regulates hundreds of genes. *P. aeruginosa* possesses at least three QS systems, two of which utilize *N*-acyl-L-homoserine lactone (AHL) (de Kievit & Iglewski, 2000), while the other utilizes 2-heptyl-3-hydroxy-4-quinolone, referred to as the *Pseudomonas* quinolone signal (PQS) (Pesci et al., 1999). AHLs are produced in a number of Gram-negative bacteria, in which signal specificity is conferred by variations in side chains (Schuster, Sexton, Diggle, & Greenberg, 2013).

P. aeruginosa produces *N*-(3-oxododecanoyl)-L-homoserine lactone (3-oxo-C12-HSL) and *N*-butyryl-L-homoserine lactone (C4-HSL). These AHL signals are received by their cognate receptors LasR and RhlR, respectively (Gambello & Iglewski, 1991; Ochsner, Koch, Fiechter, & Reiser, 1994). PQS is received by PqsR, which can also receive the direct precursor 2-heptyl-4-quinolone, but at lower affinity compared to PQS (Xiao et al., 2006). All of these QS molecules, C4-HSL, 3-oxo-C12-HSL, and PQS, repress denitrification in *P. aeruginosa* (Toyofuku et al., 2007, 2008; Yoon et al., 2002), suggesting that denitrification is repressed when the cell density is high. Several different mechanisms underlie how QS molecules influence denitrification. AHLs influence the transcription of denitrification genes, whereas PQS is indicated to impact the activity of denitrification enzymes posttranslationally. Interestingly, the influence of AHLs and PQS on denitrification has been suggested to alter NO accumulation. This was first described in a *rhlR* mutant in which NIR activity was greatly increased compared to NOR activity in the presence of nitrate and absence of oxygen, resulting in accumulation of NO and rapid cell death in a biofilm of the *rhlR* mutant (Yoon et al., 2002). Most of the cells were alive when *nirS* was deleted in addition to *rhlR*, indicating the involvement of NO in anaerobic cell death. Using planktonic cells, it was further shown that C4-HSL and 3-oxo-C12-HSL repress the expression of denitrification genes through their receptors, RhlR and LasR (Toyofuku et al., 2007). In this regulation, the *las* QS system regulates denitrification through the *rhl* QS system. In contrast to the AHLs, PQS mainly influences denitrification independently through its cognate receptor PqsR and has a direct influence on the respiratory chain (Toyofuku et al., 2008). The PQS-mediated control of denitrification represses NOR activity while increasing NIR activity. This mode of action can lead to the accumulation of NO, similar to what has been observed in the *rhlR* mutant (Yoon et al., 2002). Because PQS requires oxygen in its final step of biosynthesis, the PQS control of denitrification would be relevant under conditions where both PQS biosynthesis and denitrification take place (Toyofuku et al., 2008). Biofilms are one such environment where microenvironments can be created by the metabolic activity of the bacteria (Xu, Stewart, Xia, Huang, & McFeters, 1998). While NO generated by denitrification can induce biofilm dispersal (Barraud et al., 2006), how NO accumulates and induces dispersal at specific times is not understood. The control of denitrification by the QS systems suggests that they could be one of the factors that determine the timing of dispersal by controlling the level of NO when the biofilm is mature.

4. NO, AN IMPORTANT MOLECULE THAT INFLUENCES *P. AERUGINOSA* BEHAVIOUR

4.1 Impact of NO in *P. aeruginosa* Gene Regulation

Depending on the dose, NO can function as a cue, signal, or stress in bacteria. At low concentrations, NO alters gene expression through sensor proteins or regulatory proteins such as DNR; at higher concentrations, it induces a stress response. Recent findings of widespread NO-sensor proteins suggest that the influence of NO on gene regulation is a general feature in bacteria (Cutruzzolà & Frankenberg-Dinkel, 2016). As was first shown in *P. aeruginosa*, NO can induce biofilm dispersal, indicating that genes other than those of denitrification and NO detoxification are controlled by NO, which can act as a cue that alters the behaviour of cells (Barraud et al., 2006). Led by the discovery in *P. aeruginosa*, biofilm dispersal has also been shown to be regulated by NO in many other bacterial species. One of the key signals involved in this NO-regulated biofilm dispersal is a second messenger, bis-(3′-5′)-cyclic dimeric GMP (c-di-GMP) (Barraud et al., 2009). In general, the intracellular c-di-GMP level controls motility and biofilm matrix production in opposing manners (Jenal, Reinders, & Lori, 2017). High levels of c-di-GMP inhibit motility and promote the formation of biofilms, whereas low c-di-GMP levels upregulate motility and support a planktonic lifestyle (Boyd & O'Toole, 2012). The intracellular c-di-GMP level is controlled by proteins with diguanylate cyclase (DGC) and phosphodiesterase (PDE) activity that are involved in the synthesis or degradation of c-di-GMP. Enzymes with DGC activity can be identified by a GGDEF domain, whereas PDE activity can be identified by the EAL or HD-GYP domains. These domains are often found to be associated with sensing domains that respond to a certain stimulus, suggesting that specific stimuli regulate the activity of DGC and PDE.

In other species than *P. aeruginosa*, NO sensors with an NO-binding domain called H-NOX (haem-nitric oxide or oxygen-binding domains) are suggested to be involved in the NO regulation of c-di-GMP (Plate & Marletta, 2013). H-NOX genes are encoded in more than 300 species of sequenced bacteria and are adjacent to genes encoding GGDEF DGCs, EAL and HD-GYP c-di-GMP PDEs, or two-component system histidine kinases, suggesting the role of H-NOX in signal relay pathways (Arora, Hossain, Xu, & Boon, 2015). For example, in *Legionella pneumophila*, NO-bound H-NOX inhibits the synthesis of c-di-GMP and plays an essential role in biofilm formation (Carlson, Vance, & Marletta, 2010). Many bacteria including *P. aeruginosa* do not possess H-NOX but are still able to respond to NO and regulate c-di-GMP level, suggesting that they possess alternative NO sensors (Arora et al., 2015). BdlA is one of the key proteins involved in NO-induced biofilm dispersal. Although the NO sensory mechanism of BdlA is not understood, the deletion of the *bdlA* gene leads to impaired biofilm dispersal even in the presence of NO (Barraud, Schleheck, et al., 2009; Morgan, Kohn, Hwang, Hassett, & Sauer, 2006). BdlA lacks DGC or PDE activity, but it appears that BdlA modulates the c-di-GMP level by interacting with at least with two PDEs, DipA and RbdA (Petrova & Sauer, 2012). BdlA is also involved in biofilm dispersion in response to other environmental cues than NO and might be a general sensor (Morgan et al., 2006). A recent study has identified a new class of NO sensors that are called NosP (nitric oxide-sensing proteins) (Hossain & Boon, 2017). NosP is a haemoprotein through which haem ligates to NO. The NO-bound NosP is able to regulate the phosphorelay activity of a hybrid histidine kinase (NahK) that is encoded in the same operon and involved in biofilm regulation. When NO is present, NosP inhibits NahK activity, further leading to inhibition of the phosphotransfer to HptB. This leads to biofilm dispersal through mechanisms that are not yet identified. Importantly, NosP proteins are widely conserved among bacteria and found in ~800 sequenced genomes (Hossain, Nisbett, & Boon, 2017), suggesting the general role of these proteins in NO sensing.

4.2 Role of NO in *P. aeruginosa* Cell Morphology

LB supplemented with NO_3^- (LBN) has long been used as a standard medium for anaerobic cultivation of *P. aeruginosa*. *P. aeruginosa* is capable of robust anaerobic growth in LBN, although not as high as aerobic growth. However, recent findings (Lee et al., 2012; Yoon et al., 2011) demonstrate that anaerobic growth by NO_3^- respiration results in unique bacterial growth. *P. aeruginosa* cells grown under such conditions were found to undergo a significant degree of cell elongation (Fig. 1). What was more surprising was that this unexpected growth-associated phenotype was observed even in wild-type cells. Subsequent studies have demonstrated that this abnormal cell shape change is associated with incomplete septum formation (Yoon et al., 2011) and/or defective DNA replication (Lee et al., 2012). Cell elongation was not detected in PAO1 mutants of Δ*anr* and Δ*nirS*. As described earlier, the gene *anr* encodes an FNR (fumarate/nitrate regulator/CRP)-like transcriptional factor that plays an essential role in supporting anaerobic growth (Arai et al., 1997). Because the Δ*anr* mutant does not grow by anaerobic respiration, it is clear that cell elongation is caused by active anaerobic growth rather than by simple exposure to anaerobic environments. Cell elongation was found to be associated with anaerobiosis-induced NO production. The Δ*nirS* mutant, which does not produce NO, is not elongated, and cell elongation was significantly reduced when NO was removed by an intracellular NO scavenger (Yoon et al., 2011). Taken together, these results suggest that (i) anaerobic growth of *P. aeruginosa* by denitrification is not a biologically flawless process, as has been believed for a long time, and (ii) NO is actually produced at a higher level than *P. aeruginosa* cells can tolerate.

Of note, the above-mentioned abnormal phenotypes were not observed in a PAO1 mutant of Δ*ccoN1*Δ*ccoN2* defective in synthesizing two cbb_3-type cytochrome *c* oxidases (Hamada, Toyofuku, Miyano, & Nomura, 2014). It was counterintuitive that these two enzymes, known to be functional under aerobic conditions, participate in regulating anaerobiosis-induced changes in cell shape. In this mutant strain, NO production during anaerobic NO_3^- respiration was decreased, and cell elongation was concomitantly minimized. Accumulation of N_2O, a product of NOR, was remarkably increased in the mutant, suggesting that cbb_3-type cytochrome *c* oxidases promote NO production by suppressing NOR activity. It will be important to understand how functions of anaerobic respiratory enzymes are modulated by cytochrome *c* oxidases and aerobic enzymes.

4.3 Role of NO in *P. aeruginosa* Virulence

Since the mucus layer of the respiratory tract of CF patients has been found to be anaerobic (Worlitzsch et al., 2002), the virulence features of *P. aeruginosa* under anaerobic conditions have been actively explored. However, there are conflicting results, and additional studies are necessary to precisely understand anaerobiosis-induced modulation of *P. aeruginosa* virulence.

Δ*narGH* and Δ*nirS* mutants of the PA14 strain were found to be less capable of eliciting cytotoxicity towards a human monocyte cell line, THP-1 (Van Alst, Wellington, Clark, Haidaris, & Iglewski, 2009). This reduced virulence was due to suppressed production of type III secretion system (T3SS) components and other virulence determinants, such as extracellular protease and elastase. Interestingly, bacterial virulence was monitored aerobically in that study. Therefore, these results likely suggest that disruption of NAR and NIR, localized in the inner membrane and periplasm, can affect the bacterial cell envelope structure, which in turn prevents secretion of virulence factors.

Several reports have revealed that anaerobiosis-induced virulence alteration occurs at the transcriptional level. O'Callaghan et al. reported that the transcription of T3SS genes was upregulated by low oxygen tension (O'Callaghan, Reen, Adams, & O'Gara, 2011). Likewise, the expression of T3SS and anaerobic respiration genes was simultaneously induced in a *P. aeruginosa* strain grown in a mucin-containing artificial medium that mimics anaerobic CF lung sputum (Fung et al., 2010). Recently, it was found that AtvR, an atypical two-component response regulator, plays an important role in *P. aeruginosa* virulence in vitro and in vivo (Kaihami et al., 2017). Of particular interest is that AtvR-activated transcription of genes coding for anaerobic respiratory enzymes was induced during anaerobic NO_3^- respiration, along with its own expression (Kaihami et al., 2017). Together, these results suggest that anaerobiosis positively regulates the expression of virulence-associated genes.

Some papers, on the other hand, reported opposing findings on the virulence of *P. aeruginosa* under anaerobic conditions. Production of elastase, one of the strongest virulence determinants, was significantly downregulated when PAO1 was grown by anaerobic respiration (Fang et al., 2013; Lee, Yoon, Park, Lee, & Yoon, 2011). As a result, the capability to kill A549 human airway epithelial cells was accordingly decreased in anaerobically grown *P. aeruginosa* (Lee et al., 2011). Further investigation revealed that QS machineries, including HSL-based autoinducer production and transcription of QS-related genes, were substantially suppressed during anaerobic growth (Lee et al., 2011). Consistently, PQS, another critical QS extracellular signal molecule, requires oxygen for its synthesis, and PQS-regulated virulence determinants were not produced (Schertzer, Brown, & Whiteley, 2010). Future research will address (i) how anaerobiosis-induced cell shape changes affect the virulence potential of *P. aeruginosa* and (ii) how the balance between activation of T3SS and deactivation of QS affects anaerobic growth to control *P. aeruginosa* pathogenic mechanisms.

4.4 Role of NO in *P. aeruginosa* Biofilms

Biofilm formation is often described as a process by which bacterial cells develop into a multicellular social community (Costerton, Stewart, & Greenberg, 1999). Biofilm formation, perceived as a developmental process of bacteria, can be divided into distinct successive steps: (i) initial attachment of planktonic cells to surfaces (O'Toole & Kolter, 1998), (ii) formation of colonies of an intermediate scale (Sriramulu, Lunsdorf, Lam, & Romling, 2005), (iii) secretion of matrix molecules and growth into larger-scale communities (Ryder, Byrd, & Wozniak, 2007), and (iv) differentiation into a mature biofilm with three-dimensional architecture (Parsek & Tolker-Nielsen, 2008).

In 2002, it was found that *P. aeruginosa* forms significantly more robust biofilms during anaerobic vs aerobic respiratory growth (Yoon et al., 2002). This finding was unexpected since *P. aeruginosa* had already been known to form relatively dense biofilms during normal aerobic growth. This result also indicated that *P. aeruginosa* can proactively respond to anaerobiosis, thereby converting to a more resistant mode of growth. Subsequent studies have shown that robust biofilm formation induced upon anaerobiosis is closely related with (i) cell elongation, a hallmark of faulty anaerobic growth of *P. aeruginosa*, and (ii) extracellular DNA (eDNA) release by self-lysis of a subpopulation of cells.

The increase in biofilm formation in association with cell elongation has been demonstrated in two independent experiments. First, nonelongated Δ*nirS* mutants did not form biofilms, while robust biofilms were detected by *P. aeruginosa* strains that underwent cell elongation during anaerobic NO_3^- respiration (Hamada et al., 2014; Yoon et al., 2011). Second, when PAO1 was anaerobically grown with vitamin B_{12} (VB_{12}), a cofactor of class II ribonucleotide reductase (RNR), DNA replication proceeded normally, and cell elongation was not observed. Under this amended growth condition, robust biofilm formation was abrogated (Lee et al., 2012). These two results strongly demonstrated that anaerobiosis-induced cell elongation is responsible for robust anaerobic biofilm formation.

eDNA has shown to play a critical role in *P. aeruginosa* biofilm formation (Whitchurch, Tolker-Nielsen, Ragas, & Mattick, 2002). When PAO1 was grown in the presence of DNase I, biofilm formation was inhibited. Furthermore, established biofilms also lost their structure upon DNase 1 treatment, demonstrating that eDNA is a core component of the *P. aeruginosa* biofilm matrix. Jennings and colleagues found that the positively charged Pel exopolysaccharide can cross-link eDNA to increase biofilm physical strength (Jennings et al., 2015). eDNA release during biofilm formation is controlled by bacterial self-lysis due to the induction of prophase genes (Webb et al., 2003; Whiteley et al., 2001). The PAO1 genome harbours two distinct loci that presumably express cryptic prophage genes. The first locus ranging from PA0614 to PA0648 is also known as a cluster coding for pyocins that structurally resemble bacteriophage tails (Michel-Briand & Baysse, 2002), while the second locus from PA0717 to PA0728 encodes a Pf1 filamentous bacteriophage (Hill, Short, Perham, & Petersen, 1991). Importantly, the transcription of genes in these two loci was remarkably induced during the biofilm mode of growth (Turnbull et al., 2016; Whiteley et al., 2001), and the induction of pyocin genes resulted in explosive cell lysis, thereby releasing eDNA (Turnbull et al., 2016). Similarly, the expression of genes belonging to the first cluster was strikingly upregulated in PAO1 grown by anaerobic NO_3^- respiration (Lee et al., 2012; Waite & Curtis, 2009). Evidence suggests that gene expression of pyocins is stimulated under SOS-inducing stress conditions, such as H_2O_2 treatment (Chang, Small, Toghrol, & Bentley, 2005) and antibiotic treatment (Brazas & Hancock, 2005). Likewise, NO, produced as an inevitable intermediate during anaerobic respiration, was shown to trigger the expression of pyocin genes (Toyofuku et al., 2014).

4.5 Role of NO in *P. aeruginosa* Membrane Vesicle Formation

High concentrations of NO can kill *P. aeruginosa*, presumably due to the cytotoxicity of NO due to its direct reaction with cellular components (Privett, Broadnax, Bauman, Riccio, & Schoenfisch, 2012; Schairer, Chouake, Nosanchuk, & Friedman, 2012). Toxic concentrations of NO can also be produced by denitrification, resulting in self-induction of the SOS response under anaerobic conditions (Lee et al., 2012; Toyofuku et al., 2014). In general, the SOS response is a DNA repair system that is induced under DNA-damaging stress conditions. NO activation of the SOS response is also responsible for the majority of MVs produced under anaerobic conditions (Toyofuku et al., 2014; Turnbull et al., 2016) (Fig. 3). MVs, typically ranging in size from 20 to 400 nm in diameter, are produced in many bacteria including Gram-positive and Gram-negative bacteria as well as in natural environments (Biller et al., 2014; Brown, Wolf, Prados-Rosales, & Casadevall, 2015; Schwechheimer & Kuehn, 2015; Toyofuku et al., 2017). These MVs carry proteins, cell-to-cell communication signals, DNA, and RNA and are involved in many important biological processes. They can also neutralize antibiotics and work as a decoy against phages, indicating their role as a defence mechanism. Under aerobic conditions, PQS can induce MV formation in *P. aeruginosa* and other bacteria mainly by interacting with the

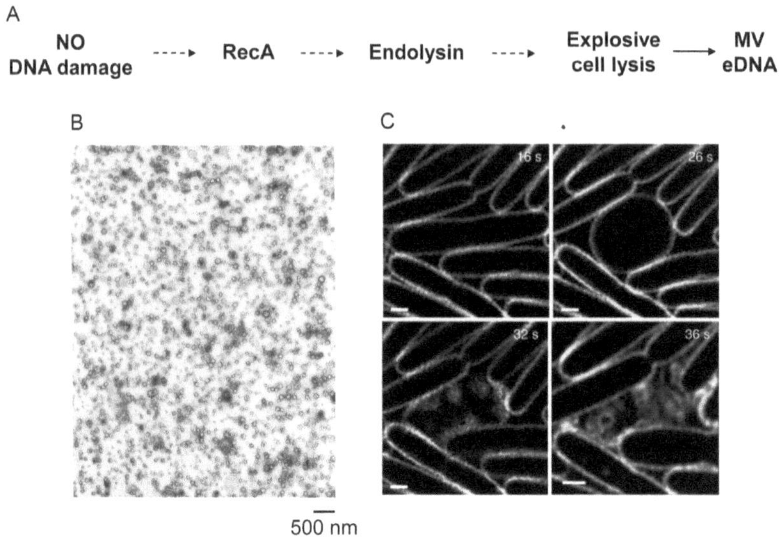

Fig. 3 Membrane vesicle formation in *P. aeruginosa*. (A) Membrane vesicle (MV) induction by DNA-damaging stress. Under anaerobic conditions, NO can trigger MV formation through explosive cell lysis. (B) MVs isolated from a *P. aeruginosa* anaerobic culture. (C) MV formation through explosive cell lysis. *Panel (B): Adapted from Toyofuku, M., Zhou, S., Sawada, I., Takaya, N., Uchiyama, H., & Nomura, N. (2014).* Environmental Microbiology, 16*(9), 2927–2938. Panel (C): Adapted and modified from Turnbull, L., Toyofuku, M., Hynen, A. L., Kurosawa, M., Pessi, G., Petty, N. K., et al. (2016).* Nature Communications, *7, 11220.*

membrane (Mashburn-Warren et al., 2008; Schertzer & Whiteley, 2012; Tashiro, Ichikawa, Nakajima-Kambe, Uchimura, & Nomura, 2010). The interaction with the membrane results in localized membrane curvature, leading to membrane budding. Because PQS is not produced under anaerobic conditions (Toyofuku et al., 2008), it is assumed that there are alternative mechanisms for MV formation (Toyofuku et al., 2014). A *nirS* mutant did not produce MVs under anaerobic conditions, indicating the involvement of NO in this process. It was further shown that pyocin production activated by RecA, a key regulatory protein in the SOS response, is involved in anaerobic MV formation. A more recent study showed that endolysins (PA0629) are key enzymes in the MV formation process. Endolysins are peptidoglycan (PG)-degrading enzymes that are typically used by dsDNA phages to disrupt their hosts and be released in the environment. In *P. aeruginosa*, an endolysin is encoded in the pyocin gene cluster and is involved in the release of pyocins (Nakayama et al., 2000). Once the endolysin degrades PG, the cells agglomerate and explode, resulting in the production of MVs from the shattered membrane (Turnbull et al., 2016). This process, named explosive cell lysis, contributes to the majority of MVs produced in biofilms where cells are exposed to higher levels of oxidative stress compared to planktonic cells (Boles & Singh, 2008). In biofilms, explosive cell lysis also contributes to the release of DNA, which is an important component of the extracellular matrix (Turnbull et al., 2016; Whitchurch et al., 2002). The simultaneous release of MVs and DNA by explosive cell lysis can explain why many proteins observed in the biofilm matrix are derived from MVs (Toyofuku, Roschitzki, Riedel, & Eberl, 2012). Although cells that undergo explosive cell lysis will eventually die, endolysins that trigger this event are only induced in a subpopulation of the cells, which allow the other cells to use the materials released from explosive cell lysis. Under aerobic conditions, less than 1% of cells express endolysins, which can be induced upon DNA-damaging stress. Under anaerobic conditions, the endolysin gene is induced significantly more than under aerobic conditions due to the accumulation of NO (Toyofuku et al., 2014). Inhibition of DNA biosynthesis by NO is the main cause of induction of the SOS response and therefore endolysin under anaerobic conditions. NO inactivates the class II ribonucleotide reductase (RNR II) that is one of the essential DNA biosynthesis proteins for *P. aeruginosa* under anaerobic conditions (Crespo, Pedraz, Astola, & Torrents, 2016; Lee et al., 2012). The inactivation of RNR II can be rescued by adding its coenzyme VB_{12} (Lee et al., 2012); in accordance with this finding, adding VB_{12} represses MV formation under anaerobic conditions (Toyofuku et al., 2014). The accumulation of NO and its inhibition of RNR II are influenced by environmental factors, which also explains the contradictory results of MV formation under different anaerobic conditions (Schertzer et al., 2010; Toyofuku et al., 2014).

5. CLINICAL RELEVANCE OF NO IN *P. AERUGINOSA* INFECTION

5.1 Bacterial Phenotypic Switching Inside Airway Mucus

Recently, many CF isolates that harbour mutations in the *lasR* gene have been recovered (D'Argenio et al., 2007; Hoffman et al., 2009). This finding is contradictory to the established knowledge that *lasR*-associated QS plays an essential role in *P. aeruginosa* virulence. A subsequent study, however, reported that mutations in *lasR* confer increased survival fitness in CF airways, where bacteria shift their energy metabolism to anaerobic respiration (Hoffman et al., 2010). This result further suggests that infection dynamics inside a patient's airway are very complicated, and the role of QS in *P. aeruginosa* pathogenesis in vivo should be reevaluated.

Interestingly, QS mutants, in which the *lasR* gene or *rhlR* gene is disrupted, have reduced viability during in vitro biofilm formation under anaerobic respiration, a phenomenon due to the overproduction of toxic NO, a by-product of anaerobic respiration (Yoon et al., 2002). This suggests that QS is required to maintain viability during anaerobic biofilm formation and supports the presence of a novel function of *P. aeruginosa* QS. Further investigation is warranted to better understand the function of QS during anaerobic growth, a mode of proliferation that occurs in the airway of CF patients.

One of the hallmarks of *P. aeruginosa* pathogenesis in the CF airway is phenotypic conversion to mucoid variants that overproduce alginate polysaccharide (Yoon et al., 2006). While mucoid conversion has been credited to the *mucA* gene mutation (Jones et al., 2010; Martin et al., 1993), the cause of this mutation is not clearly understood. The response to varying oxygen tension has been reported to trigger alginate biosynthesis (Bayer, Eftekhar, Tu, Nast, & Speert, 1990). Importantly, alginate production was promoted in nonmucoid *P. aeruginosa* cells residing in the CF mucus airway (Worlitzsch et al., 2002), further demonstrating that mucoid conversion is beneficial for the long-term survival of *P. aeruginosa*. Due to the presence of an alginate capsule (Slack & Nichols, 1981), mucoid *P. aeruginosa* strains are resistant to antibiotic treatments (Govan & Fyfe, 1978) and active neutrophils (Cabral, Loh, & Speert, 1987). It was found that mucoid *P. aeruginosa* strains showing inherent resistance are highly susceptible to NO (Yoon et al., 2006). Mucoid strains, compared with nonmucoid counterparts, exhibited considerably higher NAR activity, while the activity of NIR and NOR was nondetectable and very low, respectively. These results suggest that (i) NO_2^- accumulation is induced in the environment where NO_3^- is present, and anaerobic respiration is activated, and (ii) mucoid strains can easily be eliminated by NO, which is chemically synthesized from accumulated nitrite. Importantly, CF mucus airways are mildly acidified, promoting the formation of HNO_2, an intermediate of NO generation (Major et al., 2010; Yoon et al., 2006). Together, these results suggest that compounds capable of inducing NO production have the potential to be used as therapeutics to remove mucoid-type *P. aeruginosa* from the airway in CF patients.

5.2 Use of NO-Releasing Molecules

The antibacterial and biofilm dispersal effects of NO make this molecule attractive in the treatment of bacterial infection. Due to the cytotoxicity of NO at high doses, low-dose treatment might be preferable, especially for use in vivo. The influence of NO has been observed in a wide range of biofilms, suggesting that NO treatment of biofilms can allow novel strategies to control biofilms in environmental, industrial, and clinical settings (Barraud, Kelso, Rice, & Kjelleberg, 2015). For example, it has been shown that addition of a nanomolar dose of NO causes dispersal of multispecies biofilms formed in water distribution and treatment systems (Barraud et al., 2009). Recently, a clinical study was carried out showing that inhaling NO gas at low doses can reduce the presence of *P. aeruginosa* biofilms in CF patients treated with antibiotics (Howlin et al., 2017). Because the NO dose is one of the keys to effectively treating bacteria, there are challenges in applying NO to bacterial treatment in terms of how to supply and control the concentrations of this reactive and relatively unstable molecule in different settings where gaseous NO cannot be directly used.

To this end, NO-releasing molecules and materials have been designed that could be used in different forms. Sodium nitroprusside (SNP) is an FDA-approved NO donor that spontaneously releases NO and has been shown to induce biofilm dispersal in various species including *P. aeruginosa* (Arora et al., 2015; Barraud et al., 2015). SNP treatment of *P. aeruginosa* biofilms increases the susceptibility of the bacteria to various antimicrobial treatments, including tobramycin and H_2O_2 (Barraud et al., 2006). Some of the key NO donors are the highly versatile *N*-diazeniumdiolates (NONOates) (Keefer, 2003). NONOates can have different chemical backbones that modify their NO-releasing rates and physical properties. They can be incorporated into nanoparticles and polymers or conjugated with different moieties that allow the release of NO under specific conditions (Carpenter, Slomberg, Rao, & Schoenfisch, 2011; Duong, Adnan, et al., 2014; Duong, Jung, et al., 2014; Hetrick, Shin, Paul, & Schoenfisch, 2009; Hetrick et al., 2008; Lu, Slomberg, Shah, & Schoenfisch, 2013). These modifications can help change the physicochemical properties of NONOates to affect local concentrations of NO. NONOates embedded in macromolecules can also be used to coat surfaces to prevent bacterial adhesion. A NONOate-embedded sol–gel has been developed and was shown to reduce the adhesion of *P. aeruginosa* compared to a control sol–gel (Nablo & Schoenfisch, 2003). Biocompatible macromolecules such as chitosan-based oligosaccharides that are biodegradable have also been used as a scaffold for NONOates, which inhibits aerobic and anaerobic growth of *P. aeruginosa* nonmucoid and mucoid strains (Lu, Slomberg, & Schoenfisch, 2014; Reighard, Hill, Dixon, Worley, & Schoenfisch, 2015; Reighard & Schoenfisch, 2015). NONOates spontaneously release NO at physiological pH and temperature, but can be modified to only release NO under certain conditions. For example, NONOate-based NO donors that release NO upon cleavage by a bacteria-specific enzyme β-lactamase have been developed (Barraud et al., 2012; Yepuri et al., 2013). These molecules consist of NONOates attached to β-lactam analogs, cephalosporins, which can be cleaved by β-lactamase. Upon cleavage, cephalosporins spontaneously release NONOates that will further release NO. These complexes are stable and can store NONOates until the β-lactam rings of cephalosporins are cleaved, specifically targeting the bacteria.

A probiotic approach for generating NO in a patch has also been developed (Jones, Ganopolsky, Labbe, & Prakash, 2010). In this patch, NO production depends on the metabolic activity of immobilized lactic acid bacteria. Fermentation by lactic acid bacteria produces protons that cause a nonenzymatic reduction of nitrite to NO.

6. CONCLUSIONS

Although patient airways are exposed to diverse bacterial pathogens, *P. aeruginosa* has been a major microorganism that successfully colonizes and establishes persistent airway infections. *P. aeruginosa* airway infection should now be approached as an anaerobic lung disease, and this new idea necessitates further research directed at identifying new targets, inhibition of which will decrease bacterial virulence or survival under anaerobic conditions. Because biofilms and QS are two major virulence mechanisms of this clinically important pathogen, future therapeutic strategies for the treatment of airway infection should include a molecular-level understanding of anaerobiosis-induced biofilm formation and QS.

Our knowledge about the influence of NO on *P. aeruginosa* has expanded rapidly in the last decade. Although the detailed molecular mechanisms are not fully understood, several proteins that alter gene expression in response to NO have been identified, indicating the need for further studies on how they network together. Studies in *P. aeruginosa* have shown that NO can work as a cue or signal that alters bacterial behaviour without causing bacterial death and induces a stress response at high concentrations that can stimulate biofilm formation and release of MVs (Fig. 4). The influence of NO can also change depending on the conditions. Under anaerobic

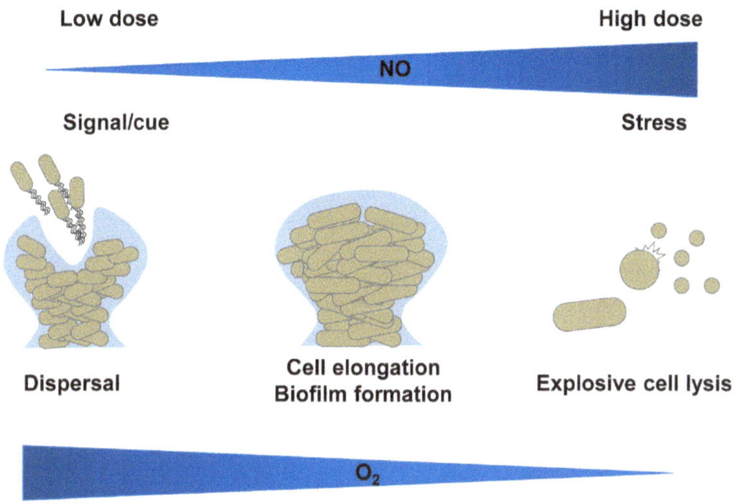

Fig. 4 Schematic model showing how NO alters the behaviour of *P. aeruginosa* depending on the dose and other environmental conditions. NO can act as a signal, cue, or stress depending on its dose and induces different responses in bacteria. Environmental factors such as oxygen can also influence the bacterial response.

conditions, biofilms show distinct structures different from those of aerobic conditions, and biofilm dispersal under such conditions is not well understood. It has been shown that NO stimulates anaerobic biofilm formation through cell elongation (Yoon et al., 2011) and presumably through increased eDNA release (Turnbull et al., 2016), whereas NO triggers dispersal under aerobic conditions (Barraud et al., 2006). Furthermore, *P. aeruginosa* shows a wide range of strain-dependent phenotypic variations under anaerobic conditions, including biofilm formation (Fang et al., 2013), and the impact of NO on anaerobiosis might not be simple. To apply NO treatment to bacteria, it is also important to understand the consequences of long-term exposure to low concentrations of NO. While toxic amounts of NO did not select for NO resistance mutants (Privett et al., 2012), it is not clear whether cells that do not disperse in response to a high level of NO would eventually emerge or not.

Overall, the impact of NO against bacteria largely depends on the NO level and the bacterial growing conditions, indicating that understanding the role of NO under different conditions will be important. Furthermore, microenvironments are naturally formed within biofilms (Xu et al., 1998). One of the next steps would be to integrate the observations studied under different conditions and develop a better understanding of the spatiotemporal patterns of the influence of NO in a heterogeneous environment.

REFERENCES

Adak, S., Aulak, K. S., & Stuehr, D. J. (2002). *Journal of Biological Chemistry*, *277*(18), 16167–16171.

Arai, H. (2011). *Frontiers in Microbiology*, *2*, 103.

Arai, H., Hayashi, M., Kuroi, A., Ishii, M., & Igarashi, Y. (2005). *Journal of Bacteriology*, *187*(12), 3960–3968.

Arai, H., & Iiyama, K. (2013). *Bioscience, Biotechnology and Biochemistry*, 77(1), 198–200.

Arai, H., Kawakami, T., Osamura, T., Hirai, T., Sakai, Y., & Ishii, M. (2014). *Journal of Bacteriology*, *196*(24), 4206–4215.

Arai, H., Kodama, T., & Igarashi, Y. (1997). *Molecular Microbiology*, *25*(6), 1141–1148.

Arora, D. P., Hossain, S., Xu, Y., & Boon, E. M. (2015). *Biochemistry*, *54*(24), 3717–3728.

Barraud, N., Hassett, D. J., Hwang, S. H., Rice, S. A., Kjelleberg, S., & Webb, J. S. (2006). *Journal of Bacteriology*, *188*(21), 7344–7353.

Barraud, N., Kardak, B. G., Yepuri, N. R., Howlin, R. P., Webb, J. S., Faust, S. N., et al. (2012). *Angewandte Chemie (International Ed. in English)*, *51*(36), 9057–9060.

Barraud, N., Kelso, M. J., Rice, S. A., & Kjelleberg, S. (2015). *Current Pharmaceutical Design*, *21*(1), 31–42.

Barraud, N., Schleheck, D., Klebensberger, J., Webb, J. S., Hassett, D. J., Rice, S. A., et al. (2009). *Journal of Bacteriology*, *191*(23), 7333–7342.

Barraud, N., Storey, M. V., Moore, Z. P., Webb, J. S., Rice, S. A., & Kjelleberg, S. (2009). *Microbial Biotechnology*, *2*(3), 370–378.

Bayer, A. S., Eftekhar, F., Tu, J., Nast, C. C., & Speert, D. P. (1990). *Infection and Immunity*, *58*(5), 1344–1349.

Benkert, B., Quack, N., Schreiber, K., Jaensch, L., Jahn, D., & Schobert, M. (2008). *Microbiology (Reading, England)*, *154*(Pt. 10), 3053–3060.

Biller, S. J., Schubotz, F., Roggensack, S. E., Thompson, A. W., Summons, R. E., & Chisholm, S. W. (2014). *Science (New York, NY)*, *343*(6167), 183–186.

Boles, B. R., & Singh, P. K. (2008). *Proceedings of the National Academy of Sciences of the United States of America*, *105*(34), 12503–12508.

Borrero-de Acuña, J. M., Molinari, G., Rohde, M., Dammeyer, T., Wissing, J., Jänsch, L., et al. (2015). *Journal of Bacteriology*, *197*(19), 3066–3075.

Borrero-de Acuña, J. M., Rohde, M., Wissing, J., Jänsch, L., Schobert, M., Molinari, G., et al. (2016). *Journal of Bacteriology*, *198*(9), 1401–1413.

Boyd, C. D., & O'Toole, G. A. (2012). *Annual Review of Cell and Developmental Biology*, *28*, 439–462.

Brazas, M. D., & Hancock, R. E. (2005). *Antimicrobial Agents and Chemotherapy*, *49*(8), 3222–3227.

Brown, L., Wolf, J. M., Prados-Rosales, R., & Casadevall, A. (2015). *Nature Reviews. Microbiology*, *13*(10), 620–630.

Cabral, D. A., Loh, B. A., & Speert, D. P. (1987). *Pediatric Research*, *22*(4), 429–431.

Carlson, H. K., Vance, R. E., & Marletta, M. A. (2010). *Molecular Microbiology*, 77(4), 930–942.

Carlsson, S., Wiklund, N. P., Engstrand, L., Weitzberg, E., & Lundberg, J. O. (2001). *Nitric Oxide: Biology and Chemistry*, *5*(6), 580–586.

Carpenter, A. W., Slomberg, D. L., Rao, K. S., & Schoenfisch, M. H. (2011). *ACS Nano*, *5*(9), 7235–7244.

Castiglione, N., Rinaldo, S., Giardina, G., & Cutruzzolà, F. (2009). *Microbiology (Reading, England)*, *155*(Pt. 9), 2838–2844.

Chang, W., Small, D. A., Toghrol, F., & Bentley, W. E. (2005). *BMC Genomics*, *6*, 115.

Chen, J., & Strous, M. (2013). *Biochimica et Biophysica Acta*, *1827*(2), 136–144.

Chen, F., Xia, Q., & Ju, L. K. (2003). *Applied and Environmental Microbiology*, *69*(11), 6715–6722.

Clarke, P. H. (1982). *Antonie van Leeuwenhoek, 48*(2), 105–130.
Clarke, S. W. (1987). *European Journal of Respiratory Diseases. Supplement, 153*, 136–144.
Costerton, J. W., Stewart, P. S., & Greenberg, E. P. (1999). *Science (New York, NY), 284*(5418), 1318–1322.
Crespo, A., Pedraz, L., Astola, J., & Torrents, E. (2016). *Frontiers in Microbiology, 7*, 688.
Cutruzzola, F., Arese, M., Ranghino, G., van Pouderoyen, G., Canters, G., & Brunori, M. (2002). *Journal of Inorganic Biochemistry, 88*(3–4), 353–361.
Cutruzzolà, F., & Frankenberg-Dinkel, N. (2016). *Journal of Bacteriology, 198*(1), 55–65.
D'Argenio, D. A., Wu, M., Hoffman, L. R., Kulasekara, H. D., Deziel, E., Smith, E. E., et al. (2007). *Molecular Microbiology, 64*(2), 512–533.
de Kievit, T. R., & Iglewski, B. H. (2000). *Infection and Immunity, 68*(9), 4839–4849.
Duong, H. T. T., Adnan, N. N. M., Barraud, N., Basuki, J. S., Kutty, S. K., Jung, K., et al. (2014). *Journal of Materials Chemistry. B, Materials for Biology and Medicine, 2*(31), 5003–5011.
Duong, H. T., Jung, K., Kutty, S. K., Agustina, S., Adnan, N. N., Basuki, J. S., et al. (2014). *Biomacromolecules, 15*(7), 2583–2589.
Fang, F. C. (2004). *Nature Reviews. Microbiology, 2*(10), 820–832.
Fang, H., Toyofuku, M., Kiyokawa, T., Ichihashi, A., Tateda, K., & Nomura, N. (2013). *Bioscience, Biotechnology and Biochemistry, 77*(8), 1747–1752.
Filiatrault, M. J., Picardo, K. F., Ngai, H., Passador, L., & Iglewski, B. H. (2006). *Infection and Immunity, 74*(7), 4237–4245.
Frey, H. R., & Russi, E. W. (1997). *Schweizerische Medizinische Wochenschrift, 127*(6), 219–230.
Frimmersdorf, E., Horatzek, S., Pelnikevich, A., Wiehlmann, L., & Schomburg, D. (2010). *Environmental Microbiology, 12*(6), 1734–1747.
Fung, C., Naughton, S., Turnbull, L., Tingpej, P., Rose, B., Arthur, J., et al. (2010). *Journal of Medical Microbiology, 59*(Pt. 9), 1089–1100.
Gambello, M. J., & Iglewski, B. H. (1991). *Journal of Bacteriology, 173*(9), 3000–3009.
Giardina, G., Rinaldo, S., Johnson, K. A., Di Matteo, A., Brunori, M., & Cutruzzolà, F. (2008). *Journal of Molecular Biology, 378*(5), 1002–1015.
Govan, J. R., & Fyfe, J. A. (1978). *Journal of Antimicrobial Chemotherapy, 4*(3), 233–240.
Gusarov, I., & Nudler, E. (2005). *Proceedings of the National Academy of Sciences of the United States of America, 102*(39), 13855–13860.
Hamada, M., Toyofuku, M., Miyano, T., & Nomura, N. (2014). *Journal of Bacteriology, 196*(22), 3881–3889.
Hayashi, N. R., Arai, H., Kodama, T., & Igarashi, Y. (1998). *Biochimica et Biophysica Acta, 1381*(3), 347–350.
Hetrick, E. M., Shin, J. H., Paul, H. S., & Schoenfisch, M. H. (2009). *Biomaterials, 30*(14), 2782–2789.
Hetrick, E. M., Shin, J. H., Stasko, N. A., Johnson, C. B., Wespe, D. A., Holmuhamedov, E., et al. (2008). *ACS Nano, 2*(2), 235–246.
Hill, D. F., Short, N. J., Perham, R. N., & Petersen, G. B. (1991). *Journal of Molecular Biology, 218*(2), 349–364.
Hino, T., Matsumoto, Y., Nagano, S., Sugimoto, H., Fukumori, Y., Murata, T., et al. (2010). *Science (New York, NY), 330*(6011), 1666–1670.
Hirai, T., Osamura, T., Ishii, M., & Arai, H. (2016). *Proceedings of the National Academy of Sciences of the United States of America, 113*(45), 12815–12819.
Hoffman, L. R., Kulasekara, H. D., Emerson, J., Houston, L. S., Burns, J. L., Ramsey, B. W., et al. (2009). *Journal of Cystic Fibrosis, 8*(1), 66–70.
Hoffman, L. R., Richardson, A. R., Houston, L. S., Kulasekara, H. D., Martens-Habbena, W., Klausen, M., et al. (2010). *PLoS Pathogens, 6*(1), e1000712.
Hossain, S., & Boon, E. M. (2017). *ACS Infectious Diseases, 3*(6), 454–461.
Hossain, S., Nisbett, L. M., & Boon, E. M. (2017). *Accounts of Chemical Research, 50*(7), 1633–1639.

Howlin, R. P., Cathie, K., Hall-Stoodley, L., Cornelius, V., Duignan, C., Allan, R. N., et al. (2017). *Molecular Therapy*, *25*(9), 2104–2116.
Jackson, A. A., Gross, M. J., Daniels, E. F., Hampton, T. H., Hammond, J. H., Vallet-Gely, I., et al. (2013). *Journal of Bacteriology*, *195*(13), 3093–3104.
Jenal, U., Reinders, A., & Lori, C. (2017). *Nature Reviews. Microbiology*, *15*(5), 271–284.
Jennings, L. K., Storek, K. M., Ledvina, H. E., Coulon, C., Marmont, L. S., Sadovskaya, I., et al. (2015). *Proceedings of the National Academy of Sciences of the United States of America*, *112*(36), 11353–11358.
Jones, A. K., Fulcher, N. B., Balzer, G. J., Urbanowski, M. L., Pritchett, C. L., Schurr, M. J., et al. (2010). *Journal of Bacteriology*, *192*(21), 5709–5717.
Jones, M. L., Ganopolsky, J. G., Labbe, A., & Prakash, S. (2010). *Applied Microbiology and Biotechnology*, *87*(2), 509–516.
Jones, K. L., Hegab, A. H., Hillman, B. C., Simpson, K. L., Jinkins, P. A., Grisham, M. B., et al. (2000). *Pediatric Pulmonology*, *30*(2), 79–85.
Jones, R. W., Lamont, A., & Garland, P. B. (1980). *Biochemical Journal*, *190*(1), 79–94.
Kaihami, G. H., Breda, L. C. D., de Almeida, J. R. F., de Oliveira Pereira, T., Nicastro, G. G., Boechat, A. L., et al. (2017). *Infection and Immunity*, *85*(8), e00207–17.
Kakishima, K., Shiratsuchi, A., Taoka, A., Nakanishi, Y., & Fukumori, Y. (2007). *Biochemical and Biophysical Research Communications*, *355*(2), 587–591.
Kalkowski, I., & Conrad, R. (1991). *FEMS Microbiology Letters*, *66*(1), 107–111.
Kawakami, T., Kuroki, M., Ishii, M., Igarashi, Y., & Arai, H. (2010). *Environmental Microbiology*, *12*(6), 1399–1412.
Keefer, L. K. (2003). *Annual Review of Pharmacology and Toxicology*, *43*, 585–607.
Kinkel, T. L., Ramos-Montañez, S., Pando, J. M., Tadeo, D. V., Strom, E. N., Libby, S. J., et al. (2016). *Nature Microbiology*, *2*, 16224.
Kolpen, M., Kühl, M., Bjarnsholt, T., Moser, C., Hansen, C. R., Liengaard, L., et al. (2014). *PLoS One*, *9*(1), e84353.
Koskenkorva, T., Aro-Kärkkäinen, N., Bachmann, D., Arai, H., Frey, A. D., & Kallio, P. T. (2008). *Archives of Microbiology*, *189*(4), 385–396.
Lee, K. M., Go, J., Yoon, M. Y., Park, Y., Kim, S. C., Yong, D. E., et al. (2012). *Infection and Immunity*, *80*(5), 1639–1649.
Lee, K. M., Yoon, M. Y., Park, Y., Lee, J. H., & Yoon, S. S. (2011). *Infection and Immunity*, *79*(7), 2792–2800.
Line, L., Alhede, M., Kolpen, M., Kuhl, M., Ciofu, O., Bjarnsholt, T., et al. (2014). *Frontiers in Microbiology*, *5*, 554.
Linnane, S. J., Keatings, V. M., Costello, C. M., Moynihan, J. B., O'Connor, C. M., Fitzgerald, M. X., et al. (1998). *American Journal of Respiratory and Critical Care Medicine*, *158*(1), 207–212.
Lu, Y., Slomberg, D. L., & Schoenfisch, M. H. (2014). *Biomaterials*, *35*(5), 1716–1724.
Lu, Y., Slomberg, D. L., Shah, A., & Schoenfisch, M. H. (2013). *Biomacromolecules*, *14*(10), 3589–3598.
Luthi, E., Mercenier, A., & Haas, D. (1986). *Journal of General Microbiology*, *132*(10), 2667–2675.
MacMicking, J., Xie, Q. W., & Nathan, C. (1997). *Annual Review of Immunology*, *15*, 323–350.
Major, T. A., Panmanee, W., Mortensen, J. E., Gray, L. D., Hoglen, N., & Hassett, D. J. (2010). *Antimicrobial Agents and Chemotherapy*, *54*(11), 4671–4677.
Martin, D. W., Schurr, M. J., Mudd, M. H., Govan, J. R., Holloway, B. W., & Deretic, V. (1993). *Proceedings of the National Academy of Sciences of the United States of America*, *90*(18), 8377–8381.
Mashburn-Warren, L., Howe, J., Garidel, P., Richter, W., Steiniger, F., Roessle, M., et al. (2008). *Molecular Microbiology*, *69*(2), 491–502.
Michel-Briand, Y., & Baysse, C. (2002). *Biochimie*, *84*(5–6), 499–510.

Moenne-Loccoz, P. (2007). *Natural Product Reports*, *24*(3), 610–620.
Morgan, R., Kohn, S., Hwang, S. H., Hassett, D. J., & Sauer, K. (2006). *Journal of Bacteriology*, *188*(21), 7335–7343.
Nablo, B. J., & Schoenfisch, M. H. (2003). *Journal of Biomedical Materials Research. Part A*, *67*(4), 1276–1283.
Nakayama, K., Takashima, K., Ishihara, H., Shinomiya, T., Kageyama, M., Kanaya, S., et al. (2000). *Molecular Microbiology*, *38*(2), 213–231.
O'Callaghan, J., Reen, F. J., Adams, C., & O'Gara, F. (2011). *Microbiology (Reading, England)*, *157*(Pt. 12), 3417–3428.
Ochsner, U. A., Koch, A. K., Fiechter, A., & Reiser, J. (1994). *Journal of Bacteriology*, *176*(7), 2044–2054.
O'Toole, G. A., & Kolter, R. (1998). *Molecular Microbiology*, *30*(2), 295–304.
Parr, S. R., Barber, D., Greenwood, C., & Brunori, M. (1977). *Biochemical Journal*, *167*(2), 447–455.
Parsek, M. R., & Tolker-Nielsen, T. (2008). *Current Opinion in Microbiology*, *11*(6), 560–566.
Pesci, E. C., Milbank, J. B., Pearson, J. P., McKnight, S., Kende, A. S., Greenberg, E. P., et al. (1999). *Proceedings of the National Academy of Sciences of the United States of America*, *96*(20), 11229–11234.
Petrova, O. E., & Sauer, K. (2012). *Proceedings of the National Academy of Sciences of the United States of America*, *109*(41), 16690–16695.
Plate, L., & Marletta, M. A. (2013). *Trends in Biochemical Sciences*, *38*(11), 566–575.
Privett, B. J., Broadnax, A. D., Bauman, S. J., Riccio, D. A., & Schoenfisch, M. H. (2012). *Nitric Oxide: Biology and Chemistry*, *26*(3), 169–173.
Reighard, K. P., Hill, D. B., Dixon, G. A., Worley, B. V., & Schoenfisch, M. H. (2015). *Biofouling*, *31*(9–10), 775–787.
Reighard, K. P., & Schoenfisch, M. H. (2015). *Antimicrobial Agents and Chemotherapy*, *59*(10), 6506–6513.
Reynolds, H. Y. (1987). *Clinics in Chest Medicine*, *8*(3), 339–358.
Rinaldo, S., Arcovito, A., Brunori, M., & Cutruzzola, F. (2007). *Journal of Biological Chemistry*, *282*(20), 14761–14767.
Rinaldo, S., Arcovito, A., Giardina, G., Castiglione, N., Brunori, M., & Cutruzzola, F. (2008). *Biochemical Society Transactions*, *36*(Pt. 6), 1155–1159.
Rinaldo, S., Brunori, M., & Cutruzzola, F. (2007). *Biochemical and Biophysical Research Communications*, *363*(3), 662–666.
Ryder, C., Byrd, M., & Wozniak, D. J. (2007). *Current Opinion in Microbiology*, *10*(6), 644–648.
Schairer, D. O., Chouake, J. S., Nosanchuk, J. D., & Friedman, A. J. (2012). *Virulence*, *3*(3), 271–279.
Schertzer, J. W., Brown, S. A., & Whiteley, M. (2010). *Molecular Microbiology*, *77*(6), 1527–1538.
Schertzer, J. W., & Whiteley, M. (2012). *MBio*, *3*(2), e00297–11.
Schreiber, K., Boes, N., Eschbach, M., Jaensch, L., Wehland, J., Bjarnsholt, T., et al. (2006). *Journal of Bacteriology*, *188*(2), 659–668.
Schreiber, K., Krieger, R., Benkert, B., Eschbach, M., Arai, H., Schobert, M., et al. (2007). *Journal of Bacteriology*, *189*(11), 4310–4314.
Schuster, M., Sexton, D. J., Diggle, S. P., & Greenberg, E. P. (2013). *Annual Review of Microbiology*, *67*, 43–63.
Schwechheimer, C., & Kuehn, M. J. (2015). *Nature Reviews. Microbiology*, *13*(10), 605–619.
Sharma, V., Noriega, C. E., & Rowe, J. J. (2006). *Applied and Environmental Microbiology*, *72*(1), 695–701.
Slack, M. P., & Nichols, W. W. (1981). *Lancet*, *2*(8245), 502–503.
Sonnleitner, E., Gonzalez, N., Sorger-Domenigg, T., Heeb, S., Richter, A. S., Backofen, R., et al. (2011). *Molecular Microbiology*, *80*(4), 868–885.

Sriramulu, D. D., Lunsdorf, H., Lam, J. S., & Romling, U. (2005). *Journal of Medical Microbiology*, *54*(Pt. 7), 667–676.
Tashiro, Y., Ichikawa, S., Nakajima-Kambe, T., Uchimura, H., & Nomura, N. (2010). *Microbes and Environments*, *25*(2), 120–125.
Tiddens, H. A., Donaldson, S. H., Rosenfeld, M., & Pare, P. D. (2010). *Pediatric Pulmonology*, *45*(2), 107–117.
Toledo-Cuevas, E. M. (1997). *Revista Latinoamericana de Microbiologia*, *39*(3–4), 167–186.
Toyofuku, M., Morinaga, K., Hashimoto, Y., Uhl, J., Shimamura, H., Inaba, H., et al. (2017). *The ISME Journal*, *11*, 1504–1509.
Toyofuku, M., Nomura, N., Fujii, T., Takaya, N., Maseda, H., Sawada, I., et al. (2007). *Journal of Bacteriology*, *189*(13), 4969–4972.
Toyofuku, M., Nomura, N., Kuno, E., Tashiro, Y., Nakajima, T., & Uchiyama, H. (2008). *Journal of Bacteriology*, *190*, 7947–7956.
Toyofuku, M., O'toole, G. A., & Nomura, N. (2015). Anaerobic life style of *Pseudomonas aeruginosa*. In J. Ramos, J. B. Goldberg, & A. Filloux (Eds.), *Pseudomonas: Vol. 7.*: Springer.
Toyofuku, M., Roschitzki, B., Riedel, K., & Eberl, L. (2012). *Journal of Proteome Research*, *11*(10), 4906–4915.
Toyofuku, M., Zhou, S., Sawada, I., Takaya, N., Uchiyama, H., & Nomura, N. (2014). *Environmental Microbiology*, *16*(9), 2927–2938.
Trunk, K., Benkert, B., Quäck, N., Münch, R., Scheer, M., Garbe, J., et al. (2010). *Environmental Microbiology*, *12*(6), 1719–1733.
Turnbull, L., Toyofuku, M., Hynen, A. L., Kurosawa, M., Pessi, G., Petty, N. K., et al. (2016). *Nature Communications*, 7, 11220.
Van Alst, N. E., Sherrill, L. A., Iglewski, B. H., & Haidaris, C. G. (2009). *Canadian Journal of Microbiology*, *55*(10), 1133–1144.
Van Alst, N. E., Wellington, M., Clark, V. L., Haidaris, C. G., & Iglewski, B. H. (2009). *Infection and Immunity*, 77(10), 4446–4454.
van de Kamp, M., Silvestrini, M. C., Brunori, M., Van Beeumen, J., Hali, F. C., & Canters, G. W. (1990). *European Journal of Biochemistry*, *194*(1), 109–118.
Vander Wauven, C., Pierard, A., Kley-Raymann, M., & Haas, D. (1984). *Journal of Bacteriology*, *160*(3), 928–934.
Waite, R. D., & Curtis, M. A. (2009). *Journal of Bacteriology*, *191*(4), 1349–1354.
Webb, J. S., Thompson, L. S., James, S., Charlton, T., Tolker-Nielsen, T., Koch, B., et al. (2003). *Journal of Bacteriology*, *185*(15), 4585–4592.
Whitchurch, C. B., Tolker-Nielsen, T., Ragas, P. C., & Mattick, J. S. (2002). *Science (New York, NY)*, *295*(5559), 1487.
Whiteley, M., Bangera, M. G., Bumgarner, R. E., Parsek, M. R., Teitzel, G. M., Lory, S., et al. (2001). *Nature*, *413*(6858), 860–864.
Williams, D. R., Rowe, J. J., Romero, P., & Eagon, R. G. (1978). *Applied and Environmental Microbiology*, *36*(2), 257–263.
Williams, H. D., Zlosnik, J. E., & Ryall, B. (2007). *Advances in Microbial Physiology*, *52*, 1–71.
Worlitzsch, D., Tarran, R., Ulrich, M., Schwab, U., Cekici, A., Meyer, K. C., et al. (2002). *Journal of Clinical Investigation*, *109*(3), 317–325.
Xiao, G., Déziel, E., He, J., Lépine, F., Lesic, B., Castonguay, M. H., et al. (2006). *Molecular Microbiology*, *62*(6), 1689–1699.
Xu, K. D., Stewart, P. S., Xia, F., Huang, C. T., & McFeters, G. A. (1998). *Applied and Environmental Microbiology*, *64*(10), 4035–4039.
Yepuri, N. R., Barraud, N., Mohammadi, N. S., Kardak, B. G., Kjelleberg, S., Rice, S. A., et al. (2013). *Chemical Communications (Cambridge, England)*, *49*(42), 4791–4793.
Yin, S., Fuangthong, M., Laratta, W. P., & Shapleigh, J. P. (2003). *Applied and Environmental Microbiology*, *69*(7), 3938–3944.

Yoon, S. S., Coakley, R., Lau, G. W., Lymar, S. V., Gaston, B., Karabulut, A. C., et al. (2006). *Journal of Clinical Investigation, 116*(2), 436–446.
Yoon, S. S., & Hassett, D. J. (2004). *Expert Review of Anti-Infective Therapy, 2*(4), 611–623.
Yoon, S. S., Hennigan, R. F., Hilliard, G. M., Ochsner, U. A., Parvatiyar, K., Kamani, M. C., et al. (2002). *Developmental Cell, 3*(4), 593–603.
Yoon, S. S., Karabulut, A. C., Lipscomb, J. D., Hennigan, R. F., Lymar, S. V., Groce, S. L., et al. (2007). *The EMBO Journal, 26*(15), 3662–3672.
Yoon, M. Y., Lee, K. M., Park, Y., & Yoon, S. S. (2011). *PLoS One, 6*(1), e16105.
Zimmermann, A., Reimmann, C., Galimand, M., & Haas, D. (1991). *Molecular Microbiology, 5*(6), 1483–1490.

CHAPTER FIVE

Emerging Roles of Nitric Oxide Synthase in Bacterial Physiology

Elizabeth H. Hutfless, Sujata S. Chaudhari, Vinai C. Thomas[1]
University of Nebraska Medical Center, Omaha, NE, United States
[1]Corresponding author: e-mail address: vinai.thomas@unmc.edu

Contents

Abstract

Nitric oxide (NO) is a potent inhibitor of diverse cellular processes in bacteria. Therefore, it was surprising to discover that several bacterial species, primarily Gram-positive organisms, harboured a gene encoding nitric oxide synthase (NOS). Recent attempts to characterize bacterial NOS (bNOS) have resulted in the discovery of structural features that may allow it to function as a NO dioxygenase and produce nitrate in addition to NO. Consistent with this characterization, investigations into the biological function of bNOS have also emphasized a role for NOS-dependent nitrate and nitrite production in aerobic and microaerobic respiration. In this review, we aim to compare, contrast, and summarize the structure, biochemistry, and biological role of bNOS with mammalian NOS and discuss how recent advances in our understanding of bNOS have enabled efforts at designing inhibitors against it.

Advances in Microbial Physiology, Volume 72
ISSN 0065-2911
https://doi.org/10.1016/bs.ampbs.2018.01.006

ABBREVIATIONS

BH_4 6*R*-tetrahydrobiopterin
bNOS bacterial nitric oxide synthase
bsNOS *Bacillus subtilis* nitric oxide synthase
eNOS endothelial NOS
FH_4 tetrahydrofolic acid
gsNOS *Geobacillus stearothermophilus* nitric oxide synthase
iNOS inducible NOS
mNOS mammalian nitric oxide synthase
nNOS neuronal NOS
NOS nitric oxide synthase
NOS_{oxy} nitric oxide synthase oxygenase domain
NOS_{red} nitric oxide synthase reductase domain
saNOS *Staphylococcus aureus* nitric oxide synthase
stNOS *Streptomyces turgidiscabies* nitric oxide synthase

1. EVOLUTIONARY SIGNIFICANCE OF NOS

The generation of NO can be traced back to earth's earliest ages during the Hadean eon (~4 billion years back) (Chyba & Sagan, 1991; Nisbet & Sleep, 2001). The prebiotic atmosphere then was primarily composed of dinitrogen ($N{\equiv}N$), carbon dioxide (CO_2), and water vapour (H_2O) (Feelisch & Martin, 1995; Hunten, 1993; Sleep, 2010; Trainer, 2013). Due to high atmospheric temperatures and frequent energy discharges during lightning storms, passing comets, and meteorite impacts, these molecules reacted to form NO (Feelisch & Martin, 1995; Walker, 1977). Once produced, the reducing environments prevalent at the time allowed rapid conversion of NO to nitroxyl (HNO), a chemical species that readily dissolves in oceans due to its high water solubility (Rothschild & Lister, 2003). Some estimates suggest that as much as 10^{10} kg of NO and 10^9 kg of HNO were produced yearly during this period (Mancinelli & McKay, 1988; Moroz & Kohn, 2011). In aqueous environments, nitroxyl dissociates into NO^- and H^+ (protons). Multiple abiotic oxidation reactions involving NO^- and oxidized intermediates, like $N_2O_2^-$, $N_3O_3^-$, and N_2O, are thought to be the source of stable nitrates (NO_3^-) and nitrites (NO_2^-) in early earth's crust and oceans (Rothschild & Lister, 2003). Also, an abundance of minerals like iron (Fe^{2+}) in the oceans during this period is hypothesized to reduce the oxidized forms of nitrogen to ammonium ions (NH_4^+) (Summers & Chang, 1993). As atmospheric N_2 is not readily bioavailable due to the high energy required to break its bonds, the abiotic pool of oxidized (primarily

NO_3^- and NO_2^-) and reduced nitrogen species (NH_4^+) in the earth's crust and oceans formed the 'fixed' nitrogen pool that catalysed the evolution of biological life (Moroz & Kohn, 2011).

Apart from abiotic NO generation, life itself evolved to produce and utilize NO for various purposes. The first living inhabitants of the earth were unicellular microbes that arose during the Archean eon (Schopf, 2006). Given the abundance of NH_4^+ in the environment, microbes evolved the ability to utilize NH_4^+ as a source of nitrogen and incorporate it into various biological molecules like nucleotides and amino acids. However, when NH_4^+ became limiting in some environments, NO_3^- and NO_2^- may have become the preferred nitrogen source (Rothschild & Lister, 2003). Both NO_3^- and NO_2^- can be converted into ammonia through the dissimilatory nitrate ammonification process (Eq. 1). Also, NO_3^- and NO_2^- are important components of the denitrification process (Eq. 2) that microbes also evolved during this period:

$$NO_3^- \rightarrow NO_2^- \rightarrow NH_4^+ \tag{1}$$

$$2NO_3^- \rightarrow 2NO_2^- \rightarrow 2[NO] \rightarrow N_2O \rightarrow N_2 \tag{2}$$

Biological denitrification involves the dissimilatory reduction of NO_3^- to N_2 through a series of intermediates that include NO_2^-, NO, and N_2O (Ye, Averill, & Tiedje, 1994). The involvement of NO as an intermediate of denitrification not only allowed a conduit for the disposal of high-energy electrons that arose during cellular metabolism but also allowed for the efficient extraction of energy from those electrons when coupled to ATP production (Ye et al., 1994).

Towards the end of the Archean period, a catastrophic oxygenation event (also called the Great Oxygenation Event, GOE) occurred due to cyanobacterial metabolism (Lyons, Reinhard, & Planavsky, 2014). Oxygen along with reactive ozone (O_3) is toxic and adversely affects various biological macromolecules (Feelisch & Martin, 1995). A decrease in atmospheric CO_2 that accompanied GOE may have not only resulted in a decrease of abiotic NO synthesis (by as much as a 100-fold) but also adversely affected biological denitrification as some of the reductases involved in denitrification optimally function under anaerobic conditions (Navarro-Gonzalez, McKay, & Mvondo, 2001). Additionally, various minerals (iron, sulphate, etc.) in the oceans could have also been oxidized due to the rise in oxygen levels during the GOE and thus significantly disrupt maintenance of an abiotic nitrogen cycle (Navarro-Gonzalez et al., 2001). An abundance of minerals like reduced iron and sulphides probably shielded cyanobacteria from toxic effects of self-produced oxygen by acting as a sink (Shaw, 2016). These minerals may also have afforded these microbes enough time to evolve novel antioxidant capabilities (Shaw, 2016). Herein, perhaps, lies the impetus for the evolution of nitric oxide synthases (NOSs) in some prokaryotes, as a defence mechanism against oxidative stress (Feelisch & Martin, 1995). Indeed, NOS-like enzymes are present in diverse microbial species including cyanobacteria (Rőszer, 2012). NOS utilizes L-arginine (Arg) as a substrate and produce L-citrulline (Cit) and NO in the presence of oxygen (Rőszer, 2012). The NO produced thus may be a way to protect cells from the harmful effects of oxygen either directly by reacting with oxygen radicals and ozone or indirectly by blocking the exposed targets of oxidation in vulnerable biomolecules (Feelisch & Martin, 1995). Alternatively, NO_3^- and NO_2^-, both oxidized derivatives of NO, may positively influence cellular adaptation to oxygen by modulating cellular respiration (Chaudhari et al., 2017).

Although a systematic and detailed evolutionary study of NOS has not been undertaken, some studies have offered initial clues as to its origin and diversity. In a study by Gusarov et al. homologs of NOS were identified in all kingdoms of life, attesting to the conservation of this enzyme through evolution (Gusarov et al., 2008). Phylogenetic analysis of these homologs led to the proposal that the multidomain protein architecture of eukaryotic NOS evolved from simpler single-domain homologs found in bacteria. Consistent with this argument, these authors point out that lower eukaryotes like snails and fruit flies have various truncations of NOS that match both primitive and evolved lineages of this enzyme found in bacteria and higher eukaryotes, respectively (Gusarov et al., 2008). In a separate study, Sudhamsu and Crane describe a simplified phylogenetic tree of NOS using sequences from bacteria, eukarya, and archaea (Sudhamsu & Crane, 2009). Comparison of the *nos* gene-specific phylogenetic tree to the corresponding species-specific tree (derived from 16S ribosomal RNA sequence) from the same organisms revealed marked differences in tree topology (Sudhamsu & Crane, 2009). Although the resolution of the phylogenetic analyses in this study was low due to a small sample size, the authors concluded that the evolution of NOS appeared to be distinct from the linear evolution of species and suggested horizontal gene transfer of NOS between species.

2. DOMAIN ORGANIZATION AND COFACTORS OF NOS

The mammalian NOS (mNOS) homologs were the first to be biochemically and structurally characterized. Mammals have three distinct NOS isoforms encoded by three separate genes. These isoforms are hypothesized to have evolutionarily risen from multiple gene and genome duplications in lower invertebrates (Agapie et al., 2009; Andreakis et al., 2011; Ghosh & Salerno, 2003; Marletta, 1994; Nathan & Xie, 1994). Additionally, unique changes in protein architecture specific to each NOS isoform appear to have contributed to their functional specialization. These isoforms are classified as endothelial NOS (eNOS), neuronal NOS (nNOS), and inducible NOS (iNOS) based on tissue specificity and differences in regulation. Just as these isoforms are expressed in different cell types, the production of NO by these three NOS isoforms has quite diverse roles, some physiological, some pathological (Bryan, Bian, & Murad, 2009; Feron et al., 1996; Guo et al., 1995; Michel & Feron, 1997; Schulz, Nava, & Moncada, 1992; Xu, Huso, Dawson, Bredt, & Becker, 1999). Physiological roles for NO include vasodilation and vascular homeostasis, inhibition of platelet aggregation, neuronal signalling, and immune defence. Pathological roles for NO include hypotension during septic shock, inflammatory tissue injury, mutagenesis, and *N*-nitrosamine formation (Alderton, Cooper, & Knowles, 2001; Griffith & Stuehr, 1995; Kiechle & Malinski, 1993; Nathan & Xie, 1994).

All isoforms of mNOS are homodimers. Each monomer consists of an isozyme-specific N-terminal domain, a β-hairpin hook (also called N-terminal hook), a Cys-X_4-Cys site that coordinates a zinc atom, and two additional domains that follow (Fig. 1)—an oxygenase domain (NOS_{oxy}) and a reductase domain (NOS_{red}). The latter two domains (NOS_{oxy} and NOS_{red}) are linked by a calmodulin-binding site (Fig. 1). Both NOS_{oxy} and NOS_{red} can independently fold and have been separately expressed. Located within the NOS_{oxy} domain are substrate (Arg)- and cofactor-binding sites. The cofactors for the NOS_{oxy} domain include haem and 6*R*-tetrahydrobiopterin (BH_4). The NOS_{red} domain, on the other hand, is closely related to cytochrome P450 reductase and contains binding sites for FAD, FMN, and NADPH (Fig. 1). Stoichiometrically, each monomer of mNOS binds one molecule of haem, BH_4, FMN, and FAD. However, a

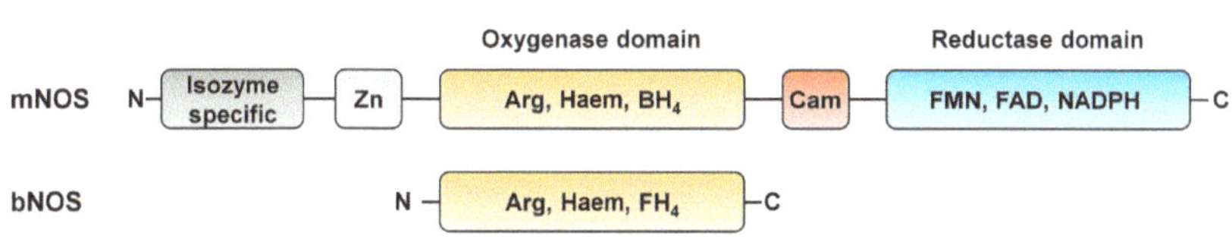

Fig. 1 Protein domain organization of mNOS and bNOS. Mammalian NOS contains an isoform-specific N-terminal domain, a Cys-X_4-Cys/zinc-binding domain (Zn), an oxygenase domain, a calmodulin-binding site, and a reductase domain. Most bacterial NOSs on the other hand lack all domains except the oxygenase domain.

single atom of zinc is shared between the functional NOS dimers (Fleming, Bauersachs, Fisslthaler, & Busse, 1998; Ghosh & Salerno, 2003; Griffith & Stuehr, 1995; Schulz et al., 1992).

Although bacterial NOSs (bNOSs) are homodimers, they exhibit significant differences compared to their eukaryotic counterparts. All bNOSs, except the one produced by *Sorangium cellulosum*, lack a reductase domain (Agapie et al., 2009). The lack of a linked reductase domain initially shed doubt as to whether NOS in bacteria truly possessed genuine NOS function, i.e., the oxygen-dependent conversion of Arg to Cit and NO. However, several groups have purified bNOS and confirmed its activity in multiple bacterial species including *Nocardia*, *Bacillus cereus*, *Bacillus subtilis*, *S. cellulosum*, *Staphylococcus aureus*, *Streptomyces turgidiscabies*, *Deinococcus radiodurans*, and *Geobacillus stearothermophilus* (Agapie et al., 2009; Bird et al., 2002; Chen & Rosazza, 1994; Jeong & Choi, 2002; Montgomery, Dupont, Leivo, & Guillemette, 2010; Pant, Bilwes, Adak, Stuehr, & Crane, 2002; Reece, Woodward, & Marletta, 2009; Sudhamsu & Crane, 2006). Additionally, bNOS also lacks the N-terminal Cys-X_4-Cys zinc- and calmodulin-binding sites (Fig. 1). The absence of a calmodulin-binding site in bNOS is not surprising since bacteria do not synthesize calmodulin. Similarly, most bacteria cannot produce the pterin cofactor, BH_4. However, there is some consensus that tetrahydrofolate (FH_4), a molecule that is closely related to BH_4, may be used instead as a cofactor for bNOS.

3. STRUCTURAL ASPECTS OF BACTERIAL NOS

Till date, the structures of three different bacterial NOS proteins have been solved (Bird et al., 2002; Pant et al., 2002; Sudhamsu & Crane, 2006). The NOSs isolated from *B. subtilis* (bsNOS), *S. aureus* (saNOS), and *G. stearothermophilus* (gsNOS) were structurally resolved to 2.1, 2.4, and 3.2 Å, respectively (Bird et al., 2002; Pant et al., 2002; Sudhamsu & Crane, 2006). All three bNOS structures are like the well-studied oxygenase domain of mNOS ($mNOS_{oxy}$) (Alderton et al., 2001; Campbell, Smith, Potter, Carragher, & Marletta, 2014; Stuehr, 1999). At its very basic form, each monomer of NOS has a winged β-sheet core and resembles a baseball catcher's mitt (Crane et al., 1997). Each of these monomers also incorporates a molecule of haem. Upon homodimerization, a dimer interface is created. A set of perpendicular helices from each monomer that form a 'T-shaped' structure surround the interface (Crane et al., 1998). Additionally, each monomer contributes a hydrophobic 'helical lariat' that protrudes into the interface (Lustig et al., 2011; Pant et al., 2002; Pant & Crane, 2005). Within the dimer interface of $mNOS_{oxy}$, each of these helical lariats binds a single BH_4 molecule, whereas bulkier FH_4 molecules are thought to associate with the 'helical lariats' in bacteria (Adak, Bilwes, et al., 2002; Ghosh & Stuehr, 1995). Studies on the structure of bNOS have provided unique insights of which some are discussed later.

3.1 How Bacterial NOS Forms a Stable Dimer

Multiple interactions within the NOS dimer interface are critical for stabilization of the NOS dimer. In common with bNOS, $mNOS_{oxy}$ has four regions within each of its monomer that make up the dimer interface (black shaded regions in Fig. 2) (Bird et al., 2002; Pant et al., 2002). However, in addition to the interactions occurring within these four regions, the N-terminal hook and adjacent Cys-X_4-Cys site are critical for dimerization of $mNOS_{oxy}$ (Crane et al., 1998). Since bNOS lack these structures, it is evident that compensatory structural changes to stabilize the dimer must exist.

Bird et al. describe many compensatory changes in saNOS that could increase dimer stability (Bird et al., 2002). First, aromatic residues in α-12 helices (Fig. 2, residues shown in blue; Phe262, Tyr273, and Tyr276) of saNOS region II allow strong intersubunit π-stacking interactions. Since the corresponding residues in $mNOS_{oxy}$ are made up of leucines and an isoleucine, the dimer interface of $mNOS_{oxy}$ in this region may not be as strongly bound as it is in saNOS (Bird et al., 2002). Second, interactions between region IV of one NOS subunit and regions I, II, and IV of the second subunit within the dimer interface are possibly strengthened due to a conserved Pro323 found within region IV of saNOS (Fig. 2, residue shown in purple) (Bird et al., 2002; Pant et al., 2002). Since Pro323 (instead of glycine present at the same position in $mNOS_{oxy}$) could restrict conformational flexibility, its presence in region IV is predicted to strengthen dimerization by increasing hydrophobicity of the bNOS dimer interface. Third, dimer stability may also be indirectly strengthened by multiple hydrogen bonds and van der Waals interactions between residues relatively unique to bNOS (Bird et al., 2002). Finally, studies on bsNOS have shown that substrate and cofactor recognition/binding can positively influence dimer stabilization (Kabir, Sudhamsu, Crane, Yeh, & Rousseau, 2008; Pant et al., 2002). Hence, multiple factors contribute to dimer stabilization in bNOS.

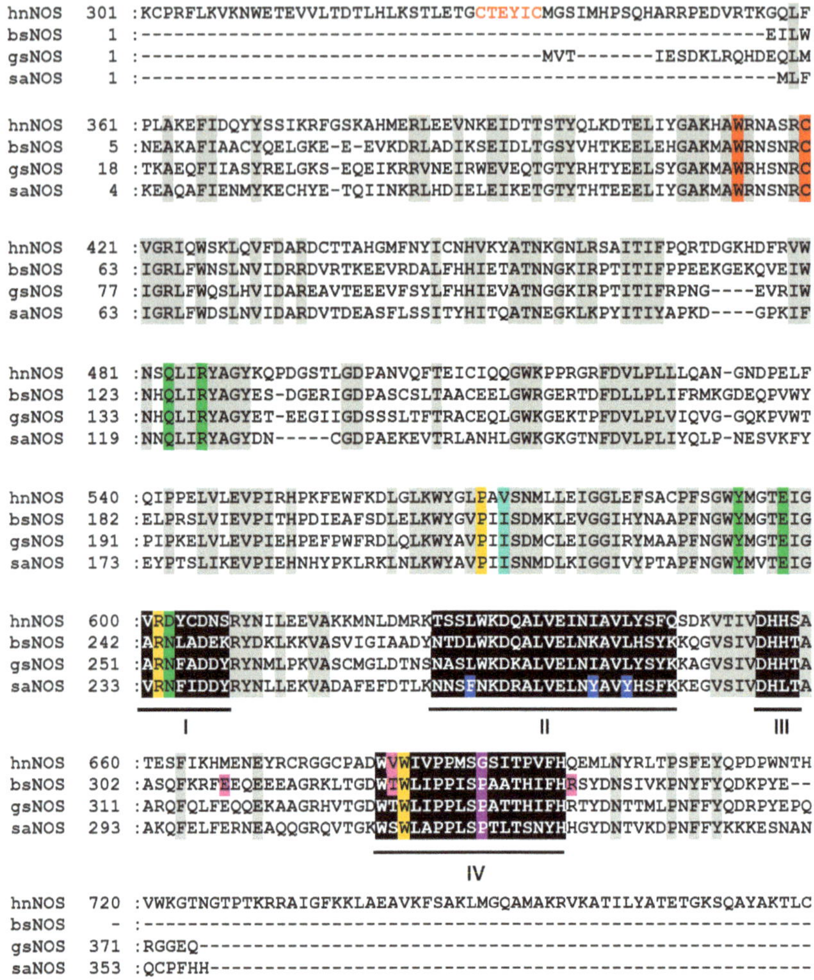

Fig. 2 Multiple sequence alignment of mNOS and bNOS. Amino acid sequences of NOS from *Bacillus subtilis*, *Staphylococcus aureus*, and *Geobacillus stearothermophilus* were compared to the human neuronal NOS isoform. The four regions of the alignment shaded in *black* (labelled I, II, III, and IV) constitute the NOS dimer interface. Residues shaded in *grey* are identical. Additional essential residues are shaded in *different colours* and are described in the text.

3.2 Pterins in Bacterial NOS

Both pterins, BH_4 and FH_4, are reducing cofactors as they can act as a source of electrons during redox reactions. Both these cofactors are associated with the folate biosynthesis pathway. Reduced folate cofactors like FH_4 can serve as a donor of one-carbon units for biosynthesis of purines, methionine, and thiamine (Bermingham & Derrick, 2002). While bacteria have a complete machinery to synthesize necessary folates and FH_4, they usually cannot synthesize BH_4 due to the lack of sepiapterin reductase, the final enzyme of the BH_4 biosynthetic pathway. The potential exception is *B. subtilis*, which may have a homolog of sepiapterin reductase. However, it is unclear if this homolog is functionally active (Crane, Sudhamsu, & Patel, 2010).

The BH_4-binding region is remarkably conserved between the three $mNOS_{oxy}$ isoforms (Fischmann et al., 1999; Stuehr, 1999). BH_4 is tightly tucked into the dimer interface due to multiple interactions it has with residues in the 'helical lariat' (Crane et al., 1998). In particular, π-stacking interactions resulting from a conserved Trp683 of the helical lariat and hydrogen bonds formed between BH_4 and a conserved Arg601 in human nNOS are thought to be important for stabilization of pterin in the dimer interface (Ghosh, Wolan, et al., 1999; Pant et al., 2002). Additionally, $mNOS_{oxy}$ also has an N-terminal region that limits the solvent accessibility to the distal end of BH_4 (Konas et al., 2004; Tejero & Stuehr, 2013). As in nNOS, both tryptophan (Trp325 in bsNOS and Trp316 in saNOS) and arginine (Arg243 in bsNOS and Arg234 in saNOS; Fig. 2, residues shown in yellow) residues are conserved in bNOS (Crane et al., 1998; Ghosh et al., 1997; Konas et al., 2004; Tejero & Stuehr, 2013). However, the bNOS homologs lack the N-terminal residues important for restricting solvent accessibility (Ghosh et al., 1997). Studies have shown that $mNOS_{oxy}$ with 114 residues deleted from the N-terminus resulting in the loss of the N-terminal pterin-binding site among others was unable to dimerize (Ghosh, Wolan, et al., 1999). One contributing factor preventing dimerization may be the loss of pterin to the bulk solvent (Ghosh, Wolan, et al., 1999). Considering this fact, the observation that both bsNOS (which lacks the N-terminal pterin-binding site) and $mNOS_{oxy}$ are still able to maintain comparable BH_4-binding affinities ($K_m = 0.1\ \mu M$) is unexpected (Adak, Aulak, & Stuehr, 2002; Adak, Bilwes, et al., 2002; List et al., 1997; Presta et al., 1998; Tejero & Stuehr, 2013).

The lack of N-terminal features including the N-terminal hook and Zn/pterin-binding sites in bNOS increases access of bulkier molecules like FH_4 to the pterin-binding pocket (Ghosh, Crane, et al., 1999; Siddhanta et al., 1998; Weisslocker-Schaetzel et al., 2017). While bsNOS can incorporate and function with either BH_4 or FH_4, it must be noted that there is some evidence to indicate that these molecules may not necessarily be the physiological cofactors of bNOS (Adak, Aulak, et al., 2002). Pant et al. point out that there are a series of conserved residues in the bNOS pterin-binding site that is different from those observed in $mNOS_{oxy}$ (Pant et al., 2002).

In particular, Arg340 (Fig. 2, residue shown in pink) of bsNOS forms a salt bridge with Glu309 (Fig. 2, residue shown in pink) resulting in a slightly altered structure relative to $mNOS_{oxy}$ that does not necessarily favour either BH_4 or FH_4 binding (Pant et al., 2002). Similarly, the presence of Thr324 (Fig. 2, residue shown in pink) in bsNOS instead of a Val (found in $mNOS_{oxy}$) introduces a hydroxyl group that does not bind with either BH_4 or FH_4 (Pant et al., 2002). Additional evidence supporting a different cofactor was observed when NAD^+ was found in the crystalized pterin-binding pocket of saNOS (Bird et al., 2002). Due to its redox properties, it is possible that NAD^+ may have some effect on the function of saNOS (Bird et al., 2002). However, surprisingly none of these compounds (BH_4, FH_4, and NAD^+) could stimulate the activity of saNOS in vitro. Collectively, these arguments favour a different native cofactor for bNOS (Bird et al., 2002). Alternatively, it is possible that the conformational changes in bNOS following interaction with the substrate or its reductase partner are necessary for proper pterin (BH_4 or FH_4) cofactor activity in bNOS (Bird et al., 2002).

3.3 The Catalytic Pocket of Bacterial NOS

The catalysis of formation of NO and Cit from Arg and oxygen occurs within the oxygenase domain at a haem-containing catalytic pocket, located in the palm of the β-sheet core 'catcher's mitt' (Crane et al., 1997, 2010; Fischmann et al., 1999). Here, at this concave, hydrophobic face, the catalytic pocket comprises binding sites for both (1) a cysteine-ligated, pterin-associated haem and (2) a tightly hugged Arg. Detailed here are the important residues of bNOS involved in the structuring of a triangular association between these three players: haem, pterin, and Arg (Crane et al., 2010).

At the base of this triangular association is haem (Fig. 3A). The proximal binding site of haem is cysteine (saNOS: Cys62; bsNOS: Cys62; gsNOS: Cys76; nNOSoxy: Cys420; Fig. 2, residues shown in red) (Bird et al., 2002; Kabir et al., 2008; Pant et al., 2002). Cysteine contains a sulphur atom that coordinates the haem by forming a thiolate bond with haem iron. Additionally, the sulphur atom of cysteine accepts a hydrogen bond from a proximal tryptophan (Fig. 3) (saNOS: Trp56; bsNOS: Trp56; gsNOS: Trp70; nNOSoxy: Trp414; Fig. 2, residues shown in red) (Bird et al., 2002; Brunel, Wilson, Henry, Dorlet, & Santolini, 2011; Kabir et al., 2008; Pant et al., 2002; Rousseau, Li, Couture, & Yeh, 2005). The interaction between cysteine and tryptophan is thought to be important for modulating electron density that then governs oxygen chemistry for NOS-mediated catalysis

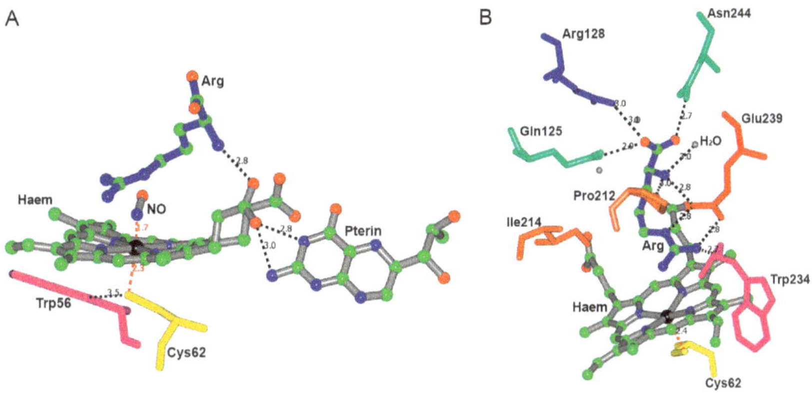

Fig. 3 (A) Triangular association between haem, pterin, and substrate Arg within the NOS catalytic site. Also shown is the proximal cysteine residue that binds to the haem iron and is hydrogen bonded to a Trp residue. NO is bound on the distal end of haem. (B) Critical residues in the NOS catalytic site that interact with the substrate Arg. Protein Data Bank (PDB #): 1M7V.

(Adak, Ghosh, Abu-Soud, & Stuehr, 1999; Adak, Wang, & Stuehr, 2000; Couture, Adak, Stuehr, & Rousseau, 2001; Rousseau et al., 2005). It is speculated that this Cys-Trp interaction increases the haem redox potential (Adak et al., 1999, 2000; Couture et al., 2001).

The second foot of the base of the triangle comes from interactions between the pterin and NOS (Fig. 3A). In contrast to mNOS, the bNOS protein only conserves a partial binding site for pterin (Pant et al., 2002). Hydrogen bonding to haem propionate D and the substrate Arg stabilizes the NOS pterin binding. Additionally, π–π stacking interactions with a conserved Trp residue (bsNOS: Trp325; Fig. 2, residue shown in yellow) also play a crucial role in pterin stabilization (Holden, Kang, Beasley, et al., 2015; Pant et al., 2002). These residues are conserved not only in different bNOS homologs but also between bacterial and mammalian homologs (Bird et al., 2002).

The apex of the triangle is the substrate, Arg, that gains access to the catalytic haem through an entry port in NOS. Once it occupies the substrate-binding pocket, Arg is nestled between tyrosine, tryptophan, and glutamate residues (Fig. 3B; bsNOS: Glu239, Tyr235, and Trp234; Fig. 2, residues shown in green) (Pant et al., 2002) of NOS via hydrogen bonding. Arg that is bound to NOS is in an extended conformation with its guanidino group facing the haem moiety. Hovering above Arg is a conserved proline residue (Fig. 4; bsNOS: Pro212) (Pant et al., 2002). The glutamate residue (bsNOS:

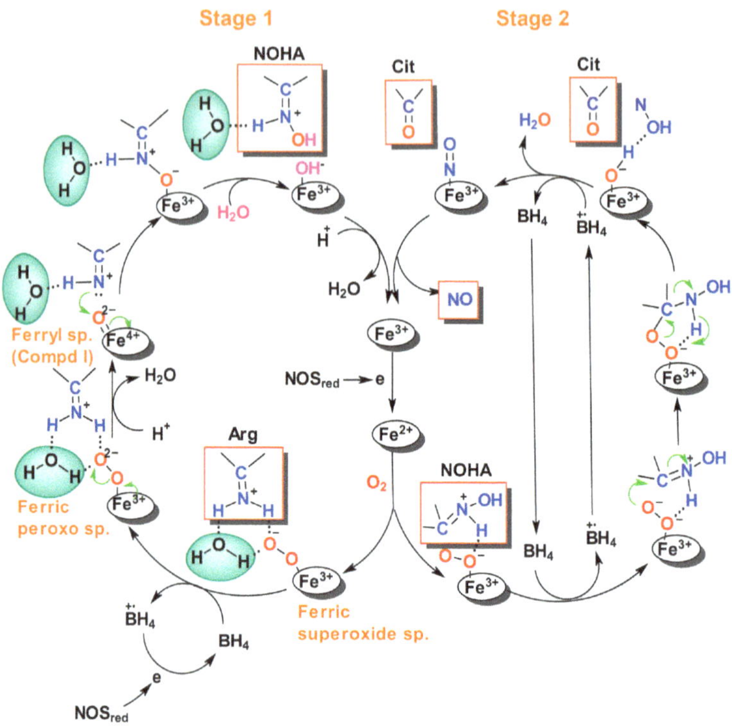

Fig. 4 Catalytic model of NOS. The conversion of Arg to Cit and NO takes place in two stages. In stage 1, the initial step involves the activation of oxygen. For this process to take place, an electron is transferred by a reductase to the ferric haem (Fe^{3+}) moiety of NOS. This results in the reduction of ferric iron to the ferrous (Fe^{2+}) state and allows dioxygen to bind to it. The resulting ferrous-dioxy ($Fe^{2+}O_2$) species may also exist as a ferric superoxide $Fe^{3+}O_2^{-}$ species. At this stage, a second electron is donated to the ferric superoxide species by a pterin (BH_4 or FH_4). A heterolytic cleavage of the O—O bond of the resulting ferric peroxo $\left(Fe^{3+}O_2^{2-}\right)$ species following protonation and water release results in the formation of the high-valent ferryl ($Fe^{4+}O^{2-}$) species (Compd 1) that is instrumental in the hydroxylation of Arg to NOHA. Throughout the first stage, a water molecule (*shaded blue*) stabilizes the substrate, Arg, and H-bonds to the distal end of the haem iron. It is also thought that the proton required for the formation of the ferryl species originates from the Arg and water molecule. During the second stage, the distal hydrogen bond network rearranges in the presence of NOHA, and the water molecule moves away. As in the first stage, oxygen is once again reactivated in stage 2. However, it is the ferric peroxo species rather than the ferryl species that perform the nucleophilic attack on NOHA resulting in the formation of Cit and nitroxyl. Following the nucleophilic attack, the nitroxyl is H-bonded to a ferric-hydroxide intermediate that is formed from the ferric peroxo species via an alkyl peroxo intermediate. Subsequently, an electron from nitroxyl is donated to the pterin radical with the simultaneous release of a water molecule resulting in the formation of NO.

Glu239) is essential for Arg binding (Fig. 3B). The amino group of substrates Arg or NOHA hydrogen bonds with Glu239, whereas the carboxyl group of Arg/NOHA interacts with Tyr235, Asn244, Gln125, and Arg128 (Pant et al., 2002). Intriguingly, Pant et al. noted that Arg128 points away from the substrate in mNOS.

One of the most prominent differences in the haem-binding pocket when comparing mammalian and bNOS isoforms arises from a valine-to-isoleucine substitution (compare Ile214 of bsNOS with Val572 of $mNOS_{oxy}$; Fig. 2, residue shown in cyan) (Pant et al., 2002). It is notable that this valine residue is conserved across the three mammalian isoforms, whereas the isoleucine residue occupying this site is conserved among bNOS homologs. Occasionally, the Ile214 is a methionine in some bNOS. Val572 in mNOS isoforms are conserved for efficient release of NO, and hence its substitution in bNOS has significant functional implications.

4. MECHANISTIC AND KINETIC ASPECTS OF BACTERIAL NOS-MEDIATED CATALYSIS

Despite the lack of a reductase domain and various structural differences outlined earlier, it is held that the catalytic reaction cycle of bNOS is very similar to that of mNOS, primarily because the catalytic site of mNOS and bNOS is remarkably well conserved (Crane et al., 2010). Multiple biochemical and biophysical studies that span over two decades have characterized various isoforms of mNOS in great detail. In this section, we will primarily focus on the general insights gleaned from these studies. However, attempts will be made to highlight mechanistic aspects of bNOS that appear to support or diverge from eukaryotic NOS.

4.1 Oxygen Activation in NOS

Catalysis of Arg to NO and Cit occurs in two stages within the catalytic pocket of NOS. The first stage involves hydroxylation of Arg to N^{ω}-hydroxy-L-arginine (NOHA) (Stuehr et al., 1991). In the second stage, NOHA is further oxidized to Cit and NO (Brunel et al., 2016; Crane et al., 2010; Papale et al., 2012; Stuehr, Pou, & Rosen, 2001). For every mole of Cit and NO produced by NOS, the reaction consumes 1 mol of Arg, 1.5 mol of NADPH, and 2 mol of oxygen (O_2). At both stages of the reaction cycle (Arg $\rightarrow$ NOHA and NOHA $\rightarrow$ Cit), NOS performs a monooxygenase reaction (Stuehr et al., 2001). These reactions involve the incorporation of a single oxygen atom from dioxygen (O_2) into organic molecules. However, for these reactions to occur, oxygen must initially be activated at the NOS catalytic site, a process that involves the formation of highly reactive haem-based oxidants (Brunel et al., 2016; Couture, Stuehr, & Rousseau, 2000; Stuehr, Wei, et al., 2004; Wang et al., 2004).

Although cleavage of the bond between oxygen is thermodynamically feasible, dioxygen does not readily react with organic molecules. The lack of reactivity is because oxygen has two unpaired electrons (triplet state) that prevent it from reacting with stable organic molecules that exist in the singlet state (molecules that do not have unpaired electrons) (Sono, Roach, Coulter, & Dawson, 1996). One way to bypass the slow reaction kinetics of dioxygen is to allow it to interact with transition metals with a suitable oxidation state, like iron (Cooper, 1999; Makris, Davydov, Denisov, Hoffman, & Sligar, 2002; Meunier, de Visser, & Shaik, 2004; Santolini, 2011). This reaction will allow the formation of appropriate metal–oxygen adducts that can then be used to oxidize organic molecules (Sono et al., 1996). In the case of mammalian and bacterial NOS, high-spin ferric iron (Fe^{3+}) is found within its haem cofactor and acts as the acceptor of dioxygen (Santolini, 2011). Once it is reduced to a high-spin ferrous state (Fe^{2+}), it readily accepts dioxygen and forms the low-spin $Fe^{2+}O_2$ (ferrous-oxo) singlet adduct which can also exist in an alternate electronic configuration as $Fe^{3+}O_2^-$ (ferric superoxide species; see Fig. 4) (Poulos, 2005a, 2005b; Santolini, 2011). Consistent with the above reaction requirements, high-spin iron necessary for dioxygen binding is found in all bacterial and mammalian NOS. Interestingly, not all iron associated with NOS necessarily exists in the high-spin state (Salard et al., 2006). Indeed, the ratio of high- to low-spin states can vary depending on the specific NOS homolog. For instance, the majority of mNOS and bsNOS iron exists in the high-spin state, but only 80% of the iron in saNOS is in the same state (Ghosh et al., 1997; Mayer et al., 1997; Presta et al., 1998; Salard et al., 2006).

The electron that is initially used to reduce the haem-associated ferric iron in $mNOS_{oxy}$ to its ferrous form is derived from the oxidation of NADPH by the mNOS reductase domain ($mNOS_{red}$) (Brunel, Santolini, & Dorlet, 2012). Since $mNOS_{red}$ is an integral part of mNOS in addition to its oxygenase domain, it is regarded as a catalytically self-sufficient enzyme (Stuehr, 1999). However, this is not necessarily the case with bNOS which lacks a dedicated reductase domain (Gusarov et al., 2008). In the case of bNOS, it is possible that NADH or NADPH may be used as the source of electrons depending on the nature of its reductase partner (Agapie et al., 2009; Gorren & Mayer, 2007). Irrespective of the source of electrons, the NOS oxygenase domain can only accept a single electron at a time (Guan, Kamatani, Kimura, & Iyanagi, 2003). How is this achieved since both NADH and NADPH release two electrons upon oxidation? The answer to this lies in the use of a flavin cofactors (flavin adenine dinucleotide, FAD, and flavin mononucleotide, FMN) within $mNOS_{red}$ that can undergo one electron redox reactions (Feng, 2012; Stuehr, Wei, et al., 2004). The electrons travel from NADPH to FAD and finally to FMN before being offloaded to the $mNOS_{oxy}$ haem. Accordingly, when the oxidized terminal FMN accepts electrons, it is reduced to a flavin hydroquinone ($FMNH_2$) and after that cycles to the semiquinone ($FMNH^{\bullet}$) form upon donating a single electron to the NOS haem (Feng, 2012).

The proximal ligand of the haem iron superoxide species is a cysteinyl sulphur, whereas the terminal oxygen on the distal end of the iron forms hydrogen bonds with the substrate (Arg) and to a water molecule in the catalytic site (Giroud et al., 2010). To cleave dioxygen within the NOS catalytic site, a 'push–pull' mechanism analogous to that observed in cytochrome P450 has been described (Cho, Carvajal, & Shaik, 2009; Giroud et al., 2010; Hurshman & Marletta, 2002; Jeong & Choi, 2002; Li, Kabir, Stuehr, Rousseau, & Yeh, 2007; McQuarters et al., 2016; Rousseau et al., 2005; Sabat et al., 2013; Stuehr, 1999; Stuehr, Wei, et al., 2004). Essentially, the strong electron-donating ability of the proximal thiolate ligand allows the strong 'push' of electrons towards the terminal oxygen at the distal end of the haem iron forming a ferric-peroxo species $Fe^{3+}O_2^{2-}$ (Rousseau et al., 2005; Woodward, Chang, Martin, & Marletta, 2009). In this case, the source of the second electron is BH_4 which gets rapidly oxidized into a BH_4 radical. Further, the hydrogen bonds associated with the oxygen at the distal end provides the 'pull' required to stabilize the resulting peroxo adduct. Notably, the 'push–pull' effect by itself is not sufficient to cause heterolytic cleavage of the O—O bond. In addition, donation of a proton is necessary to effect cleavage of dioxygen and release water in the process (Rousseau et al., 2005). The source of the proton is not entirely clear. Since the BH_4 radical ($BH_4^{\cdot +}$) still retains its proton following oxidation, it is unlikely to be the source of the proton in the first stage of reaction. Rather, cryostudies using gsNOS suggest that the substrate Arg is the source of protons that allows heterolytic cleavage of dioxygen (Davydov et al., 2014; Davydov, Sudhamsu, Lees, Crane, & Hoffman, 2009).

In the final step of oxygen activation, a highly reactive oxidizing intermediate is formed. Although the evidence as to the nature of this species is currently limited, it is commonly considered that a ferryl porphyrin cation radical (similar to compound 1) is formed immediately after heterolytic cleavage of dioxygen. This radical species is believed to hydroxylate Arg into NOHA (Rousseau et al., 2005; Stuehr et al., 2001).

4.2 One Catalytic Site, but Two Distinct Reactions

In the second stage of NOS-mediated catalysis, oxygen must be reactivated for the conversion of NOHA to Cit and NO (Marletta, Hurshman, & Rusche, 1998; Pufahl, Nanjappan, Woodard, & Marletta, 1992; Stuehr et al., 2001). Accordingly, the reductase domain donates the first electron that mediates dioxygen binding and the second electron that activates dioxygen is from BH_4 (Stuehr, Wei, et al., 2004). Despite these similarities in oxygen activation, the iron–oxygen adduct that initiates the nucleophilic attack on NOHA is thought to be the ferric-peroxo intermediate rather than the high-valent ferryl species that performs this task during the first stage (Woodward et al., 2009).

How does one catalytic centre that mainly carries out the same preliminary reactions involved in oxygen activation during both stages of NOS reaction stabilize two distinct catalytic intermediates? The answer to this question appears to be related to which of the oxygens in the ferric-peroxo species is stabilized at each stage of the reaction. Mounting evidence suggests that substrate-specific interactions and the cysteine-dependent electron 'push' are important factors that determine which of the two iron-oxo adducts is stabilized (Stuehr, Wei, et al., 2004; Woodward et al., 2009). In the first stage of the reaction, two separate hydrogen bonds from Arg and a water molecule extend to the terminal oxygen of the ferric-peroxo species and catalyse the heterolytic cleavage of the O—O bond. However, in the second stage, it is the proximal (and not the terminal) oxygen that is hydrogen bonded to NOHA. The hydrogen bond to the proximal oxygen of the ferric-peroxo species can increase electrophilicity of the terminal oxygen atom, thus allowing its reaction with NOHA to proceed (Marletta et al., 1998; Stuehr et al., 2001; Stuehr, Wei, et al., 2004; Woodward et al., 2009). In addition to hydrogen bonding, donation of protons from substrates may also play a role in stabilization of distinct oxygen atoms of the ferric-peroxo species. Whereas the guanidium group of Arg directs protons to the terminal oxygen during the first reaction stage, the N^{ω} of NOHA delivers protons to the proximal oxygen and increases substrate turnover during the second stage (Cho et al., 2009; Davydov et al., 2014; Giroud et al., 2010; Woodward et al., 2009). Finally, resonance Raman spectroscopy has revealed that the strength of the proximal iron–cysteine (Fe–Cys) bonds are distinct during both stages of the reaction (Santolini, Roman, Stuehr, & Mattioli, 2006). The Fe–Cys bond is much stronger in the NOHA-bound haem–oxo complex (stage 2) compared to the Arg-bound complex (stage 1), suggesting a stronger 'electron push' in stage 2 compared to stage 1. It is conceivable that the hydrogen bond between NOHA and oxygen strengthens following a stronger 'push' by the Cys ligand during the second stage of the reaction. Addition of the terminal oxygen of the ferric-peroxo species into NOHA results in the formation of Cit and nitroxyl via an alkyl peroxo species (Sabat et al., 2013).

Immediately following the monooxygenase reactions in both catalytic stages of NOS, a ferric-hydroxide intermediate is proposed to form on the distal end of the haem. This species has a Soret maximum of 420 nm and is too short lived to detect during the regular catalytic cycle of NOS (Tejero et al., 2008). However, substitution of the conserved tryptophan to histidine (W188H) in $iNOS_{oxy}$ resulted in the stabilization and further study of this species (Sabat et al., 2013). Importantly, it was demonstrated that the negative charge associated with the ferric-hydroxide intermediate is stabilized by the positive charge on the BH_4 radical, and hence, any process that affects turnover of the BH_4 radical would result in the decay of the ferric-hydroxide intermediate (Sabat et al., 2013). Accordingly, the reduction of the BH_4 radical in the first stage of the NOS reaction would promote the uptake of a proton to convert the negatively charged hydroxide ligand to a water molecule (Sabat et al., 2013).

The BH_4 radical generated during oxygen activation in the second stage is reduced back to the neutral form of BH_4 by an electron that is probably derived from nitroxyl (Brunel et al., 2016; Stoll et al., 2010; Wei, Wang, Hemann, Hille, & Stuehr, 2003; Wei et al., 2008). The simultaneous expulsion of a water molecule due to protonation of the ferric-hydroxide intermediate by nitroxyl would allow the formation of nitric oxide, which transiently attaches to the haem before being released into the bulk solution (Adak et al., 2000; Woodward, Nejatyjahromy, Britt, & Marletta, 2010). Thus, the kinetic coupling of the BH_4 radical turnover to the decay of the ferric-hydroxide intermediate in both stages of the NOS catalytic cycle results in the regeneration of the resting ferric haem species.

4.3 A Kinetic Model for the Catalytic Function of NOS

Stuehr and colleagues have described a kinetic model to explain various aspects of mNOS-mediated catalysis (Santolini, Adak, Curran, & Stuehr,

2001; Santolini, Meade, & Stuehr, 2001). Broadly, three kinetic parameters are fundamental to the catalytic behaviour of NOS. These parameters include the rates of (a) ferric haem reduction, (b) dissociation of ferric haem-NO complex, and (c) oxidation of ferrous haem-NO complex (Fig. 5) (Santolini, Meade, et al., 2001). The first rate-limiting step in the production of NO by NOS is the conversion of the resting ferric iron to a ferrous state (k_r) (Santolini, Meade, et al., 2001). This step is achieved by the donation of an electron to the ferric iron by the $mNOS_{red}$ domain (in eukaryotes) or a reductase partner (in bacteria) (Stuehr, Wei, et al., 2004).

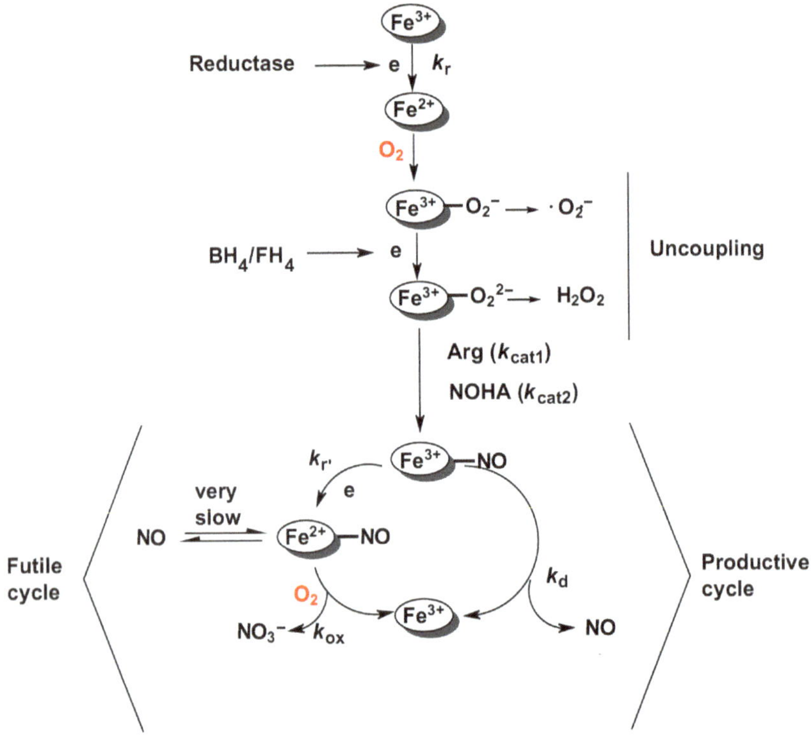

Fig. 5 Kinetic model of NOS catalysis. Following the reduction of Fe^{3+} (k_r) by a reductase to Fe^{2+}, oxygen binds reduced iron leading to its activation and successive entry into stage 1 and stage 2 of the NOS catalytic cycle (see Fig. 4). Any delay in reduction of iron due to either low BH_4 concentrations or low substrate (Arg) concentrations will result in uncoupling. The first stage involves the hydroxylation of Arg to NOHA (k_{cat1}), whereas in stage 2, further oxidation of NOHA (k_{cat2}) eventually results in a ferric-NO intermediate. At this point, NO may either dissociate from the ferric-NO intermediate and give rise to a productive cycle (k_d) or be reduced to a ferrous-NO intermediate ($k_{r'}$). As NO release from the ferrous-NO complex is very slow, NO tends to be oxidized further and released as nitrate (k_{ox}) giving rise to a futile cycle.

Once reduced, the ferrous iron takes part in two sequential cyclic reaction stages (Arg → NOHA and NOHA → Cit + NO) that are each characterized by their specific catalytic rate constants (k_{cat1} or k_{cat2}) (Fig. 5) (Santolini, Adak, et al., 2001; Santolini, Meade, et al., 2001). Both k_{cat1} and k_{cat2} are equal to the rates of BH_4 radical formation at each stage (Wang et al., 2001; Wei et al., 2001). The ferrous iron is subsequently transformed to a ferric-NO intermediate that dissociates and releases NO during the productive cycle at a rate equal to k_d (Fig. 5) (Stuehr, Wei, et al., 2004). Alternatively, reduction of ferric-NO to a ferrous-NO intermediate and further oxidation to nitrate may occur (Santolini, Adak, et al., 2001; Santolini, Meade, et al., 2001). In the latter step, the NOS reductase determines the reduction rate of ferric-NO to the ferrous-NO intermediate ($k_{r'}$) (Fig. 5). Subsequent interaction of ferrous-NO with dioxygen results in the release of nitrate (futile cycle) rather than NO at a rate equal to k_{ox} (Fig. 5). The futile cycle is favoured following the reduction of the ferric-NO species because the release of NO from the ferrous-NO intermediate is prolonged ($10^{-5}\ s^{-1}$) (Brunel et al., 2016; Santolini, Adak, et al., 2001; Santolini, Meade, et al., 2001; Stuehr, Wei, et al., 2004).

The kinetic model reveals that the efficient release of NO by NOS mandates a slow reduction of ferric haem by the NOS reductase. Typically, mNOS isoforms generate NO at a rate of 2–5 s^{-1} (k_d) and reduce the ferric iron at a relatively slower rate of 0.1–4 s^{-1} (k_r and $k_{r'}$). The slower reduction kinetics of $mNOS_{red}$ is necessary as a rapid reduction of the ferric iron in the $Fe^{3+}NO$ intermediate before NO release will lead to a futile cycle (Stuehr, Santolini, Wang, Wei, & Adak, 2004; Stuehr, Wei, Wang, & Hille, 2005).

Paradoxically, even though slow electron transfer from the NOS reductase to the ferric haem is beneficial for a productive cycle, it could retard the rate of the second step of oxygen activation (Adak et al., 2000; Rusche, Spiering, & Marletta, 1998). Specifically, constraining further reduction of $Fe^{3+}O_2^-$ in stage 1 by the NOS reductase would result in an uncoupled reaction involving the decay of $Fe^{3+}O_2^-$ species to the damaging $O_2^{\cdot -}$ radical (Fig. 5) (Abu-Soud, Gachhui, Raushel, & Stuehr, 1997; Bec, Gorren, Voelker, Mayer, & Lange, 1998; Couture et al., 2000). This leads to the question of how the NOS enzyme manages to prevent uncoupling and produce NO in the first place. It does so by utilizing BH_4 as a second electron donor that rapidly converts $Fe^{3+}O_2^-$ to $Fe^{3+}O_2^{2-}$ ($13\ s^{-1}$) (Stuehr et al., 2001). Direct evidence supporting this argument stems from the observation that at low BH_4 levels ($<2\ \mu M$) (Stuehr et al., 2001), NOS generates $O_2^{\cdot -}$. Further, data from single turnover reactions have confirmed not only that the BH_4 radical formation is coupled to the disappearance of $Fe^{3+}O_2^-$, but also that they are kinetically linked to Arg hydroxylation (Wei et al., 2001).

While BH_4-dependent reduction of $Fe^{3+}O_2^-$ species may suppress $O_2^{\cdot -}$ formation, it could potentially promote hydrogen peroxide formation due to decay of the resulting $Fe^{3+}O_2^{2-}$ species (Fig. 5). This is particularly true when the concentration of NOS substrate, Arg, is limiting ($<100\,\mu M$) (Stuehr et al., 2001). Hence, to prevent uncoupling, NOS must sufficiently be occupied by both substrate (Arg) and cofactor (BH_4). Furthermore, the kinetics of electron donation by BH_4 should be fine-tuned to maintain a coupled reaction (Wei, Wang, Arvai, et al., 2003). In this regard, it must be noted that several NOS active site residues interacting with BH_4 could influence its kinetics of radical formation. In particular, it was observed that substitution of the mNOS tryptophan residue (iNOS, Trp457) that is involved in π-stacking interaction with BH_4, with alanine or phenylalanine, significantly decreased the rate of BH_4 radical and NOHA formation (Wang et al., 2001).

Raman spectroscopic studies using NO, CO, and oxygen probes have confirmed that the interactions of substrate and cofactor with haem of bacterial NOS (bsNOS and saNOS) are similar to mNOS (Chartier, Blais, & Couture, 2006; Chartier & Couture, 2007a, 2007b; Santolini et al., 2006). Surprisingly, neither BH_4 nor FH_4 appears to affect the frequencies of the $\upsilon_{Fe\text{-}NO}$ modes (Chartier & Couture, 2004). The lack of response following pterin binding is unlike what is observed in mNOS and probably suggests that while pterins may function in vitro as an auxiliary electron donor to bNOS, they may not be its native cofactors.

4.4 Controlling Bacterial NOS Toxicity

NOS-dependent toxicity in bacteria may arise following ROS generation when NOS is uncoupled or from an excess of radical NO produced during a productive NOS cycle. Both ROS and NO can disrupt multiple physiological processes and metabolic pathways (respiration, TCA cycle, etc.) in bacteria. This is primarily expected due to the rapid reaction of these radicals with catalytic Fe—S clusters, thiols, and haem cofactors present within enzymes that participate in various physiological processes.

How do bacteria avoid toxicity resulting from NOS activity? Here, we will discuss two possible adaptations that could potentially reduce NOS toxicity and are unique to bNOS. The first involves the unusually low midpoint potential (E_m) of haem in bNOS (Cordas, Duarte, Moura, & Moura, 2013;

Field et al., 2011; Gronberg et al., 1999). Bacterial NOSs are thought to promiscuously accept electrons from multiple reductase partners in vivo. However, when concentrations of either Arg or BH_4/FH_4 become limiting, a sustained electron transfer from the reductase partner could lead to uncoupled oxidation of NADPH. Hence, the very low haem midpoint potential in bNOS may have evolved to minimize NOS uncoupling by limiting the number of possible candidate proteins that could partner with it in the capacity of a reductase. Precedence for uncoupled NADPH oxidation exists among some mNOS isoforms that harbour both reductase and oxygenase domains within the same polypeptide (Presta et al., 1998; Stuehr, 2004). For instance, in the absence of both Arg and BH_4, the midpoint potential of haem in nNOS is 108 mV higher than iNOS. This limits the transfer of electrons from NADPH (E_m, ~−340 mV) to iNOS (E_m, −347 mV), but not to nNOS (E_m, −239 mV). Consequently, nNOS activity results in increased uncoupled NADPH oxidation. Addition of BH_4 and Arg is necessary to increase the haem midpoint potential of iNOS (E_m, −263 mV) to levels that promote coupled NADPH oxidation (Presta et al., 1998; Stuehr, 2004). In contrast to mNOS isoforms, the haems within bNOS homologs like bsNOS (E_m, −361 mV) and saNOS (E_m, −334 mV) are held at a midpoint potential substantially lower than iNOS even when Arg and BH_4 are present (Hannibal, Somasundaram, Tejero, Wilson, & Stuehr, 2011; Lang, Santolini, & Couture, 2011).

However, one must note that a very low midpoint potential while beneficial in limiting uncoupled autooxidation of NADPH may lower rates of NO biosynthesis and adversely affect NO-dependent physiological processes. How then is the midpoint potential of bNOS optimized? Various studies have observed that a conserved tryptophan residue in bNOS (Trp56 in bsNOS and saNOS) is critical for tuning the haem midpoint potential by hydrogen bonding to the proximal cysteinate ligand of haem iron. The hydrogen bond that extends from the indole NH group of the conserved tryptophan to the sulphur atom of the cysteinate ligand ($NH \cdots S^-$) essentially reduces the electron density around the cysteinyl sulphur to moderate the midpoint potential (Hannibal et al., 2011; Lang et al., 2011). As the electron density around the Fe-coordinated sulphur decreases, the midpoint potential will increase. Such a modulation of electronic properties around the sulphur atom allows the magnitude of the 'push effect' to be fine-tuned to generate optimal levels of NO. Evidence to this effect was obtained in both saNOS and bsNOS following manipulation of the electronic properties of the proximal cysteinate ligand. Hannibal et al.

demonstrated that mutation of the conserved W66 residue in bsNOS to either histidine (W66H) or phenylalanine (W66F) had opposing effects on the haem midpoint potential relative to the native haem (E_m, −361 mV) (Hannibal et al., 2011). The W66H mutation resulted in an increased midpoint potential (E_m, −302 mV) due to a lower electron density around the proximal cysteine. This increase in midpoint potential was afforded by the stronger hydrogen bond between His-Cys compared to the Tpr-Cys bond of bsNOS. In contrast, the W66F variant exhibited a lower midpoint potential (E_m, −427 mV) compared to the native bsNOS due to the lack of a hydrogen bond between phenylalanine and cysteine resulting in a stronger electron density on the proximal cysteinyl residue. Consistent with the modulation of the 'push effect' and haem midpoint potential, single turnover studies using dithionite as a reductant revealed that the W66F mutation was also accompanied by an increase in oxygen activation, Arg hydroxylation, NOHA oxidation, and oxygen reactivity of the haem Fe^{2+}-NO complex relative to the wild type, while the W66H mutation had the opposite effect on the kinetics of bsNOS. Yet, both variants of W66 exhibited lower rates of NO generation when reconstituted in vitro with ferredoxins that were previously shown to act as reductase partners for bsNOS. The slower kinetics of NO generation associated with the W66H NOS variant was expected due to a compromised 'push effect' and a slower rate of oxygen activation. However, the decrease in NO generation was surprising for the W66F variant given the faster kinetic transitions that characterized this mutant. This decrease may result from an inefficient transfer of electrons from the ferredoxin reductase partners due to the very low midpoint potential of this variant. In general, similar conclusions have been made by Lang et al. using W56H and W56F variants of saNOS (Lang et al., 2011).

The second potential adaptation that could reduce NOS toxicity involves a structural feature specific to bNOS. Within the haem pocket of bNOS, a bulkier isoleucine residue (Ile214 in bsNOS) guards the opening of the haem pocket and replaces the valine residue (Val281 in $mNOS_{oxy}$) that is characteristically found within mNOS isoforms (Wang et al., 2004; Wang, Wei, & Stuehr, 2010; Weisslocker-Schaetzel et al., 2017). In bsNOS, this switch impedes the escape of synthesized NO from the catalytic site. What are the functional and catalytic consequences of such a change? Due to the high affinity of NO for haem, any structural hindrance of its release from the catalytic pocket will increase the frequency of NO binding with haem and hence decrease the rate of NO dissociation (k_d) from haem. The structural differences may in part account for the 10- to 20-fold slower release of NO by bsNOS compared to its mammalian isoforms. Evidence supporting this conclusion was reached following stopped-flow spectroscopic analysis of complementary bsNOS and $mNOS_{oxy}$ variants incorporating I224V or V346I substitutions, respectively. Whereas the $bsNOS_{I224V}$ variant exhibited a 3.6-fold increase in k_d compared to the native bsNOS, the $mNOS_{oxyV346I}$ variant exhibited a threefold slower NO dissociation rate relative to native $mNOS_{oxy}$ protein (Wang et al., 2004, 2010).

In principle, a decreased rate of NO dissociation from haem will subject the ferric-NO complex to further reduction by the NOS reductase partner and promote the production of nitrate (futile cycle) rather than NO (Fig. 5). Computational simulations using the measured kinetic constants have essentially confirmed these predictions (Fernandez, Marti, Crespo, & Estrin, 2005). Furthermore, Wang et al. have experimentally confirmed these predictions using various chimeric mNOS molecules (Wang et al., 2004, 2010). Accordingly, a nNOS reductase/$iNOS_{oxyV346I}$ chimera was estimated to form up to two to four molecules of nitrate per NO released. Thus, one may expect that the Val to Ile switch in bNOS will allow it to function as a NO dioxygenase and potentially limit NO radical toxicity.

5. BIOLOGICAL RELEVANCE OF NOS IN BACTERIA

NO is a small, lipophilic gas that rapidly diffuses within cells. As it is a radical species, the targets of NO are mostly limited to those biological molecules that also contain unpaired electrons in their outer orbitals. The most common targets of NO in vivo include free radicals and transition metals found in many cellular proteins (Ding & Demple, 2000; Ignarro, Degnan, Baricos, Kadowitz, & Wolin, 1982; Kennedy, Antholine, & Beinert, 1997; Stone & Marletta, 1994; Yukl, Elbaz, Nakano, & Moenne-Loccoz, 2008). The latter class of molecules includes metalloproteins containing iron, zinc, or copper that is usually coordinated by a haem prosthetic group or bound to sulphur atoms (e.g. iron–sulphur clusters). Additionally, various cellular and protein thiols also react with NO due to the two lone pair electrons found within thiols (Stamler et al., 1992). As multiple bacterial pathways that support DNA replication, respiration, and TCA cycle are targets of NO, it is not surprising why eukaryotic hosts have adapted NO as an antimicrobial agent.

The immediate outcome of a NO assault on bacteria is the inability to maintain intracellular redox status due to inactivation of proteins within the respiratory pathway. In this regard, NO has well-established roles in interacting with the haem centres of respiratory cytochromes resulting in a compromised electrochemical gradient (Butler et al., 2002). Indeed, NO has been shown to inhibit respiration in mitochondria (Erusalimsky & Moncada, 2007) and bacteria (Stevanin et al., 2000). However, some bacterial pathogens like *S. aureus* can overcome the bacteriostatic effects of NO by enzymatically detoxifying NO to nitrate using the flavohaemoglobin, Hmp, and induce lactic acid fermentation to compensate for the dysbalanced cellular redox status (Gardner & Gardner, 2002; Gardner, Gardner, Martin, & Salzman, 1998; Kinkel et al., 2016; Poole & Hughes, 2000; Richardson, Dunman, & Fang, 2006).

Given that NO induces bacteriostasis, it was initially surprising and intriguing to find that many bacterial species including staphylococci harbour NOS (Crane et al., 2010). In the case of staphylococci, if the saNOS activity produces significant concentrations of endogenous NO, existing dogma would dictate at least traces of lactate to be excreted. Furthermore, another prediction would be that cells unable to detoxify endogenous NO (e.g. *hmp* mutant) should also exhibit growth defects associated with NO toxicity. However, *S. aureus* produces very little lactate during growth under aerobic conditions (Richardson, Libby, & Fang, 2008). Neither does a *S. aureus hmp* mutant exhibit altered growth rates due to endogenous NO toxicity, assuming that NOS is still active in the *hmp* mutant (Lewis et al., 2015). Furthermore, since multiple studies have confirmed that saNOS can produce NO in vivo using NO-specific copper-fluorescein probes (van Sorge et al., 2013), the collective evidence suggests that endogenous NO production may either not accumulate during growth or may only be induced in excess under select circumstances and for specific purposes. Support for the latter hypothesis was first to emerge following the discovery that bNOS is induced to combat oxidative stress not only in staphylococci but also in *Bacillus* species.

5.1 Role of NOS in Countering Oxidative Stress

While NO is often thought to be cytotoxic, multiple studies performed in eukaryotic systems have described an antioxidant function for this radical when present at physiological concentrations. For instance, one study observed that brain astrocytes loaded with iron were less susceptible to intracellular oxidative stress when treated with cytokines that increased steady-state levels of NO (Robb, Gaspers, Wright, Thomas, & Connor, 1999). In a separate and more recent study, Dikalov et al. demonstrated that physiological concentrations of NO in cells diminished mitochondrial superoxide production (Dikalov, Mayorov, & Panov, 2017). Similarly, various other studies have described cytoprotective effects of NO in response to hydrogen peroxide and alkyl peroxide-dependent oxidative stress (Gusarov & Nudler, 2005; Wink et al., 1995). However, a coherent mechanism by which NO limits oxidative stress is yet to emerge.

A common source of oxidative stress within cells arises from the short lived, and highly reactive hydroxyl radicals ($OH^{\cdot}$) generated from the Haber Weiss/Fenton chemistry (Liochev, 1996; Wardman & Candeias, 1996). Cellular damage associated with hydroxyl radicals primarily involves DNA base modifications and DNA strand breaks (Balasubramanian, Pogozelski, & Tullius, 1998; Dizdaroglu & Jaruga, 2012; Kehrer, 2000). While the Haber Weiss reaction involves the formation of hydroxyl radicals (Eq. 3) following chemical reaction of hydrogen peroxide with superoxide, it is in itself not considered to be a biologically relevant source of hydroxyl radicals as it is thermodynamically unfavourable (Kehrer, 2000):

$$H_2O_2 + O_2^{-} \rightarrow O_2 + OH^{-} + OH^{\cdot} \quad (3)$$

However, reduced iron can act as a catalyst (Fenton reaction, Eq. 4) and allow the reaction to achieve physiological significance. In that case, the reduced iron necessary for the Fenton reaction may be obtained following reduction of Fe^{3+} by superoxide (Eq. 5) or other cellular reductants like $FADH_2$ and reduced cysteine (Kehrer, 2000):

$$Fe^{2+} + H_2O_2 \rightarrow \left([FeOH]^{3+} \leftrightarrow [FeO]^{2+}\right) \rightarrow Fe^{3+} + OH^{-} + OH^{\cdot} \quad (4)$$

$$Fe^{3+} + O_2^{-} \rightarrow Fe^{2+} + O_2 \quad (5)$$

How does NO function as an antioxidant? Evidence till date suggests that NO may counter oxidative stress by directly scavenging hydroxyl radicals or reacting with ferryl intermediates ($[FeOH]^{3+}$, $[FeO]^{2+}$) proposed to be part of the Fenton chemistry (Sharpe, Robb, & Clark, 2003). Note here that the reaction of hydroxyl radicals with biological macromolecular targets within cells can also proceed with second-order rate constants similar to that observed for NO and hydroxyl radicals. Hence, NO may act as a good scavenger of hydroxyl radicals only if sufficient concentrations are present within cells. Additionally, NO may nitrosate proteins and lipid radicals to prevent radical propagation.

The antioxidant nature of NO may explain why some bacteria possess NOS. Gusarov et al. were the first to systematically test whether NO derived from bNOS activity could reduce oxidative damage in bacteria (Gusarov &

Nudler, 2005). They initially observed that pretreatment of *B. subtilis* with a low concentration of exogenous NO (30 μM) was sufficient to significantly retard oxidative damage resulting from subsequent exposure to multiple oxidants including hydrogen peroxide, *t*-butyl hydroperoxide, and cumene hydroperoxide. Significantly, NO-mediated cytoprotection occurred even if NO pretreatment was administered 5 s before challenge with hydrogen peroxide, suggesting that the protection afforded by NO against this oxidant was immediate and independent of protein expression. Interestingly, the antioxidant capacity of NO was not evident when NO was added at the same time as hydrogen peroxide or after peroxide treatment, presumably because excess radicals generated from H_2O_2 treatment scavenged NO. Furthermore, they also observed that NO-mediated cytoprotection was transient and readily reversible upon depletion of exogenously supplied NO (Gusarov & Nudler, 2005).

Additional investigations revealed that the antioxidant mechanism associated with NO in bacteria appeared to be distinct from those described in mammalian systems. Importantly, exogenously added NO protected *B. subtilis* from radicals of the Fenton chemistry by two mechanisms (Gusarov & Nudler, 2005). First, NO activates *B. subtilis* catalase activity. In vitro studies revealed that free cysteine could form a complex with *B. subtilis* catalase and partially inhibit its activity. However, NO activates catalase by releasing it from this complex presumably through an *S*-nitrosylation mechanism. The activation of catalase by NO does not appear to be universal. For instance, *Escherichia coli* catalase was shown to be inhibited rather than activated by NO. Second, NO prevents rereduction of oxidized cysteine by inhibiting thioredoxin. In *B. subtilis*, thioredoxins are crucial for maintaining sizable concentrations of reduced intracellular cysteine at the expense of NADPH. When free iron levels in the cell increase due to H_2O_2 stress, reduced cysteines can be toxic as they convert Fe^{3+} to Fe^{2+} and promote hydroxyl radical formation via the Fenton chemistry (Gusarov & Nudler, 2005). Thus, inhibition of thioredoxins by NO can limit Fenton chemistry and promote its antioxidant potential in bacteria.

While the above described antioxidant mechanisms were mostly elucidated following exposure of *B. subtilis* with exogenous NO, Gusarov et al. also demonstrated that endogenous NO resulting from the activity of bsNOS was sensitive to increased intracellular concentrations of cysteine (Gusarov & Nudler, 2005). Accordingly, a *B. subtilis nos* mutant that had higher intracellular levels of cysteine was more sensitive to hydrogen peroxide challenge than its isogenic wild-type strain, thus confirming a role for endogenous NO in regulating the redox status of intracellular cysteines. Additionally, the antioxidant functions of bNOS were also shown to be crucial in *S. aureus* and *Bacillus anthracis* where cysteine is a critical low-molecular-weight thiol (Gusarov & Nudler, 2005; Gusarov, Shatalin, Starodubtseva, & Nudler, 2009; van Sorge et al., 2013). Mutation of *nos* in both *S. aureus* and *B. anthracis* sensitizes these cells to peroxide challenge.

The antioxidant mechanisms of bNOS were also demonstrated to be essential for bacterial virulence. In *B. anthracis*, generation of endogenous NO occurs at the early stages of macrophage infection in response to superoxides produced during an oxidative burst (Shatalin et al., 2008). Superoxides generated during an oxidative burst are prone to dismutation into hydrogen peroxide which can readily breach the bacterial cell membrane and feed into the cytotoxic Fenton reaction. Mutation of *nos* in *B. anthracis* resulted in a strain that was attenuated in a mouse model of systemic infection. Additionally, spores of the *B. anthracis nos* mutant were unable to germinate efficiently and exhibited poor survival within macrophages (Shatalin et al., 2008). Mechanistically, the increased killing of the *nos* mutant within macrophages was demonstrated to result from its inability to activate catalase and suppress the Fenton reaction in the absence of endogenous NO. Interestingly, NO derived from bNOS activity also protects bacteria from a variety of antibiotics (Gusarov et al., 2009; van Sorge et al., 2013). The NO-mediated cytoprotection is proposed to result either from chemical modification of the antimicrobial compound or by suppression of antibiotic-induced oxidative stress in bacteria. While the antioxidant activity of endogenously produced NO and the effect that bNOS has on virulence appear to be functionally conserved in *S. aureus*, the importance of saNOS in resisting antibiotics was selective and restricted to a few antibiotics that targeted the cell envelope (e.g. vancomycin and daptomycin). Significantly, mutation of saNOS increased resistance to aminoglycosides (Gusarov et al., 2009; van Sorge et al., 2013). These studies in *S. aureus* not only counter the notion that bNOS is universally required for antibiotic resistance but also suggest that bNOS may have additional functions to that of an antioxidant.

5.2 Physiological Significance of Bacterial NOS

Significant progress has recently been made regarding a role for bNOS in growth and physiological processes. The initial clues about the physiological function of saNOS came from studies carried out by Kinkel et al. (2016). They observed that a *nos* mutant had an elevated rate of oxygen consumption when compared to its isogenic wild-type strain. Additionally, this phenotype was exacerbated under microaerobic conditions and also observed across different species. The evidence argued for a functional role for saNOS in inhibiting bacterial aerobic respiration presumably due to endogenous NO-mediated inhibition of cytochrome activity. NO has well-established roles in interacting with haem centres that are incorporated within cytochromes. Upon binding NO, cytochromes that translocate protons across the membrane are inhibited and would thus stall the generation of an electrochemical gradient (Butler et al., 2002) as well as the transfer of electrons to the terminal electron acceptor. Therefore, NO generated by bNOS may function to contribute to respiratory inhibition. Indeed, NO has been shown to inhibit respiration in mitochondria (Brown, 2001; Erusalimsky & Moncada, 2007). Consistent with this interpretation, a *S. aureus hmp* mutant that is predicted to have increased endogenous NO levels exhibited a slower rate of oxygen uptake (Kinkel et al., 2016).

Why do cells harbour a mechanism to inhibit their respiration? Evidence put forth by Kinkel et al. suggests that inhibition of aerobic respiration by endogenous NO may aid the transition of cells from oxygen to nitrate-dependent respiration (Kinkel et al., 2016). Bacteria are adept in their ability to use a variety of terminal electron acceptors (e.g. oxygen and nitrate) to generate the energy-conferring electrochemical gradient produced by respiration. Therefore, when oxygen is limited or absent, tuning of respiratory machinery is in order. It was observed that the saNOS mutant, unlike its isogenic parent strain, was unable to consume nitrate until all oxygen was exhausted from the media (Kinkel et al., 2016). The delayed kinetics of nitrate respiration in the saNOS mutant did not result from a decrease in nitrate reductase as its expression was not altered. Further, the saNOS-dependent respiratory transition was demonstrated to be necessary for the maintenance of membrane potential. Accordingly, the saNOS mutant exhibited a more dissipated membrane potential compared to its isogenic wild-type strain. The exact cause for the dissipated membrane potential in the saNOS mutant is unclear since this mutant can compensate for the delay in nitrate respiration with increased aerobic respiration (Kinkel et al., 2016). However, it was speculated that since nitrate respiration is electrogenic, it may contribute to the maintenance of membrane potential in the wild-type strain when oxygen levels are low. Consistent with this interpretation, the saNOS mutant was more susceptible to daptomycin (a membrane-depolarizing drug) and failed to recover from a daptomycin challenge even after nitrate supplementation in the media (Kinkel et al., 2016).

In addition to inhibiting respiration, excess concentrations of endogenous NO may also be converted to nitrate via Hmp activity (Gardner & Gardner, 2002; Gardner et al., 1998; Poole & Hughes, 2000), thus providing an alternate electron acceptor to cells whose respiration is inhibited by NO. Thus the model generated from the studies of Kinkel et al. supports a role for bNOS in microaerobic growth wherein endogenous NO produced as a result of bNOS activity not only inhibits respiration but can to some capacity also constitute the source of the alternate electron acceptor, nitrate (Kinkel et al., 2016).

While Kinkel et al. advanced an elegant model for bNOS function under microaerobic conditions, subsequent studies have uncovered important roles for bNOS even during aerobic growth. Using antibodies specific for saNOS, Chaudhari et al. demonstrated that saNOS is typically expressed during aerobic postexponential and stationary phases of *S. aureus* (Chaudhari et al., 2017). Furthermore, mutation of saNOS adversely affected the growth of *S. aureus* during the postexponential phase, a phenotype that was also confirmed in a separate study. Consistent with a slower growth rate during the postexponential phase, consumption of acetate as a secondary carbon source was also delayed in the saNOS mutant. Transcriptomic and metabolomic analyses from two separate studies confirmed a decrease in TCA cycle activity of the saNOS mutant when compared to its isogenic wild-type strain (Chaudhari et al., 2017). Unexpectedly, redox indicators pointed to an increase in $NADH/NAD^+$ ratio despite a decrease in TCA cycle activity, suggesting that the inhibition of TCA cycle activity possibly resulted from the altered redox status of the cell. Consistent with this interpretation, Chaudhari et al. observed a decrease in Pdh activity that is also thought to be regulated by altered redox status of the cell. Interestingly, although the TCA cycle and Pdh activities were lower in the saNOS mutant relative to the parent strain, the observed decrease in postexponential growth of the saNOS mutant could only be partially attributed to these enzymatic deficiencies (Chaudhari et al., 2017). Instead, Chaudhari et al. determined that the primary cause for the decrease in postexponential growth results from a lower respiratory potential of the saNOS mutant relative to the isogenic wild-type strain. Thus unlike the microaerobic conditions described by Kinkel et al. wherein saNOS inhibited respiration, under aerobic conditions saNOS was required for the maintenance of optimal respiratory activity (Chaudhari et al., 2017).

How does saNOS activate respiration under aerobic conditions? Chaudhari et al. determined that nitrite generated from either enzymatic/nonenzymatic oxidation of endogenous NO or the respiratory nitrate reductase activity was responsible for activation of aerobic respiration (Chaudhari et al., 2017). Accordingly, it was observed that supplementation of nitrite in the media rescued the postexponential growth defect of the saNOS mutant. Interestingly, the target of nitrite appeared to be cytochrome *bo* within the respiratory chain. It is not mechanistically clear how nitrite activates cytochrome *bo*, but it is more than likely to be an indirect mechanism. These authors also observed that compensatory SrrAB activity significantly masked the growth defect associated with the *nos* mutation. The SrrAB two-component system is thought to activate quinol oxidase expression among others and is well poised to compensate for the lack of saNOS-dependent activation of respiration. Consistent with a role for saNOS in activating respiration, the wild-type strain but not the saNOS mutant was more sensitive to the antibiotic, streptomycin. Earlier work in *B. subtilis* had confirmed that uptake of streptomycin requires active respiration. Hence, the increased resistance observed for the saNOS mutant to streptomycin is reflective of a lower respiratory potential for this mutant. Furthermore, nitrite supplementation fully sensitized the saNOS mutant to streptomycin, once again confirming that nitrite can bypass the need for saNOS concerning the activation of respiration. Hence physiological studies have revealed that bNOS may have opposing effects on bacterial respiration. The concentration of oxygen appears to be the crucial factor that determines whether bNOS activates or inhibits respiration (Chaudhari et al., 2017). Interestingly, regardless of the mechanism by which bNOS affects cell physiology, it appears that the mutation of saNOS is detrimental in vivo as the saNOS mutant was attenuated in a mouse nasal colonization model and a mouse sepsis model (Kinkel et al., 2016). The physiological phenotypes associated with bNOS are in good agreement with the biochemical studies that characterize bNOS as both a NO and nitrate producer.

5.3 Bacterial NOS-Dependent *S*-Nitrosylation and Nitration

Endogenous NO derived from bNOS activity has been linked to two post-translational modifications, viz., *S*-nitrosylation and nitration. *S*-nitrosylation involves the addition of NO to a cysteine residue (Hess, Matsumoto, Nudelman, & Stamler, 2001). More specifically, a covalent bond between a thiol (R-SH, reduced cysteine) and a nitrosonium equivalent (NO^+, derived from NO) is formed resulting in an *S*-nitrosocysteine modification (Eq. (6). *S*-nitrosylation occurs nonenzymatically and is reversible:

$$\text{R-SH} + \text{NO}^+ \rightarrow \text{R-SNO} \tag{6}$$

Although a target protein may possess multiple cysteine residues, only a few are subject to *S*-nitrosylation. The exact mechanism of this selectivity is unknown, but the immediate environment of the target cysteine is thought to play a major role in this selectivity. Accordingly, acid–base and hydrophobic motifs may enhance *S*-nitrosylation of certain cysteines within a protein (Hess et al., 2001).

Endogenous NO from *Bacillus* sp. is thought to *S*-nitrosylate bacterial and host proteins (Gusarov & Nudler, 2005). Gusarov et al. demonstrated that endogenous NO could activate *B. subtilis* catalase. Interestingly, in vitro studies suggested that NO does not directly activate purified catalase. Hence, an indirect mechanism was suspected of taking place in the NO-mediated activation of catalase. They speculated that the activity of catalase is potentially inhibited by free cysteine within cells and demonstrated that either diamide-mediated oxidation or NO-mediated *S*-nitrosylation of free cysteines could relieve thiol inhibition of catalase. Furthermore, consistent with *S*-nitrosylation, NO was able to activate catalase within seconds after challenge in a transient (reversible) manner (Gusarov & Nudler, 2005).

Recent evidence from nitrosoproteomic analysis suggests that multiple proteins present within human small airway epithelial cells are targets of NO produced from bNOS. Among these proteins, peroxiredoxins (Prx1) appear to be significantly susceptible to *S*-nitrosylation by *B. anthracis* NOS (Chung et al., 2017). Mutation or inhibition of baNOS resulted in decreased *S*-nitrosylation. Prx1 has both peroxidase and chaperone activities. Interestingly, *S*-nitrosylation of Prx1 by bNOS-derived NO enhanced its chaperone activity and diminished its peroxidase function, thus demonstrating that bNOS may also influence host biology in addition to its bacterial physiology (Chung et al., 2017).

In addition to *S*-nitrosylation, protein or peptide nitration appears to be a common modification carried out by bacterial NO. Nitration occurs preferentially at tyrosine residues with tryptophan and methionine residues less commonly observed. Importantly, NO-mediated nitration appears to be mostly irreversible although denitrating enzymes have been found in some organisms. The first indication of nitration by bNOS was described in *S. turgidiscabies* (Kers et al., 2004), a plant pathogen that is known to cause potato scab disease. *Streptomyces* strains are generally nonpathogenic to humans, commonly found in soil, strict aerobes, and are notable for their production of antibiotics, such as streptomycin, chloramphenicol, and tetracycline. *S. turgidiscabies* NOS (stNOS) is involved in the production of thaxtomin which is a nitrated nonribosomally produced dipeptide toxin that inhibits cellulose biosynthesis. Speculations on whether stNOS nitrated thaxtomin arose following the observation that its *nos* gene resided near toxin biosynthetic genes (*txtA* and *txtB*) (Johnson et al., 2008). Surprisingly, mutation of *nos* in *S. turgidiscabies* abrogated thaxtomin production entirely and did not result in the accumulation of nonnitrated toxin in culture media. However, convincing evidence that stNOS nitrated thaxtomin came after it was experimentally demonstrated that the nitrate modification on thaxtomin resulted from the terminal guanidine group of ^{15}N-labelled Arg which is unique to NOS activity. Consistent with a role for bNOS in nitrated-toxin production, *Streptomyces atratus* NOS is also proposed to be the source of NO required for the nitration of rufomycin (Tomita, Katsuyama, Minami, & Ohnishi, 2017).

bNOS-dependent nitration has also been reported to occur in the UV-tolerant bacteria, *D. radiodurans* (Buddha, Keery, & Crane, 2004). *D. radiodurans* has two tryptophanyl-transfer RNA (tRNA) synthetases—TrpRS I and TrpRS II. While both TrpRS I and TrpRS II can aminoacylate tRNA^{Trp}, the latter protein can also incorporate different analogs of tryptophan, including 4-nitro-Trp and 5-hydroxy-Trp. Consistent with its low specificity, TrpRS II is significantly different from canonical TrpRS (only 29% sequence identity) and is up to three times less active that TrpRS I. Interestingly, drNOS was shown to form a stable complex with TrpRS II and together catalyse the synthesis of 4-nitro-Trp-tRNA^{Trp}. It is however not currently known whether 4-nitro-Trp is incorporated into any proteins or have any physiological functions (Buddha et al., 2004).

6. BACTERIAL NOS AS A THERAPEUTIC TARGET

The recent interest in bNOS-specific inhibitors arose from observations that endogenous NO could protect bacteria against antibiotic-induced oxidative injury. Studies have demonstrated that NOS inhibitors (e.g., L-N^G-nitroarginine methyl ester, L-NAME) can potentiate the antibacterial effects of certain antibiotics like β-lactams and acriflavine (Gusarov et al., 2009; Gusarov & Nudler, 2005; Kohanski, Dwyer, Hayete, Lawrence, & Collins, 2007; van Sorge et al., 2013). Additionally, bNOS inhibitors were also demonstrated to synergize with oxidants like hydrogen peroxide and cause cell death (Shatalin et al., 2008). Hence, bacterial NOS inhibitors may be used alongside traditional antibiotics.

The design of bNOS inhibitors can be challenging given the high degree of similarity between the catalytic pocket of mNOS and bNOS homologs.

Any inhibitor that is designed against bNOS should be specific and limited in its ability to inhibit various mNOS isoforms.

Poulos and colleagues have recently undertaken structure-based approaches to designing inhibitors specific to bacterial NOS (Holden, Dejam, et al., 2015; Holden, Kang, Beasley, et al., 2015; Holden, Kang, Hollingsworth, et al., 2015; Holden et al., 2016, 2013). A compound library available as part of a more extensive drug discovery program was screened against *B. subtilis* and *S. aureus* to identify inhibitors (Holden, Kang, Beasley, et al., 2015; Holden et al., 2013; Poulos & Li, 2013; Silverman, 2009). Additionally, bacterial NOS-inhibitor structures were solved to identify structural features that were selective for bNOS. Based on those studies several critical structural features were identified in bNOS that set them apart from most mNOS isoforms. The first selective feature involved the pterin-binding site of bNOS. Since the Zn-binding motif that is present within all mNOS isoforms was absent in bNOS, the pterin-binding site is more solvent exposed within bNOS homologs. Consequently, bNOS homologs have a lower affinity for pterins when compared to their mammalian counterparts. This had led to the speculation that bulky inhibitors against pterin-binding sites were more likely to target bNOS than mNOS. In agreement with this hypothesis, various inhibitors with two aminopyridine groups were found to displace the pterin associated with bNOS and H-bond with haem propionate A (Holden, Dejam, et al., 2015).

In addition to the pterin-binding site, the lack of a Zn-binding motif also allowed the formation of a second binding pocket that was unique to bNOS. It was demonstrated that this binding pocket afforded some NOS inhibitors (thiophenecarboximidamide-based inhibitors) to adopt a unique and active inhibitory conformation upon binding to bNOS (Holden et al., 2013; Jing et al., 2014).

The third selective feature that is specific to bNOS involves the presence of an isoleucine residue (I214) instead of the conserved valine in mNOS isoforms. As described earlier in this chapter, the isoleucine substitution in bNOS prevented the rapid release of NO from the active site of bNOS. However, in addition to this effect, I214 also allowed efficient binding of thiophenecarboximidamide inhibitors to the active site of bNOS (Holden et al., 2013).

Finally, some bNOS inhibitors introduced a rotameric positional change of a NOS tyrosine residue (Tyr361) upon binding and allowed inhibitors better access to haem propionate A (Holden, Dejam, et al., 2015). Although this feature is not unique for bNOS and overlaps with nNOS

(Li et al., 2014), the inability of most of these inhibitors to pass through the blood–brain barrier is thought to lower the risk of nNOS inhibition.

In summary, the importance of NOS in pathogen physiology underscores the need for the development of new inhibitors of bNOS. The observations made by Poulos and colleagues on the selective features of bNOS open the possibility for further improvements in the rational design of potent NOS inhibitors that are specific to bNOS homologs and can be used in conjunction with existing antibiotics.

7. CONCLUSIONS

It is increasingly becoming clear that bNOS plays a vital role in prokaryotic physiology. While eukaryotic NOS possesses an oxygenase and reductase domain within the same polypeptide, the bacterial NOS is truncated (missing the reductase domain). Despite this, they have been demonstrated to function as mammalian NOSs and can convert Arg to Cit and NO. Furthermore, similar to eukaryotes, the latter molecule (NO) or its derivatives (NO_2^- and NO_3^-) appear to play the most critical functions in bacterial physiology. Emerging functions of endogenous NO include its ability to respond as a cellular antioxidant, modulate bacterial respiration, and posttranscriptionally regulate the function of several bacterial and host proteins. Often, these functions have significant consequences on host–pathogen interactions in addition to affecting bacterial physiology.

The biochemical characterization of bNOS has provided some surprising insights and highlights multiple differences between mNOS and bNOS. First, although BH_4 can act as an electron donor for NOS, most bacteria lack the BH_4 biosynthetic pathway. Hence, NOS activity in vivo mandates an alternate electron donor. In this regard, although FH_4 can to some extent substitute for the role of BH_4 in vitro, it is still not clear whether this pterin is essential for NOS function in vivo. Second, the lack of a linked reductase domain and a very low haem midpoint potential within bNOS should significantly limit initial electron transfer to the oxygenase domain. It is not clear how NOS overcomes this limitation or what the identities of the native bNOS reductases are. Although bacterial flavodoxins are considered to be excellent candidates for bNOS reductases, studies so far have not been able to support a strong role for them, in vivo. Finally, the presence of isoleucine (I214) in bNOS instead of a corresponding valine in mNOS has been shown to affect the dissociation rate of NO adversely. It is predicted that the I214V substitution in bNOS will support the production of nitrate in addition to

NO. With regard to this last point, it must be noted that during aerobic growth, studies have highlighted an essential role for nitrite rather than NO or nitrate per se as the effector of NOS in bacteria. NOS-dependent nitrite was shown to activate cytochrome *bo*-mediated respiration in bacteria. Hence, even if both nitrate and NO are produced their efficient and rapid conversion to nitrite by nitrate reductase and Hmp will aid in the fulfilment of their biological roles. Additionally, the nitrate derived either directly from bNOS activity or through Hmp activity was shown to impact microaerobic respiration (Kinkel et al., 2016). Thus, irrespective of the biochemical and structural differences between mNOS isoforms and bNOS homologs, physiological studies so far have shown that both NOS-dependent NO and its oxidized derivatives are essential contributors to prokaryotic physiology. As clarity on the physiological functions and biochemical characteristics of bNOS emerges, it is expected to complement and further fuel the search for bNOS inhibitors.

ACKNOWLEDGEMENT

The authors gratefully acknowledge financial support from the National Institutes of Health awarded to VCT (R01 AI125588).

REFERENCES

Abu-Soud, H. M., Gachhui, R., Raushel, F. M., & Stuehr, D. J. (1997). The ferrous-dioxy complex of neuronal nitric oxide synthase. Divergent effects of L-arginine and tetrahydrobiopterin on its stability. *Journal of Biological Chemistry*, *272*(28), 17349–17353.

Adak, S., Aulak, K. S., & Stuehr, D. J. (2002). Direct evidence for nitric oxide production by a nitric-oxide synthase-like protein from *Bacillus subtilis*. *Journal of Biological Chemistry*, *277*(18), 16167–16171. https://doi.org/10.1074/jbc.M201136200.

Adak, S., Bilwes, A. M., Panda, K., Hosfield, D., Aulak, K. S., McDonald, J. F., et al. (2002). Cloning, expression, and characterization of a nitric oxide synthase protein from *Deinococcus radiodurans*. *Proceedings of the National Academy of Sciences of the United States of America*, *99*(1), 107–112. https://doi.org/10.1073/pnas.012470099.

Adak, S., Ghosh, S., Abu-Soud, H. M., & Stuehr, D. J. (1999). Role of reductase domain cluster 1 acidic residues in neuronal nitric-oxide synthase. Characterization of the FMN-FREE enzyme. *Journal of Biological Chemistry*, *274*(32), 22313–22320.

Adak, S., Wang, Q., & Stuehr, D. J. (2000). Arginine conversion to nitroxide by tetrahydrobiopterin-free neuronal nitric-oxide synthase. Implications for mechanism. *Journal of Biological Chemistry*, *275*(43), 33554–33561. https://doi.org/10.1074/jbc.M004337200.

Agapie, T., Suseno, S., Woodward, J. J., Stoll, S., Britt, R. D., & Marletta, M. A. (2009). NO formation by a catalytically self-sufficient bacterial nitric oxide synthase from *Sorangium cellulosum*. *Proceedings of the National Academy of Sciences of the United States of America*, *106*(38), 16221–16226. https://doi.org/10.1073/pnas.0908443106.

Alderton, W. K., Cooper, C. E., & Knowles, R. G. (2001). Nitric oxide synthases: Structure, function and inhibition. *Biochemical Journal*, *357*(Pt 3), 593–615.

Andreakis, N., D'Aniello, S., Albalat, R., Patti, F. P., Garcia-Fernandez, J., Procaccini, G., et al. (2011). Evolution of the nitric oxide synthase family in metazoans. *Molecular Biology and Evolution*, *28*(1), 163–179. https://doi.org/10.1093/molbev/msq179.

Balasubramanian, B., Pogozelski, W. K., & Tullius, T. D. (1998). DNA strand breaking by the hydroxyl radical is governed by the accessible surface areas of the hydrogen atoms of the DNA backbone. *Proceedings of the National Academy of Sciences of the United States of America*, *95*(17), 9738–9743.

Bec, N., Gorren, A. C., Voelker, C., Mayer, B., & Lange, R. (1998). Reaction of neuronal nitric-oxide synthase with oxygen at low temperature. Evidence for reductive activation of the oxy-ferrous complex by tetrahydrobiopterin. *Journal of Biological Chemistry*, *273*(22), 13502–13508.

Bermingham, A., & Derrick, J. P. (2002). The folic acid biosynthesis pathway in bacteria: Evaluation of potential for antibacterial drug discovery. *BioEssays*, *24*(7), 637–648. https://doi.org/10.1002/bies.10114.

Bird, L. E., Ren, J., Zhang, J., Foxwell, N., Hawkins, A. R., Charles, I. G., et al. (2002). Crystal structure of SANOS, a bacterial nitric oxide synthase oxygenase protein from *Staphylococcus aureus*. *Structure*, *10*(12), 1687–1696.

Brown, G. C. (2001). Regulation of mitochondrial respiration by nitric oxide inhibition of cytochrome c oxidase. *Biochimica et Biophysica Acta*, *1504*(1), 46–57.

Brunel, A., Lang, J., Couture, M., Boucher, J. L., Dorlet, P., & Santolini, J. (2016). Oxygen activation in NO synthases: Evidence for a direct role of the substrate. *FEBS Open Bio*, *6*(5), 386–397. https://doi.org/10.1002/2211-5463.12036.

Brunel, A., Santolini, J., & Dorlet, P. (2012). Electron paramagnetic resonance characterization of tetrahydrobiopterin radical formation in bacterial nitric oxide synthase compared to mammalian nitric oxide synthase. *Biophysical Journal*, *103*(1), 109–117. https://doi.org/10.1016/j.bpj.2012.05.032.

Brunel, A., Wilson, A., Henry, L., Dorlet, P., & Santolini, J. (2011). The proximal hydrogen bond network modulates *Bacillus subtilis* nitric-oxide synthase electronic and structural properties. *Journal of Biological Chemistry*, *286*(14), 11997–12005. https://doi.org/10.1074/jbc.M110.195446.

Bryan, N. S., Bian, K., & Murad, F. (2009). Discovery of the nitric oxide signaling pathway and targets for drug development. *Frontiers in Bioscience (Landmark Ed)*, (14), 1–18.

Buddha, M. R., Keery, K. M., & Crane, B. R. (2004). An unusual tryptophanyl tRNA synthetase interacts with nitric oxide synthase in *Deinococcus radiodurans*. *Proceedings of the National Academy of Sciences of the United States of America*, *101*(45), 15881–15886. https://doi.org/10.1073/pnas.0405483101.

Butler, C., Forte, E., Maria Scandurra, F., Arese, M., Giuffre, A., Greenwood, C., et al. (2002). Cytochrome bo(3) from Escherichia coli: The binding and turnover of nitric oxide. *Biochemical and Biophysical Research Communications*, *296*(5), 1272–1278.

Campbell, M. G., Smith, B. C., Potter, C. S., Carragher, B., & Marletta, M. A. (2014). Molecular architecture of mammalian nitric oxide synthases. *Proceedings of the National Academy of Sciences of the United States of America*, *111*(35), E3614–3623. https://doi.org/10.1073/pnas.1413763111.

Chartier, F. J., Blais, S. P., & Couture, M. (2006). A weak Fe-O bond in the oxygenated complex of the nitric-oxide synthase of *Staphylococcus aureus*. *Journal of Biological Chemistry*, *281*(15), 9953–9962. https://doi.org/10.1074/jbc.M513893200.

Chartier, F. J., & Couture, M. (2004). Stability of the heme environment of the nitric oxide synthase from *Staphylococcus aureus* in the absence of pterin cofactor. *Biophysical Journal*, *87*(3), 1939–1950. https://doi.org/10.1529/biophysj.104.042119.

Chartier, F. J., & Couture, M. (2007a). Interactions between substrates and the haem-bound nitric oxide of ferric and ferrous bacterial nitric oxide synthases. *Biochemical Journal*, *401*(1), 235–245. https://doi.org/10.1042/BJ20060913.

Chartier, F. J., & Couture, M. (2007b). Substrate-specific interactions with the heme-bound oxygen molecule of nitric-oxide synthase. *Journal of Biological Chemistry, 282*(29), 20877–20886. https://doi.org/10.1074/jbc.M701800200.

Chaudhari, S. S., Kim, M., Lei, S., Razvi, F., Alqarzaee, A. A., Hutfless, E. H., et al. (2017). Nitrite derived from endogenous bacterial nitric oxide synthase activity promotes aerobic respiration. *mBio, 8*(4). https://doi.org/10.1128/mBio.00887-17.

Chen, Y., & Rosazza, J. P. (1994). A bacterial nitric oxide synthase from a Nocardia species. *Biochemical and Biophysical Research Communications, 203*(2), 1251–1258.

Cho, K. B., Carvajal, M. A., & Shaik, S. (2009). First half-reaction mechanism of nitric oxide synthase: The role of proton and oxygen coupled electron transfer in the reaction by quantum mechanics/molecular mechanics. *Journal of Physical Chemistry B, 113*(1), 336–346. https://doi.org/10.1021/jp8073199.

Chung, M. C., Alem, F., Hamer, S. G., Narayanan, A., Shatalin, K., Bailey, C., et al. (2017). S-nitrosylation of peroxiredoxin 1 contributes to viability of lung epithelial cells during *Bacillus anthracis* infection. *Biochimica et Biophysica Acta, 1861*(1 Pt A), 3019–3029. https://doi.org/10.1016/j.bbagen.2016.09.006.

Chyba, C., & Sagan, C. (1991). Electrical energy sources for organic synthesis on the early Earth. *Origins of Life and Evolution of Biospheres, 21*, 3–17.

Cooper, C. E. (1999). Nitric oxide and iron proteins. *Biochimica et Biophysica Acta, 1411*(2–3), 290–309.

Cordas, C. M., Duarte, A. G., Moura, J. J., & Moura, I. (2013). Electrochemical behaviour of bacterial nitric oxide reductase—Evidence of low redox potential non-heme Fe(B) gives new perspectives on the catalytic mechanism. *Biochimica et Biophysica Acta, 1827*(3), 233–238. https://doi.org/10.1016/j.bbabio.2012.10.018.

Couture, M., Adak, S., Stuehr, D. J., & Rousseau, D. L. (2001). Regulation of the properties of the heme-NO complexes in nitric-oxide synthase by hydrogen bonding to the proximal cysteine. *Journal of Biological Chemistry, 276*(41), 38280–38288. https://doi.org/10.1074/jbc.M105341200.

Couture, M., Stuehr, D. J., & Rousseau, D. L. (2000). The ferrous dioxygen complex of the oxygenase domain of neuronal nitric-oxide synthase. *Journal of Biological Chemistry, 275*(5), 3201–3205.

Crane, B. R., Arvai, A. S., Gachhui, R., Wu, C., Ghosh, D. K., Getzoff, E. D., et al. (1997). The structure of nitric oxide synthase oxygenase domain and inhibitor complexes. *Science, 278*(5337), 425–431.

Crane, B. R., Arvai, A. S., Ghosh, D. K., Wu, C., Getzoff, E. D., Stuehr, D. J., et al. (1998). Structure of nitric oxide synthase oxygenase dimer with pterin and substrate. *Science, 279*(5359), 2121–2126.

Crane, B. R., Sudhamsu, J., & Patel, B. A. (2010). Bacterial nitric oxide synthases. *Annual Review of Biochemistry, 79*, 445–470. https://doi.org/10.1146/annurev-biochem-062608-103436.

Davydov, R., Labby, K. J., Chobot, S. E., Lukoyanov, D. A., Crane, B. R., Silverman, R. B., et al. (2014). Enzymatic and cryoreduction EPR studies of the hydroxylation of methylated N(omega)-hydroxy-L-arginine analogues by nitric oxide synthase from *Geobacillus stearothermophilus*. *Biochemistry, 53*(41), 6511–6519. https://doi.org/10.1021/bi500485z.

Davydov, R., Sudhamsu, J., Lees, N. S., Crane, B. R., & Hoffman, B. M. (2009). EPR and ENDOR characterization of the reactive intermediates in the generation of NO by cryoreduced oxy-nitric oxide synthase from *Geobacillus stearothermophilus*. *Journal of the American Chemical Society, 131*(40), 14493–14507. https://doi.org/10.1021/ja906133h.

Dikalov, S. I., Mayorov, V. I., & Panov, A. V. (2017). Physiological levels of nitric oxide diminish mitochondrial superoxide. Potential role of mitochondrial dinitrosyl iron complexes and nitrosothiols. *Frontiers in Physiology 8*, 907. https://doi.org/10.3389/fphys.2017.00907.

Ding, H., & Demple, B. (2000). Direct nitric oxide signal transduction via nitrosylation of iron-sulfur centers in the SoxR transcription activator. *Proceedings of the National Academy of Sciences of the United States of America*, *97*(10), 5146–5150.
Dizdaroglu, M., & Jaruga, P. (2012). Mechanisms of free radical-induced damage to DNA. *Free Radical Research*, *46*(4), 382–419. https://doi.org/10.3109/10715762.2011.653969.
Erusalimsky, J. D., & Moncada, S. (2007). Nitric oxide and mitochondrial signaling: From physiology to pathophysiology. *Arteriosclerosis, Thrombosis, and Vascular Biology*, *27*(12), 2524–2531. https://doi.org/10.1161/ATVBAHA.107.151167.
Feelisch, M., & Martin, J. F. (1995). The early role of nitric oxide in evolution. *Trends in Ecology & Evolution*, *10*(12), 496–499.
Feng, C. (2012). Mechanism of nitric oxide synthase regulation: Electron transfer and interdomain interactions. *Coordination Chemistry Reviews*, *256*(3–4), 393–411. https://doi.org/10.1016/j.ccr.2011.10.011.
Fernandez, M. L., Marti, M. A., Crespo, A., & Estrin, D. A. (2005). Proximal effects in the modulation of nitric oxide synthase reactivity: A QM-MM study. *Journal of Biological Inorganic Chemistry*, *10*(6), 595–604. https://doi.org/10.1007/s00775-005-0004-6.
Feron, O., Belhassen, L., Kobzik, L., Smith, T. W., Kelly, R. A., & Michel, T. (1996). Endothelial nitric oxide synthase targeting to caveolae. Specific interactions with caveolin isoforms in cardiac myocytes and endothelial cells. *Journal of Biological Chemistry*, *271*(37), 22810–22814.
Field, S. J., Roldan, M. D., Marritt, S. J., Butt, J. N., Richardson, D. J., & Watmough, N. J. (2011). Electron transfer to the active site of the bacterial nitric oxide reductase is controlled by ligand binding to heme b(3). *Biochimica et Biophysica Acta*, *1807*(4), 451–457. https://doi.org/10.1016/j.bbabio.2011.01.009.
Fischmann, T. O., Hruza, A., Niu, X. D., Fossetta, J. D., Lunn, C. A., Dolphin, E., et al. (1999). Structural characterization of nitric oxide synthase isoforms reveals striking active-site conservation. *Nature Structural & Molecular Biology*, *6*(3), 233–242. https://doi.org/10.1038/6675.
Fleming, I., Bauersachs, J., Fisslthaler, B., & Busse, R. (1998). Ca^{2+}-independent activation of the endothelial nitric oxide synthase in response to tyrosine phosphatase inhibitors and fluid shear stress. *Circulation Research*, *82*(6), 686–695.
Gardner, A. M., & Gardner, P. R. (2002). Flavohemoglobin detoxifies nitric oxide in aerobic, but not anaerobic, Escherichia coli. Evidence for a novel inducible anaerobic nitric oxide-scavenging activity. *Journal of Biological Chemistry*, *277*(10), 8166–8171. https://doi.org/10.1074/jbc.M110470200.
Gardner, P. R., Gardner, A. M., Martin, L. A., & Salzman, A. L. (1998). Nitric oxide dioxygenase: An enzymic function for flavohemoglobin. *Proceedings of the National Academy of Sciences of the United States of America*, *95*(18), 10378–10383.
Ghosh, D. K., Crane, B. R., Ghosh, S., Wolan, D., Gachhui, R., Crooks, C., et al. (1999). Inducible nitric oxide synthase: Role of the N-terminal beta-hairpin hook and pterin-binding segment in dimerization and tetrahydrobiopterin interaction. *The EMBO Journal*, *18*(22), 6260–6270. https://doi.org/10.1093/emboj/18.22.6260.
Ghosh, D. K., & Salerno, J. C. (2003). Nitric oxide synthases: Domain structure and alignment in enzyme function and control. *Frontiers in Bioscience*, *8*, d193–209.
Ghosh, D. K., & Stuehr, D. J. (1995). Macrophage NO synthase: Characterization of isolated oxygenase and reductase domains reveals a head-to-head subunit interaction. *Biochemistry*, *34*(3), 801–807.
Ghosh, S., Wolan, D., Adak, S., Crane, B. R., Kwon, N. S., Tainer, J. A., et al. (1999). Mutational analysis of the tetrahydrobiopterin-binding site in inducible nitric-oxide synthase. *Journal of Biological Chemistry*, *274*(34), 24100–24112.

Ghosh, D. K., Wu, C., Pitters, E., Moloney, M., Werner, E. R., Mayer, B., et al. (1997). Characterization of the inducible nitric oxide synthase oxygenase domain identifies a 49 amino acid segment required for subunit dimerization and tetrahydrobiopterin interaction. *Biochemistry*, *36*(35), 10609–10619. https://doi.org/10.1021/bi9702290.

Giroud, C., Moreau, M., Mattioli, T. A., Balland, V., Boucher, J. L., Xu-Li, Y., et al. (2010). Role of arginine guanidinium moiety in nitric-oxide synthase mechanism of oxygen activation. *Journal of Biological Chemistry*, *285*(10), 7233–7245. https://doi.org/10.1074/jbc.M109.038240.

Gorren, A. C., & Mayer, B. (2007). Nitric-oxide synthase: A cytochrome P450 family foster child. *Biochimica et Biophysica Acta*, *1770*(3), 432–445. https://doi.org/10.1016/j.bbagen.2006.08.019.

Griffith, O. W., & Stuehr, D. J. (1995). Nitric oxide synthases: Properties and catalytic mechanism. *Annual Review of Physiology*, *57*, 707–736. https://doi.org/10.1146/annurev.ph.57.030195.003423.

Gronberg, K. L., Roldan, M. D., Prior, L., Butland, G., Cheesman, M. R., Richardson, D. J., et al. (1999). A low-redox potential heme in the dinuclear center of bacterial nitric oxide reductase: Implications for the evolution of energy-conserving heme-copper oxidases. *Biochemistry*, *38*(42), 13780–13786.

Guan, Z. W., Kamatani, D., Kimura, S., & Iyanagi, T. (2003). Mechanistic studies on the intramolecular one-electron transfer between the two flavins in the human neuronal nitric-oxide synthase and inducible nitric-oxide synthase flavin domains. *Journal of Biological Chemistry*, *278*(33), 30859–30868. https://doi.org/10.1074/jbc.M301929200.

Guo, F. H., De Raeve, H. R., Rice, T. W., Stuehr, D. J., Thunnissen, F. B., & Erzurum, S. C. (1995). Continuous nitric oxide synthesis by inducible nitric oxide synthase in normal human airway epithelium in vivo. *Proceedings of the National Academy of Sciences of the United States of America*, *92*(17), 7809–7813.

Gusarov, I., & Nudler, E. (2005). NO-mediated cytoprotection: Instant adaptation to oxidative stress in bacteria. *Proceedings of the National Academy of Sciences of the United States of America*, *102*(39), 13855–13860. https://doi.org/10.1073/pnas.0504307102.

Gusarov, I., Shatalin, K., Starodubtseva, M., & Nudler, E. (2009). Endogenous nitric oxide protects bacteria against a wide spectrum of antibiotics. *Science*, *325*(5946), 1380–1384. https://doi.org/10.1126/science.1175439.

Gusarov, I., Starodubtseva, M., Wang, Z. Q., McQuade, L., Lippard, S. J., Stuehr, D. J., et al. (2008). Bacterial nitric-oxide synthases operate without a dedicated redox partner. *Journal of Biological Chemistry*, *283*(19), 13140–13147. https://doi.org/10.1074/jbc.M710178200.

Hannibal, L., Somasundaram, R., Tejero, J., Wilson, A., & Stuehr, D. J. (2011). Influence of heme-thiolate in shaping the catalytic properties of a bacterial nitric-oxide synthase. *Journal of Biological Chemistry*, *286*(45), 39224–39235. https://doi.org/10.1074/jbc.M111.286351.

Hess, D. T., Matsumoto, A., Nudelman, R., & Stamler, J. S. (2001). S-nitrosylation: Spectrum and specificity. *Nature Cell Biology*, *3*(2), E46–49. https://doi.org/10.1038/35055152.

Holden, J. K., Dejam, D., Lewis, M. C., Huang, H., Kang, S., Jing, Q., et al. (2015). Inhibitor bound crystal structures of bacterial nitric oxide synthase. *Biochemistry*, *54*(26), 4075–4082. https://doi.org/10.1021/acs.biochem.5b00431.

Holden, J. K., Kang, S., Beasley, F. C., Cinelli, M. A., Li, H., Roy, S. G., et al. (2015). Nitric oxide synthase as a target for methicillin-resistant *Staphylococcus aureus*. *Chemical Biology*, *22*(6), 785–792. https://doi.org/10.1016/j.chembiol.2015.05.013.

Holden, J. K., Kang, S., Hollingsworth, S. A., Li, H., Lim, N., Chen, S., et al. (2015). Structure-based design of bacterial nitric oxide synthase inhibitors. *Journal of Medicinal Chemistry*, *58*(2), 994–1004. https://doi.org/10.1021/jm501723p.

Holden, J. K., Lewis, M. C., Cinelli, M. A., Abdullatif, Z., Pensa, A. V., Silverman, R. B., et al. (2016). Targeting bacterial nitric oxide synthase with aminoquinoline-based inhibitors. *Biochemistry*, *55*(39), 5587–5594. https://doi.org/10.1021/acs.biochem.6b00786.

Holden, J. K., Li, H., Jing, Q., Kang, S., Richo, J., Silverman, R. B., et al. (2013). Structural and biological studies on bacterial nitric oxide synthase inhibitors. *Proceedings of the National Academy of Sciences of the United States of America*, *110*(45), 18127–18131. https://doi.org/10.1073/pnas.1314080110.

Hunten, D. M. (1993). Atmospheric evolution of the terrestrial planets. *Science*, *259*(5097), 915–920. https://doi.org/10.1126/science.259.5097.915.

Hurshman, A. R., & Marletta, M. A. (2002). Reactions catalyzed by the heme domain of inducible nitric oxide synthase: Evidence for the involvement of tetrahydrobiopterin in electron transfer. *Biochemistry*, *41*(10), 3439–3456.

Ignarro, L. J., Degnan, J. N., Baricos, W. H., Kadowitz, P. J., & Wolin, M. S. (1982). Activation of purified guanylate cyclase by nitric oxide requires heme. Comparison of heme-deficient, heme-reconstituted and heme-containing forms of soluble enzyme from bovine lung. *Biochimica et Biophysica Acta*, *718*(1), 49–59.

Jeong, H. G., & Choi, C. Y. (2002). Expression of inducible nitric oxide synthase by alpha-hederin in macrophages. *Planta Medica*, *68*(5), 392–396. https://doi.org/10.1055/s-2002-32071.

Jing, Q., Li, H., Roman, L. J., Martasek, P., Poulos, T. L., & Silverman, R. B. (2014). Combination of chiral linkers with thiophenecarboximidamide heads to improve the selectivity of inhibitors of neuronal nitric oxide synthase. *Bioorganic & Medicinal Chemistry Letters*, *24*(18), 4504–4510. https://doi.org/10.1016/j.bmcl.2014.07.079.

Johnson, E. G., Sparks, J. P., Dzikovski, B., Crane, B. R., Gibson, D. M., & Loria, R. (2008). Plant-pathogenic Streptomyces species produce nitric oxide synthase-derived nitric oxide in response to host signals. *Chemical Biology*, *15*(1), 43–50. https://doi.org/10.1016/j.chembiol.2007.11.014.

Kabir, M., Sudhamsu, J., Crane, B. R., Yeh, S. R., & Rousseau, D. L. (2008). Substrate-ligand interactions in *Geobacillus stearothermophilus* nitric oxide synthase. *Biochemistry*, *47*(47), 12389–12397. https://doi.org/10.1021/bi801491e.

Kehrer, J. P. (2000). The Haber-Weiss reaction and mechanisms of toxicity. *Toxicology*, *149*(1), 43–50.

Kennedy, M. C., Antholine, W. E., & Beinert, H. (1997). An EPR investigation of the products of the reaction of cytosolic and mitochondrial aconitases with nitric oxide. *Journal of Biological Chemistry*, *272*(33), 20340–20347.

Kers, J. A., Wach, M. J., Krasnoff, S. B., Widom, J., Cameron, K. D., Bukhalid, R. A., et al. (2004). Nitration of a peptide phytotoxin by bacterial nitric oxide synthase. *Nature*, *429*(6987), 79–82. https://doi.org/10.1038/nature02504.

Kiechle, F. L., & Malinski, T. (1993). Nitric oxide. Biochemistry, pathophysiology, and detection. *American Journal of Clinical Pathology*, *100*(5), 567–575.

Kinkel, T. L., Ramos-Montanez, S., Pando, J. M., Tadeo, D. V., Strom, E. N., Libby, S. J., et al. (2016). An essential role for bacterial nitric oxide synthase in *Staphylococcus aureus* electron transfer and colonization. *Nature Microbiology 2*, 16224. https://doi.org/10.1038/nmicrobiol.2016.224.

Kohanski, M. A., Dwyer, D. J., Hayete, B., Lawrence, C. A., & Collins, J. J. (2007). A common mechanism of cellular death induced by bactericidal antibiotics. *Cell*, *130*(5), 797–810. https://doi.org/10.1016/j.cell.2007.06.049.

Konas, D. W., Zhu, K., Sharma, M., Aulak, K. S., Brudvig, G. W., & Stuehr, D. J. (2004). The FAD-shielding residue Phe1395 regulates neuronal nitric-oxide synthase catalysis by controlling NADP+ affinity and a conformational equilibrium within the flavoprotein domain. *Journal of Biological Chemistry*, *279*(34), 35412–35425. https://doi.org/10.1074/jbc.M400872200.

Lang, J., Santolini, J., & Couture, M. (2011). The conserved Trp-Cys hydrogen bond dampens the "push effect" of the heme cysteinate proximal ligand during the first catalytic cycle of nitric oxide synthase. *Biochemistry*, *50*(46), 10069–10081. https://doi.org/10.1021/bi200965e.

Lewis, A. M., Matzdorf, S. S., Endres, J. L., Windham, I. H., Bayles, K. W., & Rice, K. C. (2015). Examination of the *Staphylococcus aureus* nitric oxide reductase (saNOR) reveals its contribution to modulating intracellular NO levels and cellular respiration. *Molecular Microbiology*, *96*(3), 651–669. https://doi.org/10.1111/mmi.12962.

Li, H., Jamal, J., Delker, S., Plaza, C., Ji, H., Jing, Q., et al. (2014). The mobility of a conserved tyrosine residue controls isoform-dependent enzyme-inhibitor interactions in nitric oxide synthases. *Biochemistry*, *53*(32), 5272–5279. https://doi.org/10.1021/bi500561h.

Li, D., Kabir, M., Stuehr, D. J., Rousseau, D. L., & Yeh, S. R. (2007). Substrate- and isoform-specific dioxygen complexes of nitric oxide synthase. *Journal of the American Chemical Society*, *129*(21), 6943–6951. https://doi.org/10.1021/ja070683j.

Liochev, S. L. (1996). The role of iron-sulfur clusters in in vivo hydroxyl radical production. *Free Radical Research*, *25*(5), 369–384.

List, B. M., Klosch, B., Volker, C., Gorren, A. C., Sessa, W. C., Werner, E. R., et al. (1997). Characterization of bovine endothelial nitric oxide synthase as a homodimer with down-regulated uncoupled NADPH oxidase activity: Tetrahydrobiopterin binding kinetics and role of haem in dimerization. *Biochemical Journal*, *323*(Pt 1), 159–165.

Lustig, D. B., Kempt, C., Alam, S., Clancy, J., Yee, J., & Rafferty, S. P. (2011). Mutation of conserved tryptophan residues at the dimer interface of *Staphylococcus aureus* nitric oxide synthase. *Archives of Biochemistry and Biophysics*, *506*(2), 165–172. https://doi.org/10.1016/j.abb.2010.11.024.

Lyons, T. W., Reinhard, C. T., & Planavsky, N. J. (2014). The rise of oxygen in Earth's early ocean and atmosphere. *Nature*, *506*(7488), 307–315. https://doi.org/10.1038/nature13068.

Makris, T. M., Davydov, R., Denisov, I. G., Hoffman, B. M., & Sligar, S. G. (2002). Mechanistic enzymology of oxygen activation by the cytochromes P450. *Drug Metabolism Reviews*, *34*(4), 691–708. https://doi.org/10.1081/DMR-120015691.

Mancinelli, R. L., & McKay, C. P. (1988). The evolution of nitrogen cycling. *Origins of Life and Evolution of Biospheres*, *18*, 311–325.

Marletta, M. A. (1994). Nitric oxide synthase: Aspects concerning structure and catalysis. *Cell*, *78*(6), 927–930.

Marletta, M. A., Hurshman, A. R., & Rusche, K. M. (1998). Catalysis by nitric oxide synthase. *Current Opinion in Chemical Biology*, *2*(5), 656–663.

Mayer, B., Wu, C., Gorren, A. C., Pfeiffer, S., Schmidt, K., Clark, P., et al. (1997). Tetrahydrobiopterin binding to macrophage inducible nitric oxide synthase: Heme spin shift and dimer stabilization by the potent pterin antagonist 4-amino-tetrahydrobiopterin. *Biochemistry*, *36*(27), 8422–8427. https://doi.org/10.1021/bi970144z.

McQuarters, A. B., Speelman, A. L., Chen, L., Elmore, B. O., Fan, W., Feng, C., et al. (2016). Exploring second coordination sphere effects in nitric oxide synthase. *Journal of Biological Inorganic Chemistry*, *21*(8), 997–1008. https://doi.org/10.1007/s00775-016-1396-1.

Meunier, B., de Visser, S. P., & Shaik, S. (2004). Mechanism of oxidation reactions catalyzed by cytochrome P450 enzymes. *Chemical Reviews*, *104*(9), 3947–3980. https://doi.org/10.1021/cr020443g.

Michel, T., & Feron, O. (1997). Nitric oxide synthases: Which, where, how, and why? *Journal of Clinical Investigation*, *100*(9), 2146–2152. https://doi.org/10.1172/JCI119750.

Montgomery, H. J., Dupont, A. L., Leivo, H. E., & Guillemette, J. G. (2010). Cloning, expression, and purification of a nitric oxide synthase-like protein from *Bacillus cereus*. *Biochemistry Research International 2010*, 489892. https://doi.org/10.1155/2010/489892.

Moroz, L. L., & Kohn, A. B. (2011). Parallel evolution of nitric oxide signaling: Diversity of synthesis and memory pathways. *Frontiers in Bioscience (Landmark Ed)*, *16*, 2008–2051.

Nathan, C., & Xie, Q. W. (1994). Nitric oxide synthases: Roles, tolls, and controls. *Cell*, *78*(6), 915–918.

Navarro-Gonzalez, R., McKay, C. P., & Mvondo, D. N. (2001). A possible nitrogen crisis for Archaean life due to reduced nitrogen fixation by lightning. *Nature*, *412*(6842), 61–64. https://doi.org/10.1038/35083537.

Nisbet, E. G., & Sleep, N. H. (2001). The habitat and nature of early life. *Nature*, *409*(6823), 1083–1091. https://doi.org/10.1038/35059210.

Pant, K., Bilwes, A. M., Adak, S., Stuehr, D. J., & Crane, B. R. (2002). Structure of a nitric oxide synthase heme protein from *Bacillus subtilis*. *Biochemistry*, *41*(37), 11071–11079.

Pant, K., & Crane, B. R. (2005). Structure of a loose dimer: An intermediate in nitric oxide synthase assembly. *Journal of Molecular Biology*, *352*(4), 932–940. https://doi.org/10.1016/j.jmb.2005.07.070.

Papale, D., Bruckmann, C., Gazur, B., Miles, C. S., Mowat, C. G., & Daff, S. (2012). Oxygen activation in neuronal NO synthase: Resolving the consecutive mono-oxygenation steps. *Biochemical Journal*, *443*(2), 505–514. https://doi.org/10.1042/BJ20111644.

Poole, R. K., & Hughes, M. N. (2000). New functions for the ancient globin family: Bacterial responses to nitric oxide and nitrosative stress. *Molecular Microbiology*, *36*(4), 775–783.

Poulos, T. L. (2005a). Structural and functional diversity in heme monooxygenases. *Drug Metabolism and Disposition*, *33*(1), 10–18. https://doi.org/10.1124/dmd.104.002071.

Poulos, T. L. (2005b). Structural biology of heme monooxygenases. *Biochemical and Biophysical Research Communications*, *338*(1), 337–345. https://doi.org/10.1016/j.bbrc.2005.07.204.

Poulos, T. L., & Li, H. (2013). Structural basis for isoform-selective inhibition in nitric oxide synthase. *Accounts of Chemical Research*, *46*(2), 390–398. https://doi.org/10.1021/ar300175n.

Presta, A., Siddhanta, U., Wu, C., Sennequier, N., Huang, L., Abu-Soud, H. M., et al. (1998). Comparative functioning of dihydro- and tetrahydropterins in supporting electron transfer, catalysis, and subunit dimerization in inducible nitric oxide synthase. *Biochemistry*, *37*(1), 298–310. https://doi.org/10.1021/bi971944c.

Pufahl, R. A., Nanjappan, P. G., Woodard, R. W., & Marletta, M. A. (1992). Mechanistic probes of N-hydroxylation of L-arginine by the inducible nitric oxide synthase from murine macrophages. *Biochemistry*, *31*(29), 6822–6828.

Reece, S. Y., Woodward, J. J., & Marletta, M. A. (2009). Synthesis of nitric oxide by the NOS-like protein from *Deinococcus radiodurans*: A direct role for tetrahydrofolate. *Biochemistry*, *48*(23), 5483–5491. https://doi.org/10.1021/bi900385g.

Richardson, A. R., Dunman, P. M., & Fang, F. C. (2006). The nitrosative stress response of *Staphylococcus aureus* is required for resistance to innate immunity. *Molecular Microbiology*, *61*(4), 927–939. https://doi.org/10.1111/j.1365-2958.2006.05290.x.

Richardson, A. R., Libby, S. J., & Fang, F. C. (2008). A nitric oxide-inducible lactate dehydrogenase enables *Staphylococcus aureus* to resist innate immunity. *Science*, *319*(5870), 1672–1676. https://doi.org/10.1126/science.1155207.

Robb, S. J., Gaspers, L. D., Wright, K. J., Thomas, A. P., & Connor, J. R. (1999). Influence of nitric oxide on cellular and mitochondrial integrity in oxidatively stressed astrocytes. *Journal of Neuroscience Research*, *56*(2), 166–176.

Rőszer, T. S. (2012). *The biology of subcellular nitric oxide*. Dordrecht; New York: Springer.

Rothschild, L. J., & Lister, A. (2003). *Evolution on planet earth: The impact of the physical environment*. Amsterdam; Boston: Academic Press.

Rousseau, D. L., Li, D., Couture, M., & Yeh, S. R. (2005). Ligand-protein interactions in nitric oxide synthase. *Journal of Inorganic Biochemistry*, *99*(1), 306–323. https://doi.org/10.1016/j.jinorgbio.2004.11.007.

Rusche, K. M., Spiering, M. M., & Marletta, M. A. (1998). Reactions catalyzed by tetrahydrobiopterin-free nitric oxide synthase. *Biochemistry*, *37*(44), 15503–15512. https://doi.org/10.1021/bi9813936.

Sabat, J., Egawa, T., Lu, C., Stuehr, D. J., Gerfen, G. J., Rousseau, D. L., et al. (2013). Catalytic intermediates of inducible nitric-oxide synthase stabilized by the W188H mutation. *Journal of Biological Chemistry*, *288*(9), 6095–6106. https://doi.org/10.1074/jbc.M112.403238.

Salard, I., Mercey, E., Rekka, E., Boucher, J. L., Nioche, P., Mikula, I., et al. (2006). Analogies and surprising differences between recombinant nitric oxide synthase-like proteins from *Staphylococcus aureus* and *Bacillus anthracis* in their interactions with l-arginine analogs and iron ligands. *Journal of Inorganic Biochemistry*, *100*(12), 2024–2033. https://doi.org/10.1016/j.jinorgbio.2006.09.017.

Santolini, J. (2011). The molecular mechanism of mammalian NO-synthases: A story of electrons and protons. *Journal of Inorganic Biochemistry*, *105*(2), 127–141. https://doi.org/10.1016/j.jinorgbio.2010.10.011.

Santolini, J., Adak, S., Curran, C. M., & Stuehr, D. J. (2001). A kinetic simulation model that describes catalysis and regulation in nitric-oxide synthase. *Journal of Biological Chemistry*, *276*(2), 1233–1243. https://doi.org/10.1074/jbc.M006858200.

Santolini, J., Meade, A. L., & Stuehr, D. J. (2001). Differences in three kinetic parameters underpin the unique catalytic profiles of nitric-oxide synthases I, II, and III. *Journal of Biological Chemistry*, *276*(52), 48887–48898. https://doi.org/10.1074/jbc.M108666200.

Santolini, J., Roman, M., Stuehr, D. J., & Mattioli, T. A. (2006). Resonance Raman study of *Bacillus subtilis* NO synthase-like protein: Similarities and differences with mammalian NO synthases. *Biochemistry*, *45*(5), 1480–1489. https://doi.org/10.1021/bi051710q.

Schopf, J. W. (2006). Fossil evidence of Archaean life. *Philosophical Transactions of the Royal Society B: Biological Sciences*, *361*(1470), 869–885. https://doi.org/10.1098/rstb.2006.1834.

Schulz, R., Nava, E., & Moncada, S. (1992). Induction and potential biological relevance of a ca(2+)-independent nitric oxide synthase in the myocardium. *British Journal of Pharmacology*, *105*(3), 575–580.

Sharpe, M. A., Robb, S. J., & Clark, J. B. (2003). Nitric oxide and Fenton/Haber-Weiss chemistry: Nitric oxide is a potent antioxidant at physiological concentrations. *Journal of Neurochemistry*, *87*(2), 386–394.

Shatalin, K., Gusarov, I., Avetissova, E., Shatalina, Y., McQuade, L. E., Lippard, S. J., et al. (2008). *Bacillus anthracis*-derived nitric oxide is essential for pathogen virulence and survival in macrophages. *Proceedings of the National Academy of Sciences of the United States of America*, *105*(3), 1009–1013. https://doi.org/10.1073/pnas.0710950105.

Shaw, G. H. (2016). *Earth's early atmosphere and oceans, and the origin of life*. Springer International Publishing.

Siddhanta, U., Presta, A., Fan, B., Wolan, D., Rousseau, D. L., & Stuehr, D. J. (1998). Domain swapping in inducible nitric-oxide synthase. Electron transfer occurs between flavin and heme groups located on adjacent subunits in the dimer. *Journal of Biological Chemistry*, *273*(30), 18950–18958.

Silverman, R. B. (2009). Design of selective neuronal nitric oxide synthase inhibitors for the prevention and treatment of neurodegenerative diseases. *Accounts of Chemical Research*, *42*(3), 439–451. https://doi.org/10.1021/ar800201v.

Sleep, N. H. (2010). The Hadean-Archaean environment. *Cold Spring Harbor Perspectives in Biology*, *2*(6). a002527. https://doi.org/10.1101/cshperspect.a002527.

Sono, M., Roach, M. P., Coulter, E. D., & Dawson, J. H. (1996). Heme-containing oxygenases. *Chemical Reviews*, *96*(7), 2841–2888.

Stamler, J. S., Simon, D. I., Osborne, J. A., Mullins, M. E., Jaraki, O., Michel, T., et al. (1992). S-nitrosylation of proteins with nitric oxide: Synthesis and characterization of biologically active compounds. *Proceedings of the National Academy of Sciences of the United States of America*, *89*(1), 444–448.

Stevanin, T. M., Ioannidis, N., Mills, C. E., Kim, S. O., Hughes, M. N., & Poole, R. K. (2000). Flavohemoglobin Hmp affords inducible protection for *Escherichia coli* respiration, catalyzed by cytochromes *bo'* or *bd*, from nitric oxide. *Journal of Biological Chemistry*, *275*(46), 35868–35875. https://doi.org/10.1074/jbc.M002471200.

Stoll, S., NejatyJahromy, Y., Woodward, J. J., Ozarowski, A., Marletta, M. A., & Britt, R. D. (2010). Nitric oxide synthase stabilizes the tetrahydrobiopterin cofactor radical by controlling its protonation state. *Journal of the American Chemical Society*, *132*(33), 11812–11823. https://doi.org/10.1021/ja105372s.

Stone, J. R., & Marletta, M. A. (1994). Soluble guanylate cyclase from bovine lung: Activation with nitric oxide and carbon monoxide and spectral characterization of the ferrous and ferric states. *Biochemistry*, *33*(18), 5636–5640.

Stuehr, D. J. (1999). Mammalian nitric oxide synthases. *Biochimica et Biophysica Acta*, *1411*(2–3), 217–230.

Stuehr, D. J. (2004). Enzymes of the L-arginine to nitric oxide pathway. *Journal of Nutrition*, *134*(10 Suppl), 2748S–2751S. discussion 2765S–2767S.

Stuehr, D. J., Kwon, N. S., Nathan, C. F., Griffith, O. W., Feldman, P. L., & Wlseman, J. (1991). N^{ω}-Hydroxy-L-arginine is an intermediate in the biosynthesis of nitric oxide from L-arginine. *Journal of Biological Chemistry*, *266*(10), 6259–6263.

Stuehr, D., Pou, S., & Rosen, G. M. (2001). Oxygen reduction by nitric-oxide synthases. *Journal of Biological Chemistry*, *276*(18), 14533–14536. https://doi.org/10.1074/jbc.R100011200.

Stuehr, D. J., Santolini, J., Wang, Z. Q., Wei, C. C., & Adak, S. (2004). Update on mechanism and catalytic regulation in the NO synthases. *Journal of Biological Chemistry*, *279*(35), 36167–36170. https://doi.org/10.1074/jbc.R400017200.

Stuehr, D. J., Wei, C. C., Santolini, J., Wang, Z., Aoyagi, M., & Getzoff, E. D. (2004). Radical reactions of nitric oxide synthases. *Biochemical Society Symposia*, *71*, 39–49.

Stuehr, D. J., Wei, C. C., Wang, Z., & Hille, R. (2005). Exploring the redox reactions between heme and tetrahydrobiopterin in the nitric oxide synthases. *Dalton Transactions*, (21), 3427–3435. https://doi.org/10.1039/b506355h.

Sudhamsu, J., & Crane, B. R. (2006). Structure and reactivity of a thermostable prokaryotic nitric-oxide synthase that forms a long-lived oxy-heme complex. *Journal of Biological Chemistry*, *281*(14), 9623–9632. https://doi.org/10.1074/jbc.M510062200.

Sudhamsu, J., & Crane, B. R. (2009). Bacterial nitric oxide synthases: What are they good for? *Trends in Microbiology*, *17*(5), 212–218. https://doi.org/10.1016/j.tim.2009.02.003.

Summers, D. P., & Chang, S. (1993). Prebiotic ammonia from reduction of nitrite by iron (II) on the early earth. *Nature*, *365*, 630–633. https://doi.org/10.1038/365630a0.

Tejero, J., Biswas, A., Wang, Z. Q., Page, R. C., Haque, M. M., Hemann, C., et al. (2008). Stabilization and characterization of a heme-oxy reaction intermediate in inducible nitric-oxide synthase. *Journal of Biological Chemistry*, *283*(48), 33498–33507. https://doi.org/10.1074/jbc.M806122200.

Tejero, J., & Stuehr, D. (2013). Tetrahydrobiopterin in nitric oxide synthase. *IUBMB Life*, *65*(4), 358–365. https://doi.org/10.1002/iub.1136.

Tomita, H., Katsuyama, Y., Minami, H., & Ohnishi, Y. (2017). Identification and characterization of a bacterial cytochrome P450 monooxygenase catalyzing the 3-nitration of tyrosine in rufomycin biosynthesis. *Journal of Biological Chemistry*, *292*(38), 15859–15869. https://doi.org/10.1074/jbc.M117.791269.

Trainer, M. G. (2013). Atmospheric prebiotic chemistry and organic hazes. *Current Organic Chemistry*, *17*, 1710–1723. https://doi.org/10.2174/13852728113179990078.

van Sorge, N. M., Beasley, F. C., Gusarov, I., Gonzalez, D. J., von Kockritz-Blickwede, M., Anik, S., et al. (2013). Methicillin-resistant *Staphylococcus aureus* bacterial nitric-oxide synthase affects antibiotic sensitivity and skin abscess development. *Journal of Biological Chemistry*, *288*(9), 6417–6426. https://doi.org/10.1074/jbc.M112.448738.

Walker, J. C. G. (1977). *Evolution of the atmosphere*. New York: Macmillan.
Wang, Z. Q., Wei, C. C., Ghosh, S., Meade, A. L., Hemann, C., Hille, R., et al. (2001). A conserved tryptophan in nitric oxide synthase regulates heme-dioxy reduction by tetrahydrobiopterin. *Biochemistry*, *40*(43), 12819–12825.
Wang, Z. Q., Wei, C. C., Sharma, M., Pant, K., Crane, B. R., & Stuehr, D. J. (2004). A conserved Val to Ile switch near the heme pocket of animal and bacterial nitric-oxide synthases helps determine their distinct catalytic profiles. *Journal of Biological Chemistry*, *279*(18), 19018–19025. https://doi.org/10.1074/jbc.M311663200.
Wang, Z. Q., Wei, C. C., & Stuehr, D. J. (2010). How does a valine residue that modulates heme-NO binding kinetics in inducible NO synthase regulate enzyme catalysis? *Journal of Inorganic Biochemistry*, *104*(3), 349–356. https://doi.org/10.1016/j.jinorgbio.2009.11.006.
Wardman, P., & Candeias, L. P. (1996). Fenton chemistry: An introduction. *Radiation Research*, *145*(5), 523–531.
Wei, C. C., Wang, Z. Q., Arvai, A. S., Hemann, C., Hille, R., Getzoff, E. D., et al. (2003). Structure of tetrahydrobiopterin tunes its electron transfer to the heme-dioxy intermediate in nitric oxide synthase. *Biochemistry*, *42*(7), 1969–1977. https://doi.org/10.1021/bi026898h.
Wei, C. C., Wang, Z. Q., Hemann, C., Hille, R., & Stuehr, D. J. (2003). A tetrahydrobiopterin radical forms and then becomes reduced during Nomega-hydroxyarginine oxidation by nitric-oxide synthase. *Journal of Biological Chemistry*, *278*(47), 46668–46673. https://doi.org/10.1074/jbc.M307682200.
Wei, C. C., Wang, Z. Q., Tejero, J., Yang, Y. P., Hemann, C., Hille, R., et al. (2008). Catalytic reduction of a tetrahydrobiopterin radical within nitric-oxide synthase. *Journal of Biological Chemistry*, *283*(17), 11734–11742. https://doi.org/10.1074/jbc.M709250200.
Wei, C. C., Wang, Z. Q., Wang, Q., Meade, A. L., Hemann, C., Hille, R., et al. (2001). Rapid kinetic studies link tetrahydrobiopterin radical formation to heme-dioxy reduction and arginine hydroxylation in inducible nitric-oxide synthase. *Journal of Biological Chemistry*, *276*(1), 315–319. https://doi.org/10.1074/jbc.M008441200.
Weisslocker-Schaetzel, M., Andre, F., Touazi, N., Foresi, N., Lembrouk, M., Dorlet, P., et al. (2017). The NOS-like protein from the microalgae Ostreococcus tauri is a genuine and ultrafast NO-producing enzyme. *Plant Science*, *265*, 100–111. https://doi.org/10.1016/j.plantsci.2017.09.019.
Wink, D. A., Cook, J. A., Krishna, M. C., Hanbauer, I., DeGraff, W., Gamson, J., et al. (1995). Nitric oxide protects against alkyl peroxide-mediated cytotoxicity: Further insights into the role nitric oxide plays in oxidative stress. *Archives of Biochemistry and Biophysics*, *319*(2), 402–407. https://doi.org/10.1006/abbi.1995.1310.
Woodward, J. J., Chang, M. M., Martin, N. I., & Marletta, M. A. (2009). The second step of the nitric oxide synthase reaction: Evidence for ferric-peroxo as the active oxidant. *Journal of the American Chemical Society*, *131*(1), 297–305. https://doi.org/10.1021/ja807299t.
Woodward, J. J., Nejatyjahromy, Y., Britt, R. D., & Marletta, M. A. (2010). Pterin-centered radical as a mechanistic probe of the second step of nitric oxide synthase. *Journal of the American Chemical Society*, *132*(14), 5105–5113. https://doi.org/10.1021/ja909378n.
Xu, K. Y., Huso, D. L., Dawson, T. M., Bredt, D. S., & Becker, L. C. (1999). Nitric oxide synthase in cardiac sarcoplasmic reticulum. *Proceedings of the National Academy of Sciences of the United States of America*, *96*(2), 657–662.
Ye, R. W., Averill, B. A., & Tiedje, J. M. (1994). Denitrification: Production and consumption of nitric oxide. *Applied and Environmental Microbiology*, *60*(4), 1053–1058.
Yukl, E. T., Elbaz, M. A., Nakano, M. M., & Moenne-Loccoz, P. (2008). Transcription factor NsrR from *Bacillus subtilis* senses nitric oxide with a 4Fe-4S cluster (dagger). *Biochemistry*, *47*(49), 13084–13092. https://doi.org/10.1021/bi801342x.

CHAPTER SIX

Anaerobic Bacterial Response to Nitrosative Stress

Jeffrey A. Cole[1]
School of Biosciences and Institute of Microbiology and Infection, University of Birmingham, Birmingham, United Kingdom
[1]Corresponding author: e-mail address: j.a.cole@bham.ac.uk

Contents

Advances in Microbial Physiology, Volume 72
ISSN 0065-2911
https://doi.org/10.1016/bs.ampbs.2018.01.001

Abstract

This chapter provides an overview of current knowledge of how anaerobic bacteria protect themselves against nitrosative stress. Nitric oxide (NO) is the primary source of this stress. Aerobically its removal is an oxidative process, whereas reduction is required anaerobically. Mechanisms required to protect aerobic and anaerobic bacteria are therefore different. Several themes recur in the review. First, how gene expression is regulated often provides clues to the physiological function of the gene products. Second, the physiological significance of reports based upon experiments under extreme conditions that bacteria do not encounter in their natural environment requires reassessment. Third, responses to the primary source of stress need to be distinguished from secondary consequences of chemical damage due to failure of repair mechanisms to cope with extreme conditions. NO is generated by many mechanisms, some of which remain undefined. An example is the recent demonstration that the hybrid cluster protein combines with YtfE (or RIC protein, for repair of iron centres damaged by nitrosative stress) in a new pathway to repair key iron–sulphur proteins damaged by nitrosative stress. The functions of many genes expressed in response to nitrosative stress remain either controversial or are completely unknown. The concentration of NO that accumulates in the bacterial cytoplasm is essentially unknown, so dogmatic statements cannot be made that damage to transcription factors (Fur, FNR, SoxRS, MelR, OxyR) occurs naturally as part of a physiologically relevant signalling mechanism. Such doubts can be resolved by simple experiments to meet six proposed criteria.

ABBREVIATIONS

GSNO S-nitrosoglutathione
NONOates NO-releasing compounds such as proli NONOate [1-(hydroxy-NNO-azoxy)-L-proline]
RIC repair of iron centres damaged by nitrosative stress
ROO rubredoxin:oxygen oxidoreductase
TMPD *N,N,N′,N′*-tetramethyl-*p*-phenylenediamine

1. INTERACTION BETWEEN OXIDATIVE AND NITROSATIVE STRESS RESPONSES

According to the Fountain of All Knowledge, "Reactive nitrogen species (RNS) act together with reactive oxygen species (ROS) to damage cells, causing nitrosative stress. Therefore, these two species are often collectively referred to as ROS/RNS" (Wikipedia). Underlying this definition is the assumption that oxygen is available. Certainly the definition is relevant to both plants and animals in which two defence mechanisms, the oxidative burst and the nitrosative burst, are induced simultaneously in response to attack by a pathogen, or stresses such as dehydration, exposure to high salt concentrations, or excessive heat. It can also be extended to aerobic microorganisms and even to those at the aerobic–anaerobic interface, for example, in oxygen-limited soils or in animals when a bacterial biofilm attaches to the epithelial cell layer. However, many microbes, especially bacteria, survive anaerobically in environments such as the lower gastrointestinal tracts of animals, in anaerobic soils or estuarine sediments, or in wastewater treatment plants. In these environments, oxidative stress is not an issue, and microbes have developed different strategies to minimise damage caused by nitrosative stress, or to reverse damage as it occurs. Key questions to be answered are how bacteria sense the danger of nitrosative damage, how they minimise the accumulation of NO, how they reverse metal-dependent nitrosative damage, and mechanisms available to reverse secondary consequences of *trans*-nitrosylation. For all four topics, current knowledge is fragmented and incomplete: the literature is therefore both controversial and confusing. The main aim of this chapter is to review current knowledge of how bacteria cope with nitrosative stress in anaerobic environments. This raises the fundamental question: why does it matter?

Microbial physiology can reasonably be defined as "structure–function relationships in microorganisms, especially how microbes respond to their environment". Protection against nitrosative stress can be effective only if the required gene products are synthesised under physiologically relevant conditions. Although obvious, this point has sometimes been ignored. This can be extrapolated to the general principle that knowledge of how gene expression is regulated can provide clues to the function of a gene product. The caveat is that the experiments must be completed under conditions encountered by the strain under study in its natural environment. Many reports in the literature that specific transcription factors do or do not regulate a response to nitrosative stress fail to meet this criterion and therefore require reexamination. Ideally differences in levels of mRNA in the presence or absence of a transcription factor determined in microarray, RNAseq, or promoter–probe reporter fusion experiments should be confirmed by experiments designed to answer the following questions.

1. Is the signal to which the transcription factor responds present at similar levels in any natural environment?
2. Does the transcription factor bind to the regulatory region of the relevant gene?
3. Is protein binding ligand dependent?
4. Can binding and gene regulation be reconstituted with pure components in vitro?
5. Are there alternative explanations not considered by the authors for the differential transcription detected?
6. Do old data require reinterpretation in light of subsequent discoveries?

It is well established that there are multiple global responses of microbial gene expression to the presence or absence of oxygen. In enteric bacteria, two of the global regulators that control adaptation between aerobic and anaerobic growth are FNR (the regulator of fumarate and nitrate reduction), and the ArcB–ArcA two-component regulatory system (Arc stands for anaerobic respiratory control). Expression of many genes whose products are required during anaerobic growth is activated by FNR or repressed by ArcA, but only during anaerobic growth. The converse is also usually true: genes repressed by FNR usually encode products required for aerobic growth. Recently published whole-genome transcriptomic (and to a lesser extent proteomic) data have guided much of the recent research into how bacteria cope with nitrosative stress. The first point is that different responses are required during aerobic or anaerobic growth, and even when oxygen is limited or available in excess. This review will highlight some of the many unanswered questions derived from whole system approaches that provide targets for future research. The primary aim of this chapter is to provide an overview of current knowledge of how anaerobic bacteria protect themselves against nitrosative stress. Attempts will be made to achieve three additional objectives: first, to distinguish between enzymes that remove the main source of damage, NO, from those that repair the damage after it has occurred; second to distinguish between enzymes involved in repairing secondary consequences of nitrosative damage from those targeting primary damage by NO; and third to distinguish between data derived from experiments under physiologically relevant conditions from those generated by chemical rather than biochemical mechanisms. Hopefully the explanations will provide motivation for the design of simple experiments to prove they are wrong!

2. NITRIC OXIDE AS THE PRIMARY SOURCE OF NITROSATIVE STRESS

In both aerobic and anaerobic environments, the free radical NO is a major source of nitrosative stress. There are two reasons why the microbial response to nitrosative stress is fundamentally different in an aerobic or anaerobic environment. First, NO removal in the presence of oxygen is likely to be an oxidative process, whereas reduction is required anaerobically. Second, NO initiates free radical chain reactions with ROS, for example, in a rapid reaction on collision with superoxide to form peroxynitrite (reviewed by Pacher, Beckman, & Liaudet, 2007), or with oxygen itself to form the nitrosating agent, N_2O_3. In anaerobic environments, hydrogen peroxide and peroxynitrite either do not occur, or due to diffusion from aerobic environments, are only a minor problem.

2.1 Chemical and Biochemical Sources of NO

Anaerobic bacteria encounter NO from essentially three main sources—but there are also other minor sources. Oxygen is severely limited in the lower regions of the gastrointestinal tract, where a wide variety of anaerobic bacteria live by fermentation of the abundant supply of carbon compounds. Many of the bacteria are in biofilms attached to the epithelial cell layer. Nevertheless, NO generated from arginine in the blood stream diffuses rapidly across membranes and through tissues. Bacteria in closest contact with the epithelium are therefore exposed to this source of NO and must be able to protect themselves accordingly.

A second source is NO released by denitrifying bacteria, which, although scarce in the gastrointestinal tract, are abundant in anaerobic soils, sediments and water treatment plants. For many years whether NO is an obligate intermediate in denitrification was itself controversial (Averill & Tiedje, 1982; Payne, Riley, & Cox, 1971). The controversy was resolved when Braun and Zumft (1991) demonstrated that a *Pseudomonas stutzeri* mutant defective in NO reductase could not survive exposure to nitrate or nitrite unless there was a secondary mutation in the genes required for nitrite reduction to NO, which in this organism is the gene *nirK* encoding a copper-containing nitrite reductase. Denitrifying bacteria differ widely in their ability to minimise NO accumulation, reflecting their different abilities to tolerate or detoxify NO (Mania, Heylen, van Spanning, & Frostegård, 2015; and references therein).

Finally, NO is generated during the anaerobic reduction of nitrite to ammonia when the supply of electron donors is abundant, but the concentrations of nitrate or nitrite are low (see Section 2.3).

Among the minor sources of NO in an anaerobic environment is the chemical reduction of nitrite by ferrous ions or the reduced forms of cytochromes and iron–sulphur centres. This has not been quantified.

Other sources include NO generated from atmospheric N_2 and O_2 at high temperatures such as those generated during a lightning strike (Levine, Augustsson, Anderson, Hoell, & Brewer, 1984). Worldwide this could account for the generation of 8.6 million tonnes of nitric oxide and nitrogen dioxide annually (Ott et al., 2010). In comparison, it has been estimated that emissions of nitrogen oxides, NO and N_2O, resulting from fossil fuel combustion are about 30 million tonnes (Schumann & Huntrieser, 2007). Although NO is a reactive-free radical gas, its reaction with O_2 is highly concentration dependent, increasing with the square of the NO concentration. At low concentrations, NO is relatively stable even in the presence of oxygen. For example, the half-life of oxidation of 10 parts per million of atmospheric NO to NO_2 is almost 6 h, but less than 1 min at 10,000 ppm. At low concentrations, the half-life is sufficiently long for some NO to be washed into oxygen-limited layers of soil by rain.

2.2 Sources of Stress From RNS Other Than NO

Although NO generated from nitrite is the dominant source of nitrosative stress, it would be a gross oversimplification to suggest it is the only source. Pacher et al. (2007) highlighted the misconception that accumulation of nitrotyrosine can reliably be used to measure the degree of nitrosative stress. Even NO-saturated water in which the NO concentration is 1.8–2.0 mM does not react significantly with tyrosine (Beckman et al., 1994). However, peroxynitrite readily reacts with tyrosine, a process that should correctly be designated nitration. As peroxynitrite is formed from NO only under conditions of oxidative stress, nitrotyrosine accumulation is more an indicator of oxidative than nitrosative stress. Another nitrating agent, N_2O_3, can be generated from the nitrogen dioxide formed by the O_2-dependent oxidation of NO. This reacts with another molecule of NO to generate N_2O_3, which in turn nitrosates primary and secondary amines in addition to thiol groups. Both peroxynitrite and N_2O_3 are formed from NO only when oxygen is also available. They are therefore unlikely to be important in the anaerobic nitrosative stress response.

Even the chemistry of NO oxidation has been controversial. It was originally believed that NO reacts with O_2 in a trimolecular reaction to generate a mixture of nitric and nitrous acids (Abel & Schmid, 1928).

$$2\ NO + O_2 + H_2O \rightarrow HNO_2 + HNO_3$$

However, this equation requires the production of equal quantities nitric and nitrous acids, which is contrary to the observation that nitrous acid is the only major product of NO oxidation (Lewis & Deen, 1994). The data suggested that nitrogen dioxide and dinitrogen trioxide are intermediates in the aerobic decomposition of NO:

$$2\ NO + O_2 \rightarrow 2\ NO_2$$

$$NO + NO_2 \rightarrow N_2O_3$$

$$N_2O_3 + H_2O \rightarrow 2\ HNO_2$$

According to Lewis and Deen (1994), "the pseudo-steady-state assumptions used for NO_2 and N_2O_3 depend on these species being present only in trace amounts. These assumptions were justified by the overall nitrogen balances, which showed that essentially all of the nitrogen added initially as NO could be accounted for by the sum of the NO and NO_2^- remaining in the reactor, together with the amount of NO calculated to have left the system (using the measured mass transfer coefficients). Previous reports did not utilise nitrogen balances to verify that NO and NO_2^- were the only major species present". As one of the intermediates is N_2O_3, nitrosation of proteins and other biological macromolecules is an inevitable side reaction of the aerobic oxidation of NO.

2.3 Compounds Used Experimentally as Surrogates for NO

The many different chemicals that have been used experimentally as sources of NO or nitrosative stress include S-nitrosoglutathione (GSNO), S-nitroso-*N*-acetylpenicillamine, sodium nitroprusside, acidified nitrite, and a range of NO-releasing compounds known collectively as NONOates. The concentration of glutathione (GSH) in cells of all types can be as high as 5 mM. Although it cannot react directly with NO, it serves as a sink for NO from other RNS, including nitrosylated metal centres. It was originally thought that GSNO releases NO by spontaneous homolytic cleavage, so GSNO has been used as a source of "NO" in many studies. However, three alternative GSNO metabolic reactions are, or in some circumstances can be, more rapid. These are donation of the nitrosium ion, NO^+, reduction to ammonia via hydroxylamine, and even donation of nitroxyl, NO^- (Arnelle & Stamler, 1995; Miao & King, 2016). It is therefore not surprising that the genome-wide responses to NO or surrogates such as GSNO or sodium nitroprusside are quite distinct (Flatley et al., 2005; Moore, Nakano, Wang, Ye, & Helmann, 2004; Pullan et al., 2007).

2.4 Metabolism of GSNO

From bacteria to mammals, high cytoplasmic concentrations of GSH serve as a sink for NO by S-*trans*-nitrosylation from nitrosylated metal ions. GSH can also be converted to GSNO by RNS such as N_2O_3 generated from NO in the presence of oxygen.

Various enzymes have been proposed to metabolise the NO group attached GSNO (summarised in Liu et al., 2001). They include a GSNO lyase that releases free NO and especially the widely distributed GSH-dependent formaldehyde dehydrogenase (Liu et al., 2001). This NADH-dependent enzyme is suggested to reduce GSNO via GSNHOH to $GSNH_2$, which then reacts with a second GSH molecule to release NH_3 and oxidised GSH, GSSG. Deletion of the gene for the yeast enzyme resulted in a strain that was highly sensitive to nitrosative stress, and similar results have been reported for "alcohol dehydrogenases" such as AdhC in *Haemophilus parainfluenzae*.

Many further reactions occur once S-nitrosothiols have been formed. They include the *trans*-nitrosation of thiol-containing amino acids, peptides and proteins. In summary, GSNO cannot be used as a reliable surrogate for NO itself.

2.5 NO Production During Nitrate Reduction to Ammonia

In bacteria that reduce nitrate to ammonia, a major source of nitrosative stress is due to the production of NO from any nitrite that accumulates during nitrate reduction. The best-documented enzyme that catalyses this one-electron reduction is the membrane-associated nitrate reductase, NarG. The active site of NarG is located in the cytoplasm, which is therefore the site of both nitrite production and its reduction to NO. In enteric bacteria the *narGHJI* operon is expressed most highly when nitrate is abundant, a situation more likely to occur in the outside world than in the human body. Many bacteria encode a periplasmic nitrate reductase that has a higher affinity than NarG for nitrate and is preferentially synthesised when electron donors are more abundant than nitrate. Less NO is produced from nitrite by the periplasmic nitrate reductase than by NarG (Rowley et al., 2012). Some studies have implicated one or both of the two nitrite reductases in NO production (Corker & Poole, 2003; Weiss, 2006). One of them, NirBD, is located in the cytoplasm: the other, NrfAB or NrfHA (for nitrite reduction by formate), is located in the periplasm. It is unknown whether there are environments in which NO production is mainly catalysed by one of these nitrite reductases rather than by nitrate reductase. However, we recently discovered that an *Escherichia coli* mutant that lacks nitrate reductase activity and both of the nitrite reductases can still generate NO from a further unidentified source (B. Balasiny & J.A. Cole, unpublished data).

3. TRANSCRIPTION FACTORS THAT REGULATE THE *E. COLI* AND *SALMONELLA ENTERICA* RESPONSE TO NITROSATIVE STRESS

There is agreement that the primary function of two *E. coli* DNA-binding proteins, NorR and NsrR, is to sense the presence of NO and regulate expression of specific genes accordingly. Controversies have arisen concerning the physiological relevance of reports of other *E. coli* transcription factors involved in the bacterial response to nitrosative stress. The problem was highlighted in the comprehensive review by Spiro (2007). They include Fur, FNR, SoxRS, OxyR, and MetR. Their roles will be reviewed in Sections 4.5 and 10.

3.1 The Nonhaem Iron Protein, NorR

NorR is an NO-binding enhancer protein required for the expression of the flavorubredoxin gene, *norV* (Gardner, Gessner, & Gardner, 2003). The *norV* promoter requires the σ^{54} form of RNA polymerase, which in turn requires positive control by enhancer proteins to initiate transcription. NorR was shown to bind specifically to the sequence, GT–N_7–AC (N is any base), which occurs as multiple direct repeats (Tucker, D'Autréaux, Studholm, Spiro, & Dixon, 2004). Various attempts to find *E. coli* genes other than *norV* that are significantly regulated by NorR have consistently returned negative results. NorR is therefore a dedicated transcription factor with the sole function of activating expression of genes for the NO reductase, flavorubredoxin, and flavorubredoxin reductase. Recombinant NorR from *E. coli* was shown to bind NO with rather low affinity, and although sufficient NO was released to activate expression of NsrR-regulated genes when *E. coli* was incubated with 500 μM GSNO, expression of *norVW* was at most only slightly induced (Jarboe, Hyduke, Tran, Chou, & Liao, 2008). Consequently the *norV* and *norW* genes are expressed only under conditions of moderately severe nitrosative stress, almost operating as an SOS response.

3.2 The Iron–Sulphur Protein, NsrR

In contrast to NorR, NsrR is a more global transcription factor. Genes expected to be regulated by NsrR were predicted bioinformatically by Rodionov, Dubchak, Arkin, Alm, and Gelfand (2005). Bodenmiller and Spiro (2006) provided the first experimental demonstration of the role of NsrR in regulating the *E. coli* response to nitrosative stress. Subsequently, about 20 transcripts in the NsrR regulon have been identified by various whole-genome analysis techniques (Efromovich, Grainger, Bodenmiller, & Spiro, 2008; Filenko et al., 2007; Karlinsey et al., 2012; Partridge, Bodenmiller, Humphrys, & Spiro, 2009). NsrR is a transcription repressor of the Rrf2 protein family that senses nitric oxide directly via an iron–sulphur cluster (Tucker et al., 2008). Binding of NO inactivates the ability of NsrR to bind to its target sequence, thus relieving repression.

3.3 IscR and the Repair of Damaged Iron–Sulphur Proteins

Like NsrR, IscR is a member of the Rrf2 protein family with cysteine residues that bind a [2Fe–2S] iron–sulphur centre (Schwartz et al., 2001). Its primary role is to regulate expression of the *isc* genes encoding proteins required to repair damaged iron–sulphur centres. As exposure to NO can be both the cause of the damage and can directly by nitrosylate IscR itself, it has been proposed that IscR is an NO sensor (Jarboe et al., 2008). What is not clear, however, is whether IscR binds NO with sufficiently high affinity to be inactivated by concentrations of NO that occur naturally in the bacterial cytoplasm.

4. REGULATION OF THE ANAEROBIC NITROSATIVE STRESS RESPONSE IN OTHER BACTERIA

4.1 Prevention of NO Accumulation in Denitrifying Bacteria

In most denitrifying bacteria, nitrite and NO reduction are regulated by mechanisms that severely limit NO accumulation. Examples of the proteins involved include the Nnr proteins in *Paracoccus denitrificans* and *Rhodobacter sphaeroides*, and DnrD in *P. stutzeri*. All of these transcription factors coordinately regulate nitrite and nitric oxide reductase synthesis in response to traces of NO (Tosques, Shi, & Shapleigh, 1996; Van Spanning et al., 1999; Vollack, Härtig, Körner, & Zumft, 1999). However, specific differences in this regulation in response to different growth conditions can also be found, for example, in the pathogenic *Neisseria*. In *Neisseria gonorrhoeae*, although expression of the nitrite reductase gene *aniA* is coordinately regulated with expression of the NO reductase genes, *norBC*, by the NO-sensing transcription repressor, DnrN, only nitrite reduction is regulated by the gonococcal FNR in response to oxygen starvation. This enables the gonococcus to reduce NO generated by the host inducible NO synthase even when oxygen is available (Overton et al., 2006).

4.2 Regulation of the Response to NO by CRP/FNR Family Proteins

Outside the *Enterobacteriaceae*, bioinformatics analysis, whole-genome sequencing, and experimental studies have provided strong evidence that proteins of the CRP–FNR family play key roles in sensing nitrosative stress and regulating gene expression accordingly (Rodionov et al., 2005). One of the first to be discovered was Anr in *Pseudomonas aeruginosa* (Haas, Camper, & Zimmermann, 1992). Other examples include NnrR in *R. sphaeroides* 2.4.3; Nnr in *P. denitrificans*; and DnrD in *P. stutzeri* (Tosques et al., 1996; Van Spanning et al., 1995; Vollack et al., 1999). These proteins all lack cysteines required to bind the [4Fe–4S] iron–sulphur centre that is typical of FNR proteins that sense oxygen. In contrast, the cysteines are present in FNR in the *Neisseria* where it is required to activate expression of the nitrite reductase gene, *aniA* (Whitehead et al., 2007).

In many bacteria, regulation of nitrate reduction and the nitrosative stress response is far more complicated than in *E. coli*, involving cascades of regulatory proteins or multiple CRP-related transcription factors. Only two of many examples will be mentioned.

In *Thermus thermophilus*, DnrS–DnrT are master regulators encoded within the nitrate respiration gene cluster (Cava et al., 2007; Cava, Zafra, da Costa, & Berenguer, 2008). The nitrite and nitric oxide reductases are encoded in a separate gene cluster that appears to be a mobile element (Alvarez, Bricio, Blesa, Hidalgo, & Berenguer, 2014; Alvarez, Bricio, Hidalgo, & Berenguer, 2014). DnrS is required for *nirSJM* transcription. DnrT is a member of the DNR subgroup of the Crp protein family that might activate transcription of the nitric oxide reductase genes, *norCBH*, but its precise role still needs to be defined.

A nitrosative stress response in the microaerophilic, food-borne pathogen *Campylobacter jejuni* is activated by the Crp–FNR family protein NssR. *Campylobacter* NssR regulates synthesis of the globin Cgb, which provides protection when oxygen is available. It also regulates synthesis of four other proteins, including a truncated haemoglobin (Elvers et al., 2005; Elvers, Wu, Gilberthorpe, Poole, & Park, 2004). An *nssR* mutant is more sensitive to nitrosative stress than a *cgb* mutant, so other proteins are required for full protection, possibly to protect itself against nitrosative stress even when oxygen is severely limited. Recombinant Cgb could complement an *E. coli hmp norR–norVW* mutant aerobically but not anaerobically (Tinajero-Trejo, Vreugdenhil, Sedelnikova, Davidge, & Poole, 2013).

The bioinformatic analysis by Rodionov et al. (2005) revealed that genes for some of the Crp–FNR homologues were located close to genes for the hybrid cluster protein (Hcp) and a gene encoding a potential [4Fe–4S] iron–sulphur protein, *frdX*, with a ferredoxin fold. These proteins were therefore designated HcpR, the regulator of expression of the *hcp* gene. Conserved residues in HcpR and the presence of inverted repeat sequences upstream of the *hcp* gene enabled consensus binding sites to be proposed. In *Desulfovibrio vulgaris*, there is a single *hcpR* gene encoding a protein with conserved arginine, glutamate, and arginine residues at positions 1, 2, and 6 in the DNA-binding helix that are identical to the *E. coli* Crp protein, and the proposed binding site for HcpR is identical to the binding site for Crp in *E. coli*. In contrast, there are two genes encoding HcpR homologues in *Desulfovibrio desulfuricans* and many other sulphate-reducing bacteria that have been isolated from the human body.

4.3 Haem as the Basis for NO Activation of Some of the CRP/FNR Family Transcription Factors

A haem group remained attached during purification of the NO-sensitive DnrD from *P. stutzeri* (Vollack & Zumft, 2001). Lee, Shearer, and Spiro (2006) reported that NNR can be activated in anaerobic cultures with μM concentrations of NO itself, or by NO generated metabolically from nitrate or nitrite. It also responds to the nitrosylating agent sodium nitroprusside. This broad specificity was consistent with a haem-based mechanism for NO sensing and explained the partial loss of NNR activity in a *hemA* mutant. Recombinant HcpR1 protein from *D. desulfuricans* was colourless when purified from *E. coli*, but was shown to bind haem stoichiometrically (Cadby et al., 2016). In contrast, HcpR2 is clearly an iron–sulphur protein. It was proposed that HcpR1 and HcpR2 respond to NO by independent mechanisms, HcpR1 by a haem-based mechanism, and HcpR2 by NO inactivation of an iron–sulphur centre. Although none of the above studies proved unequivocally that haem is the ligand that binds NO, other transcription factors of the CRP–FNR family that definitely bind haem have been reported.

The most convincing evidence for a haem-binding mechanism for sensing NO came from the Cutruzzolà group, who showed that DNR from *P. aeruginosa* could activate *E. coli norVW* expression only if haem is available (Castiglione, Rinaldo, Giardina, & Cutruzzolà, 2009). The haem-binding pocket in DNR has been identified (Cutruzzolà et al., 2014). High-affinity binding of DNR to DNA required nitrosylation of the haem by NO (Lobato et al., 2014). A recent paper characterised the transcription factor DnrF from the marine denitrifying bacterium *Dinoroseobacter shibae* DFL12^{T} (Ebert et al., 2017). NO-activated DnrF fulfils multiple roles. It coordinately induces both a nitrosative stress response and synthesis of NO reductase and also represses the periplasmic nitrate reductase operon, *nap*, thereby decreasing nitrite production, which is the source of nitrosative stress. Reporter fusions, transcriptomics, and gel retardation assays with proposed target promoters were combined with spectroscopic characterisation of the purified apo- and holo-proteins to show that DnrF binds to its DNA target sites as a homodimer with a single haem bound stoichiometrically to each subunit. The ability of DnrF to bind to DNA in response to NO increases with increasing haem content. The consensus binding site, 5′-TTGAT 4 bases-ATCAA-3′, perfectly matches the consensus binding site for *E. coli* FNR.

4.4 Widespread Distribution of the NsrR Protein

NsrR is a global transcription factor, a repressor in the Rrf2 family, that is widely distributed among different groups of bacteria. The first report of an NsrR orthologue was the protein designated NsrR from *Nitrosomonas europaea*, which was initially thought to respond to nitrite (Beaumont, Lens, Reijnders, Westerhoff, & van Spanning, 2004). It is an iron–sulphur protein that binds NO with very high affinity. Nitrosylation inactivates the DNA-binding ability of NsrR, so NsrR-regulated genes are derepressed in response to nitrosative stress (Bodenmiller & Spiro, 2006; Filenko et al., 2007; Gilberthorpe, Lee, Stevanin, Read, & Poole, 2007; Isabella, Lapek, Kennedy, & Clark, 2009; Overton et al., 2006; Wang et al., 2016).

The recombinant NsrR proteins from *N. gonorrhoeae* and the strict aerobe *Streptomyces coelicolor* were originally reported to contain a [2Fe–2S] cluster (Isabella et al., 2009; Tucker et al., 2008). In contrast, the *Bacillus subtilis* NsrR was isolated with a [4Fe–4S] centre (Yukl, Elbaz, Nakano, & Moënne-Loccoz, 2008). Recently, however, clear evidence has been presented that the *S. coelicolor* NsrR contains a [4Fe–4S] centre that on exposure to any oxygen is rapidly converted to the [2Fe–2S] form (Crack et al., 2015, 2016; Crack, Stapleton, Green, Thomson, & Le Brun, 2013). Presumably all NsrR proteins contain a labile [4Fe–4S] centre, but only the partially degraded forms were isolated initially. Purification under strictly anaerobic conditions is clearly essential to isolate the form that binds to DNA with high affinity. The *S. coelicolor* protein is of special interest for another reason. It is not a global regulator, but regulates only its own synthesis and synthesis of the two Hmp NO oxygenases encoded by the bacterium (Crack et al., 2016).

4.5 Other Transcription Factors That Regulate the Response to Nitrosative Stress

Various other transcription factors have been reported to regulate the bacterial nitrosative stress response. In most of the examples, the transcription factor senses a metabolic signal via either a metal centre or a cysteinyl –SH group. Some of the reports might be dismissed as consequences of chemical damage to the metal centre under conditions that are physiologically irrelevant. Not so other studies that meet at least some of the criteria specified in Section 1 of this review.

Reports that the iron-responsive Fur protein regulates the nitrosative stress response directly or only indirectly have been especially controversial. Virtually all of the experiments mentioned below were completed with aerobic cultures. Fur binds to its specific DNA target sequences in response to excess iron, which results in transcription repression. Fur regulates the intracellular concentration of iron and iron plays a key role in oxidative stress. Increased sensitivity to nitrosative stress might therefore be secondary consequences of increased oxidative stress in the absence of Fur. Binding of NO to iron can mimic iron starvation, so one key question is whether the concentration of NO required to relieve Fur repression ever accumulates in the bacterial cytoplasm. Given that concentrations of nitrate and nitrite in the human intestine are low (Saul, Kkabir, Cohen, Bruce, & Archer, 1981) and most bacteria are able to reduce NO as rapidly as it is formed, the concentrations of NO in the bacterial cytoplasm are likely to be in the low nM range, not in the high μM range used in most of the following experiments. Even then the question remains whether Fur regulates gene expression directly or indirectly, for example, via changes in other transcription factors that contain iron as part of their sensing mechanism.

In both *E. coli* and *B. subtilis*, *fur* mutants were reported to be more sensitive to nitrosative stress than the parent strains (Moore et al., 2004; Mukhopadhyay, Zheng, Bedzyk, LaRossa, & Storz, 2004). However, an early report that Fur represses *hmp* transcription was subsequently reassessed (Crawford & Goldberg, 1998, 2006). Although weak repression by Fur was detected in *E. coli* using an *hmp::lacZ* reporter plasmid, this repression was not relieved by the iron-chelator dipyridyl (Bang et al., 2006; Hernández-Urzúa et al., 2007). These results are more consistent with severe damage to Fur under conditions that bacteria do not encounter naturally, which would effectively generate a *fur* mutant phenotype, but have little physiological relevance. Husain et al. (2014) demonstrated that a *Salmonella fur* mutant is extremely sensitive to less than 10 μM NO released from proli NONOate [1-(hydroxy-NNO-azoxy)-L-proline] (half-life [$t_{1/2}$] = 1.8 s at 37°C). This was attributed to the secondary consequences of the inability of the mutant to synthesise the haem required for Hmp maturation. Despite hints to the contrary, no direct role was found for either Fur or PerR in regulating the NssR response of *C. jejuni* (Elvers et al., 2005). In general, there is currently little evidence that Fur is required for protection against nitrosative stress under anaerobic conditions. Nevertheless, as there is evidence that concentrations of NO below 10 μM are sufficient to relieve Fur repression in vivo and to inactivate DNA-binding activity of Fur in vitro, it remains an open question whether NO sensing by Fur is physiologically significant (D'Autréaux, Touati, Bersch, Latour, & Michaud-Soret, 2002). These points will be discussed further in Section 10.2.

Although the CRP–FNR family protein DNR is required for anaerobic expression of the nitrate reductase genes in *P. aeruginosa*, it does not regulate the expression of *fhp*, which encodes the Hmp orthologue (Arai, Hayashi, Kuroi, Ishii, & Igarashi, 2005). Upstream of *fhp* is the gene for a σ^{54}-dependent enhancer protein, FhpR. Fhp was synthesised in response to NO itself or NO from a range of RNS. Mutants that lacked *fhp* or *fhpR* were sensitive to NO-generating agents under aerobic conditions, but not during anaerobic growth. Not only is this evidence that a flavohaemoglobin provides protection aerobically, but not anaerobically, but it also illustrates how the aerobic response to NO differs from the anaerobic response.

Methionine biosynthesis is repressed by MetR in response to the availability of homocysteine. Homocysteine is nitrosylated by GSNO, so MetR repression is relieved by nitrosative stress from this source (Membrillo-Hernández, Coopamah, Channa, Hughes, & Poole, 1998). In contrast, 5 μM NO released from the donor molecules NOC-5 (3-(aminopropyl)-1-hydroxy-3-isopropyl-2-oxo-1-triazene) and NOC-7 (3-(2-hydroxy-1-methyl-2-nitrosohydrazino)-*N*-methyl-1-propanamine) failed to induce *met* gene transcription (Pullan et al., 2007). This indicates that NO does not directly nitrosylate homocysteine in short-term anaerobic experiments and that MetR regulation is disrupted by S-nitrosylation rather than by NO itself. This conclusion was supported by the systems analysis of data generated by Jarboe et al. (2008).

In response to oxidative stress caused by the accumulation of superoxide, SoxR is activated by oxidation of its [2Fe–2S] cluster. This increases the synthesis of SoxS, which then activates SoxRS target gene expression. High concentrations of NO bind to the SoxR iron–sulphur centre and partially mimic the oxidative stress response. In some of the experiments demonstrating this effect, the concentration of NO used was massive, 150 μM, so further experiments under conditions encountered naturally are still required to confirm its significance (Vasil'eva, Stupakova, Lobysheva, Mikoyan, & Vanin, 2001).

Novel components of an NO defence system were reported to be required for systemic infection by *Streptococcus pneumoniae*. The *nmlR* gene encodes a transcription factor, NmlR, that is cotranscribed with *adhC* encoding an alcohol dehydrogenase that is able to reduce GSNO with NADH as reductant. Low levels of GSNO reductase activity of mutants defective in *nmlR* and *adhC* correlated with increased killing by NO compared with the parent strain (Stroeher et al., 2007).

5. NITRIC OXIDE REDUCTION BY ENTERIC BACTERIA

Until recently, three enzymes were known to reduce NO in anaerobic cultures of *E. coli* and related bacteria. The first discovered by Poole and his colleagues in 1991 was the flavohaemoglobin, Hmp (Poole et al., 1996; Vasudevan, Armarego, Shaw, Lilley, & Poole, 1991). A *Salmonella hmp* mutant was less able to survive in macrophages than its parent, illustrating the importance of Hmp in protection against NO generated by the nitrosative burst in humans and warm-blooded animals (Bang et al., 2006). Hmp binds both O_2 and NO, but binds NO preferentially (Hausladen,

Gow, & Stamler, 2001). The primary physiological role of Hmp is to oxidise NO to nitrate by a mechanism that first involves the formation of an enzyme-bound nitroxyl equivalent. The reduced form of Hmp donates an electron to NO to generate the nitroxyl anion, NO^-, which can then be oxidised to nitrate by O_2. The equation for the reaction can therefore be written:

$$NO^- + O_2 \rightarrow NO_3^-$$

It was then realised that the purified protein can also reduce NO to N_2O, raising the possibility that Hmp might protect both aerobic and anaerobic bacteria against nitrosative stress (Kim, Orii, Lloyd, Hughes, & Poole, 1999). This led to a study that concluded that, at least in *E. coli*, Hmp is kinetically not competent to provide adequate protection anaerobically because the turnover number for NO reduction, 20 s^{-1}, is low and the K_m for NO is far too high (Gardner & Gardner, 2002).

The second NO reductase to be discovered was the flavorubredoxin, NorV, which also reduces NO to N_2O (Gardner et al., 2003; Gardner, Gardner, Martin, & Salzman, 1998; Gardner, Helmick, & Gardner, 2002; Gomes et al., 2002). Kinetically NorV is ideally suited for a protection role with a high V_{max} and a low K_m for NO. However, expression of the *norVW* operon is dependent upon the σ^{54} form of RNA polymerase plus the NO-sensing enhancer protein NorR. As NorR is essential for *norVW* transcription and NorR binds NO with low affinity, synthesis of NorV and its reductase, NorW, is induced only under conditions of severe nitrosative stress. Clearly such conditions are encountered during infection of a mammalian host because a *norV* mutant of an enterohaemorrhagic *E. coli* was severely defective in its ability to survive within macrophages (Shimizu, Tsutsuki, Matsumoto, Nakaya, & Noda, 2012).

The periplasmic nitrite reductase, NrfAB, is the third enzyme with NO reductase activity. It uses electrons from menaquinol to reduce NO to ammonia (Poock, Leach, Moir, Cole, & Richardson, 2002). In contrast to Hmp and NorV, NrfAB is synthesised only during anaerobic growth. Although the V_{max} is sufficiently high for it to reduce high concentrations of NO that enter the periplasm, the K_m for NO is also too high to prevent NO from diffusing into the cytoplasm (van Wonderen, Burlat, Richardson, Cheesman, & Butt, 2008). Consequently, NrfA at best provides initial protection against attack by exogenous NO generated, for example, by the host inducible nitric oxide synthase.

A PhD student, Claire Vine, constructed isogenic mutants defective in all combinations of three enzymes, Hmp, NorVW, and NrfAB, as well as a quadruple mutant that also lacked NirBD. This quadruple mutant was therefore unable to reduce nitrite to ammonia (Vine & Cole, 2011). Although it lacks all three previously known NO reductases, it retains NarG and hence the ability to reduce nitrite to NO. It therefore reduced nitrate quantitatively to nitrite, some of which was slowly reduced to NO. All of the mutants, including the quadruple mutant, surprisingly still reduced high concentrations of NO at rates similar to the parent strain. Although it was more sensitive to nitrosative stress than the parent, it was still able to grow in the presence of repeated additions of 1 μM NO, but not in 5 μM NO. It was concluded from these experiments that another physiologically important NO reductase remained to be discovered. Similar data have been obtained for a *Salmonella enterica* triple mutant defective in Hmp, NrfA, and NorV (Mills, Rowley, Spiro, Hinton, & Richardson, 2008). Deletion of the *hmp* gene alone or in combination with *nrfA* or *norV* had very little effect on the ability to resist a massive dose of 40 μM NO during fermentative growth or during anaerobic growth with a high concentration of nitrate. In contrast to deletion of either *nrfA* or *norV*, which only slightly decreased the ability to recover from addition of NO, the triple mutant or a double mutant defective in both *norV* and *nrfA* was more sensitive to NO. However, they were still able to resist concentrations of NO well above those expected to accumulate in any natural environment.

6. *E. COLI* GENES OF UNKNOWN FUNCTION EXPRESSED UNDER CONDITIONS OF NITROSATIVE STRESS

Various bacteria have been the subject of studies of the global transcriptomic response to nitrosative stress, but inevitably far more is known about the underlying regulatory mechanisms and functions of the products of the regulated genes in *E. coli* strain K-12 than in other bacteria. This and the next two sections will therefore summarise current knowledge of how *E. coli* K-12 and its close relatives respond to NO; Section 9 will then provide an overview of the nitrosative stress responses and their regulation in other bacteria.

Sources of nitrosative stress used in different studies ranged from NO-saturated water to various compounds that release NO such as DEANO or S-nitrosoglutathione. In almost all of the *E. coli* and *Salmonella* experiments, four transcripts were highly expressed in response to nitrosative stress:

hmp encoding the flavohaemoglobin, Hmp; *ytfE* encoding a diiron protein; the *hcp–hcr* operon encoding the Hcp and its reductase, Hcr; and the two-gene *yeaR–yoaG* operon. In some but not all studies, several other genes of unknown function such as *ygbA* and *yibIH* were differentially expressed, as were genes for the periplasmic nitrite reductase. In enteric bacteria all of these genes are part of the NsrR regulon.

The transcriptional regulation of the *yeaR–yoaG* operon has been studied in detail, but the function of the gene products is unknown (Lin, Bledsoe, & Stewart, 2007; Squire, Xu, Cole, Busby, & Browning, 2009). Nothing is known about the functions of YgbA, YibI, or YibH. Mutations in these genes of unknown function were introduced into the quadruple mutant described later, but the resulting strains with five mutations were as resistant to nitrosative stress as the parent. Introduction of a *ytfE* mutation even improved growth in the presence of nitrate, but the effect was so small that initially it was ignored. Section 7 will review current ideas about the physiological roles of YtfE. The functions of YgbA, YeaR, YoaG, YibI, and YibH in the bacterial defence to nitrosative stress remain to be determined.

The importance of maintaining respiration for bacterial survival during infection was demonstrated recently in experiments with uropathogenic *E. coli* strain EC953. A mutant defective in the high-affinity cytochrome oxidase, cytochrome *bd*-1, was as sensitive as an *hmp* mutant to killing by neutrophils and murine macrophages. Loss of NorV and YtfE also increased sensitivity to killing, but loss of NrfAB had no effect (Shepherd et al., 2016).

7. THE ENIGMATIC HCP

Discovery of a protein with a unique iron–sulphur cluster in anaerobic sulphate-reducing bacteria was first reported by Pierik, Wolbert, Mutsaers, Hagen, and Veeger (1992). As the protein was originally thought to bind a [6Fe–6S] iron–sulphur centre with the shape of a regular triangular prism, it was called the "prismane protein". Analysis of its crystal structure revealed the presence of a hybrid [4Fe–2S–2O] cluster, so the protein was renamed the Hcp (Arendsen et al., 1998). Subsequently, Hcp was found in a range of anaerobic and facultatively anaerobic gamma, delta, and epsilon proteobacteria as well as in archea. The Hcp protein is present in most bacteria that reduce nitrate or nitrite anaerobically to ammonia, but absent from denitrifying α or β proteobacteria. In enteric bacteria *hcp* is the first gene in the two-gene *hcp–hcr* operon, which is expressed only during anaerobic growth under the positive control of FNR (Constantinidou et al., 2006; van den

Berg, Hagen, & van Dongen, 2000). The *hcr* product is a [2Fe–2S] iron–sulphur protein. In other bacterial groups, *hcp* is commonly located close to a *frdX* gene encoding a [4Fe–4S] protein with a ferredoxin fold.

The amino acid sequence of Hcp resembles that of a carbon monoxide dehydrogenase that can also reduce hydroxylamine (Heo, Wolfe, Staples, & Ludden, 2002; Wolfe, Heo, Garavelli, & Ludden, 2002). Amino acid substitutions were introduced into the *E. coli* Hcp by site-directed mutagenesis in successful attempts to generate a modified protein with hydroxylamine reductase activity. However, the rate of hydroxylamine reduction was too low and the K_m for hydroxylamine too high for Hcp to provide protection against hydroxylamine toxicity. The authors were careful to state that hydroxylamine reduction was unlikely to be the function of Hcp (Wolfe et al., 2002). Nevertheless, basic tenets of microbial physiology have subsequently been ignored, and Hcp is still often annotated in whole-genome databases and even research papers as a hydroxylamine reductase. The crystal structure revealed the presence of a narrow channel that might accommodate a small linear molecule such as NO, but early attempts to demonstrate NO reduction by Hcp were unsuccessful (van den Berg et al., 2000).

7.1 Identification of the Hcp as the Missing High-Affinity NO Reductase

The construction of the quadruple mutant that lacked both nitrite reductase and known NO reductase activities provided the essential starting point to determine the physiological role of the Hcp. First, it was shown that deletion of the *hcp–hcr* operon resulted in a strain that was unable to grow in the presence of 1 mM nitrite or when challenged with very low concentrations of NO (Wang et al., 2016). The NO reductase activity of the quadruple mutant was abolished by deletion of either the *hcp* gene alone, or both *hcp* and *hcr*. This suggested that Hcp and Hcr might form a short electron transfer chain as shown in Fig. 1 and proposed previously by van den Berg et al. (2000). If this is correct, a mutation in *hcr* might also be expected to result in total loss of NO reductase activity.

Evidence from initial complementation experiments indicated that this assumption was likely to be correct. In experiments with 5 μM or higher concentrations of NO, a plasmid encoding only Hcp was unable to complement the loss of NO reduction by the *hcp–hcr* deletion mutant. However, at concentrations of NO below 2 μM and contrary to expectation, Hcp alone could reduce NO even in the absence of Hcr. Hcp and Hcr were shown to interact, forming a sufficiently tight complex for them to be copurified by

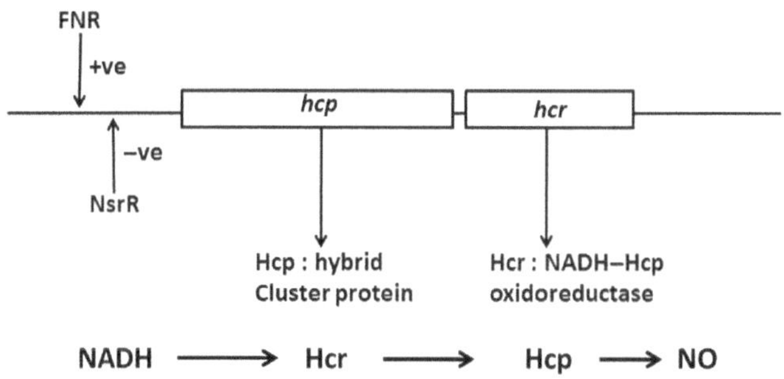

Fig. 1 Scheme originally proposed for electron transfer from NADH via the Hcp reductase, Hcr, and the hybrid cluster protein, Hcp. The *upper part* of the figure shows the gene organisation and regulation of transcription by FNR (+ for positive control) and NsrR (− for repression). Optimal levels of transcription therefore require anaerobic conditions to activate FNR and the presence of NO to inactivate NsrR repression.

Ni^{2+}-affinity chromatography (Wang et al., 2016). We then demonstrated high-affinity NO reductase activity in vitro with the copurified proteins, but not with Hcp that has been purified in the absence of Hcr. We therefore proposed that the primary role of Hcr is to protect Hcp from inactivation by its reactive substrate. These experiments resolved the long-standing controversy of the physiological role of the enigmatic Hcp, which is to protect cytoplasmic proteins against nitrosative stress.

8. THE ROLE OF THE YtfE (RIC) PROTEIN IN THE REPAIR OF NITROSYLATED IRON–SULPHUR PROTEINS

That the YtfE protein is a key player in protection against nitrosative stress was first proposed on the basis of microarray data reported by Justino, Vicente, Teixeira, and Saraiva (2005). Transcription of the *ytfE* gene was induced dramatically when cultures were challenged by addition of NO. A *ytfE* mutant was then constructed and shown to be more sensitive than the parent to nitrosative stress. This initial mutant was also defective in anaerobic growth under a variety of conditions. Dehydratases such as aconitase and fumarase that contain surface-exposed [4Fe–4S] centres are especially sensitive to inactivation by NO (Gardner, Costantino, Szabo, & Salzman, 1997; Hyduke, Jarboe, Tran, Chou, & Liao, 2007; Ren, Zhang, Yang, & Ding, 2008). Recombinant YtfE protein was shown to be a diiron protein that could restore activity to aconitase or fumarase proteins damaged by nitrosylation (Justino, Almeida, Goncalves, Teixeira, & Saraiva, 2007). Two subsequent papers revealed that iron atoms could be transferred from YtfE to iron-deficient enzymes, but the rate of Fe transfer was extremely slow, and required stoichiometric rather than catalytic quantities of YtfE (Justino, Baptista, & Saraiva, 2009; Nobre et al., 2014).

Three lines of evidence alerted us to the possibility that there might be an alternative, more relevant function for YtfE. First, the growth defects of the original *ytfE* mutant were due to a secondary 124-gene deletion that included genes required for the synthesis of the molybdopterin cofactor that is essential for anaerobic growth (Vine, Justino, Saraiva, & Cole, 2010). Critically, the deletion also included the *hcp–hcr* operon, which alone could account for the sensitivity of the mutant to nitrosative stress.

The publication of the crystal structure of YtfE provided further strong evidence that the molecular mechanisms underlying the repair of nitrosylated [Fe–S] clusters by YtfE required further analysis (Lo et al., 2016). As expected, the structure coupled with a range of spectroscopic techniques revealed how the conserved histidine and glutamate residues bind to the oxo-bridged Fe^{II}–Fe^{III} diiron core. Unexpectedly, it also showed that "there are two solvent-accessible tunnels, both of which converge to the diiron center and are critical for capturing substrates. The reactivity of the reduced-form Fe^{II}–Fe^{II} YtfE toward nitric oxide demonstrates that the prerequisite for N_2O production requires the two iron sites to be nitrosylated simultaneously. Specifically, the nitrosylation of the two iron sites prior to their reductive coupling to produce N_2O is cooperative. This result suggests that, in addition to any repair of iron centers (RIC) activity, YtfE acts as an NO-trapping scavenger to promote the NO to N_2O transformation under low NO flux, which precedes nitrosative stress" (Lo et al., 2016). In summary, structural data suggest that the restoration of aconitase and fumarase activity by YtfE is more likely to be due to NO absorption rather than to an iron donation role. Reduction of NO to N_2O was also detected, but the rate of NO reduction was very low: in the presence of ascorbate + TMPD to maintain a reducing environment, 100 μM YtfE had converted less than 50% of the available NO to N_2O after incubation for 30 min. In contrast, abundant evidence was presented by Wang et al. (2016) that a *ytfE*$^+$ strain was completely unable to reduce NO in the absence of Hcp.

The conclusion that the primary role of YtfE is likely to be to release NO rather than reduce NO was strongly supported by a series of in vivo experiments with a set of isogenic strains derived from the quadruple mutant described later. The mutant defective in *hcp* and *hcr*, both of the nitrite reductases and the previously characterised NO reductases (NorVW and Hmp), was extremely sensitive even to very low concentrations of nitrite or NO. This raised the question why YtfE was unable to protect this strain from nitrosative stress. The answer came from introduction of an additional *ytfE* mutation, which restored growth in the presence of nitrite (Fig. 2). This suggested that YtfE generates the toxic product, NO, that accumulates in the cytoplasm in the absence of a functional nitric oxide reductase, Hcp (Balasiny et al., 2018). Supporting evidence for this alternative role for YtfE includes: demonstration that nitrosative stress can be relieved by deletion of the nitrate reductase genes, thus eliminating the major source of NO generation from nitrite; by adding an NO-degrading chemical, PTIO (2-phenyl-4,4,5,5-tetramethylimidazoline-1-oxyl 3-oxide), to the growth medium; or by restoring a functional *hcp–hcr* operon so that any NO produced can be reduced to N_2O. Aconitase and fumarase were totally inactivated only in the $YtfE^+$ Hcp^- strain; and relief of NsrR repression of the *hcp* promoter was dependent upon cytoplasmic NO generated by YtfE. The combined data lead to the proposal that there is a coordinated pathway to minimise damage to cytoplasmic proteins by NO (Fig. 3). According to this pathway, NO generated in the cytoplasm by the nitrate reductase NarG is either trapped by sensitive iron–sulphur proteins or reduced directly to a harmless product. The primary role of YtfE is to reverse immediate damage to key metabolic enzymes such as aconitase, fumarase, and IlvD, the dihydroxy-acid dehydratase required for isoleucine and valine synthesis, before these and other proteins are more severely damaged by *trans*-nitrosylation to protein –SH groups. If so, this leads to the major current

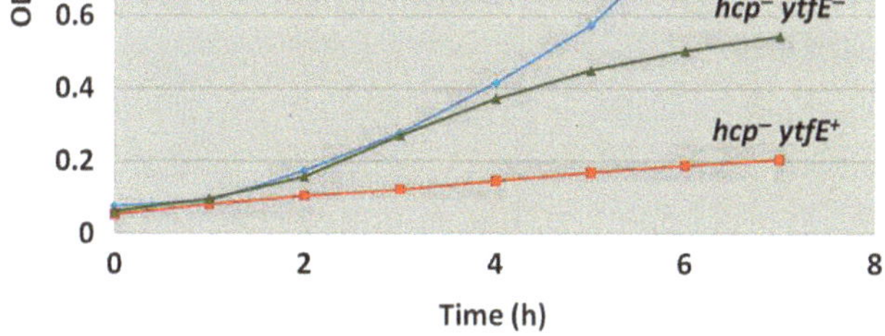

Fig. 2 Suppression of the inhibition of anaerobic growth of an *hcp* mutant by deletion of *ytfE*. Note that all three strains carry deletions in genes for previously reported nitrite and NO reductases: *nirBD, nrfAB; hmp;* and *norVW*.

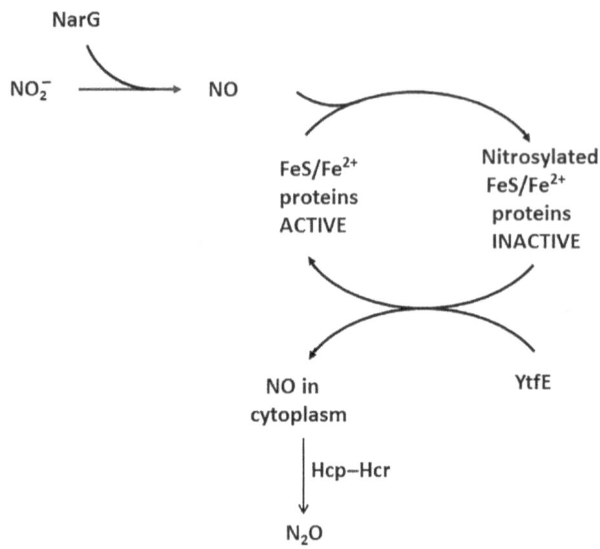

Fig. 3 Proposed pathway for reversing nitrosative damage to iron–sulphur proteins by recycling NO into the cytoplasm for its reduction to N_2O by the hybrid cluster protein.

challenge: can the repair process be reconstituted in vitro? Preliminary attempts by us, by the Saraiva group and others, have so far been unsuccessful. Possible reasons for lack of success include the possibility that no one has yet successfully isolated the active form of YtfE, that additional proteins are required, that assay conditions were inappropriate, that a small molecule such as an electron donor or a labile ligand is required, or simply that the ideas expressed earlier are wrong.

9. PROTEINS INVOLVED IN THE ANAEROBIC NITROSATIVE STRESS RESPONSE IN BACTERIA OTHER THAN ENTEROBACTERIACEAE

9.1 The Bacterial Flavohaemoglobin, Hmp, and Truncated Globins

There is a vast literature showing that Hmp proteins are widely distributed and undoubtedly confer resistance to nitrosative stress in aerobic environments. They also play important roles during infection or persistence within mammalian cells (Stevanin, Read, & Poole, 2007; and references cited therein). Oxygen was available in most of the reported experiments, so it is less clear whether Hmp is sufficient to protect some bacteria against nitrosative stress even in an anaerobic environment.

The *B. subtilis* ResD–ResE two-component system regulates adaptation to anaerobic growth (Nakano, Geng, Nakano, & Kobayashi, 2006; Rogstam, Larsson, Kjelgaard, & von Wachenfeldt, 2007). There is also an *nsrR* gene in *B. subtilis*. Both *hmp* and especially *nasDE* require the ResD–ResE system for full induction, and both are regulated by NsrR. Although cultures used by Rogstam et al. (2007) were described as "aerobic", bacteria for RNA analysis were harvested at a cell density sufficient for oxygen to be limiting (OD 0.6 or much higher in stationary phase). In response to the presence of NO or NO^+, NsrR was shown to induce synthesis of ResD, ResE, Hmp, and a truncated haemoglobin, YjbI, as well as the nitrite reductase system, NasDEF. Mutants defective in YjbI and the product of the next gene, YjbH, were extremely sensitive to nitrosative stress, suggesting that the truncated haemoglobin plays a critical defence role (Rogstam et al., 2007). Subsequently, Kommineni, Lama, Popescu, and Nakano (2012) confirmed that NsrR represses both *hmp* and *yjbI* during anaerobic growth with glucose and pyruvate, but repression is relieved in the presence of nitrate or NO. This global study revealed further genes that appeared to be repressed by NsrR, but no consensus NsrR binding sites were found in their regulatory regions. Although this "class II" NsrR regulon included genes regulated by Fur, it was concluded that Fur plays a major role only in repression in the presence of excess iron. Further experiments are still required to reveal whether other proteins play key roles in the anaerobic nitrosative stress response of Gram-positive bacteria, or whether Hmp is kinetically competent to reduce NO under anaerobic conditions.

9.2 The Flavorubredoxin, NorV, and Its Reductase, NorW

Orthologues of the flavorubredoxin NorV are widely distributed among both aerobic and anaerobic bacteria. The first to be characterised was the protein from *Desulfovibrio gigas*, which was shown to reduce oxygen to water with electron transfer from NADH catalysed by flavorubredoxin reductase (Chen et al., 1993; Wildschut, Lang, Voordouw, & Voordouw, 2006). The protein was initially called ROO, for rubredoxin:oxygen oxidoreductase. Discovery of the NO reductase activity of the *E. coli* protein and its higher affinity for NO than oxygen led to reassessment of its role. It is synthesised even in the absence of nitrosative stress, but levels of *roo* mRNA increased in response to challenge by the nitrosating agent, GSNO. As the *roo* mutant was more sensitive to NO toxicity than the parent, its assignment as an NO reductase is now accepted (Rodrigues et al., 2006). Nevertheless, ROO is able to protect the strict anaerobes like *D. gigas* and *D. vulgaris* against oxygen toxicity, so a dual role was proposed in affording protection against both oxidative and nitrosative stress (Rodrigues et al., 2006; Yurkiw, Voordouw, & Voordouw, 2012).

The discovery of NorVW orthologues in the anaerobe *Moorella thermoacetica* followed almost the same path as for the *D. gigas* proteins. The genes *frpA* (now known to encode flavorubredoxin) and *hrb* (encoding an NADH-dependent flavorubredoxin reductase) were first assigned as part of an oxidative stress response network (Das, Coulter, Kurtz, & Lungdahl, 2001). It was soon realised that these proteins are part of an NO reduction pathway to protect against severe nitrosative stress (Silaghi-Dumitrescu et al., 2003). Significantly the K_m for NO was determined to be 4 μM, lower than the K_m for oxygen, but still higher than the concentration of NO that would inactivate *E. coli* Hcp.

9.3 Hybrid Cluster Proteins

Hcps are widely distributed from extremely thermophilic archea such as *Pyrococcus furiosus* to Spirochaetes, *Treponema*, and the photosynthetic *Rhodobacter capsulatus* (Cabello et al., 2004; Overeijnder, Hagen, & Hagedoorn, 2009). It is present in most γ, δ, and ε proteobacteria, but absent from methanogens and most denitrifying bacteria.

The high-resolution structure of the Hcp from *D. vulgaris* (Hildenborough) revealed three domains: an N-terminal mainly α-helical domain and two similar domains comprising a central β-sheet flanked by α-helices. There are two iron–sulphur clusters located within 11 Å of each other. The conventional N-terminal [4Fe–4S] cubane cluster is close to an indentation on the surface of the protein and can also be approached on the opposite side by a long solvent channel. The hybrid [4Fe–2S–2O] cluster is buried in the central hydrophobic region. A large hydrophobic cavity in domain 3 is consistent with it functioning as a channel for NO diffusion. There is a second more hydrophilic channel that links the hybrid cluster to the outside of the protein. If like the *E. coli* protein the *Desulfovibrio* Hcp is a high-affinity NO reductase, one can speculate that the surface cavity might be a binding site for an electron donor protein or a protein that protects the [4Fe–4S] centre from substrate inactivation. The hydrophilic channel might function as the electron transfer pathway to the hybrid cluster, and the hydrophobic channel would provide access to the site of NO reduction by the hybrid cluster.

D. vulgaris Hildenborough is especially interesting because it encodes two copies of both ROO and Hcp. One set of genes, *hcp1* and *roo1*, is located on a 47 kb mobile genetic element that can be lost spontaneously. The *hcp1* and *roo1* genes are adjacent to each other, but are they cotranscribed? The second copies of the genes are distributed elsewhere on the chromosome (Johnston et al., 2009). The rate of growth in a normal lactate plus sulphate medium is lower when the genomic island is present than when it has been lost, but the island provides protection against both oxidative stress and nitrosative stress induced by the addition of 2 mM nitrite. Yurkiw et al. (2012) then showed that the genomic island is required for electron transfer from lactate to nitrite, but not for electron transfer from hydrogen to nitrite. They also confirmed that the genomic island (ROO1?) and ROO2 confer increased resistance to oxygen and proposed that both Hcp1 and Hcp2 contribute to resistance to nitrite stress. Subsequently, Figueiredo et al. (2013) demonstrated that mutants defective in all four *hcp* and *roo* genes lost viability when challenged with macrophages. The *hcp2* gene is most highly induced in response to NO released from 100 μM dipropylenetriamine NONOate, and the *hcp2* mutant was least able to survive nitrosative stress.

Many questions remain unanswered. Although growth inhibition by nitrite was originally attributed to the chemical formation of toxic polysulphide, in view of the reassigned function for ROO, it is more likely due to NO generated from nitrite. This assumption has not been checked experimentally. It would be interesting to determine whether growth in the presence of nitrite can be restored by adding PTIO, which stoichiometrically destroys NO, to the growth medium (see, for example, Wang et al., 2016, for details). Expression of the *hcp2* gene was reported to be induced 100-fold only in the presence of the genomic island, and nitrite addition induced both *hcp1* and *hcp2* expression. Further experiments are now required to determine whether one or both Hcp1 and Hcp2 can function as NO reductases. It is also unknown whether the genomic island generates an induction signal or encodes a transcription factor required for increased *hcp2* expression. Even if NO is the inducer to which *hcp1* and *hcp2* respond, it is unknown how NO is generated from nitrite in a strain that lacks both NrfA and genes for any nitrate reductase.

9.4 YtfE in Other Bacteria

Justino et al. (2009) highlighted the widespread distribution of YtfE homologues not only in different bacterial phyla but also in some protozoa and fungi. These proteins were identified not on the basis of their amino acid sequence similarity, which can fall below 10%, but from conservation of residues in the ligand diiron centre. We recently used this criterion to identify Ddes_1165 as the structural gene for the *D. desulfuricans* orthologue (Cadby et al., 2017). If the model proposed for YtfE function in Fig. 3 is correct, all bacteria with NO-responsive *hcp* genes should also encode either a YtfE homologue or another protein that can fulfil the same NO-releasing function.

In the denitrifying bacterium, *Ralstonia eutropha* H16, NorR regulates synthesis of NorA and NorB in which NorB is the nitric oxide reductase (Strube, De Vries, & Cramm, 2007). NorA is clearly an orthologue of YtfE with a role in defence against nitrosative stress. This would make physiological sense if, as proposed earlier, the role of YtfE is to release NO from nitrosylated iron–sulphur centres so it can be reduced to the relatively harmless product, N_2O, by a high-affinity NO reductase—in this case NorB rather than Hcp.

Harrington et al. (2009) reported that both YtfE and NrfA protect *Haemophilus influenzae* against nitrosative stress generated by macrophages and that the FNR protein is required for their gene expression. In *E. coli*, although expression of *ytfE* increases during anaerobic growth, no direct regulation by FNR has been detected (Constantinidou et al., 2006).

9.5 Nitric Oxide Reduction by *Helicobacter pylori*

The gene Hp0013 in *Helicobacter pylori* encodes a previously undescribed type of NO reductase. The recombinant protein increased the NO reductase activity of *E. coli*, and deletion of the gene results in a strain that is less able to reduce NO or survive nitrosative stress in vivo (Justino, Ecobichon, Fernandes, Boneca, & Saraiva, 2012).

10. REASONS WHY DIFFERENT CONCLUSIONS HAVE RESULTED FROM DIFFERENT STUDIES

Three major factors have resulted in confusion, and even errors, in the literature in attempts to identify biological sources of NO or how anaerobic bacteria protect themselves against nitrosative stress. Sometimes different groups have reached conflicting conclusions because different growth conditions were used that lead to activation or repression of key genes by different transcription factors. This reinforces the point made above that knowledge of how genes are regulated provides important clues about the functions of the encoded products. Second, bacteria evolve to survive in their natural environment, not in a pure culture in a laboratory shake flask where nutrients are supplied in concentrations well above those that occur naturally. The third point follows from the second. Exposing bacteria to high concentrations of reactive chemicals such as NO results in chemical modification and damage unrelated to the primary physiological role of the process under study. Further examples of all three phenomena will briefly be summarised in the following paragraphs.

10.1 Effects of Different Transcription Factors on NO Production in Response to Different Growth Conditions

A prime example of the first source of controversy concerns which enteric bacterial protein is responsible for the conversion of nitrite to NO. Corker and Poole (2003) concluded that NO is generated primarily by the periplasmic cytochrome *c* nitrite reductase, NrfAB. In contrast, Weiss (2006) concluded that the cytoplasmic nitrite reductase was a major source. Both of these conclusions challenged an extensive and growing literature showing that most of the NO is generated by the membrane-associated NarGHI complex, a conclusion also reached subsequently at least for *S. enterica* serovar Typhimurium by the Poole group (Calmels, Oshima, & Bartsch, 1988; Gilberthorpe & Poole, 2008; Metheringham & Cole, 1997; Ralt, Wishnok, Fitts, & Tannenbaum, 1988; Seth, Hausladen, Wang, & Stamler, 2012; Smith, 1983). Fig. 4 illustrates why there is no longer any doubt that NarG plays a major role in causing nitrosative stress. The growth experiments shown in part A were completed with a *narGHJI*$^{+}$ strain that is defective in synthesis of both Hcp and YtfE as well as NorVW, Hmp, NirBD, and NrfAB. Note the inverse dose-dependent effect of nitrate concentration on inhibition of growth. The explanation is that low concentrations of nitrate are rapidly reduced to nitrite, which is slowly reduced by NarG to NO. Higher concentrations of nitrate saturate NarG, effectively preventing the side reaction that generates NO from the nitrite formed until most of the nitrate has been reduced. Fig. 4B shows that growth inhibition is almost completely reversed by further deletion of *narG*.

Despite the fact that many of the studies involved the same strains and mutants supplied by the author of this chapter, the explanation lies in the growth conditions used in the various studies. Synthesis of the nitrate reductase NarG is induced during growth with relatively high concentrations of nitrate, conditions that repress synthesis of the alternative periplasmic nitrate reductase, Nap (Potter, Millington, Thomas, & Cole, 1999). Conversely,

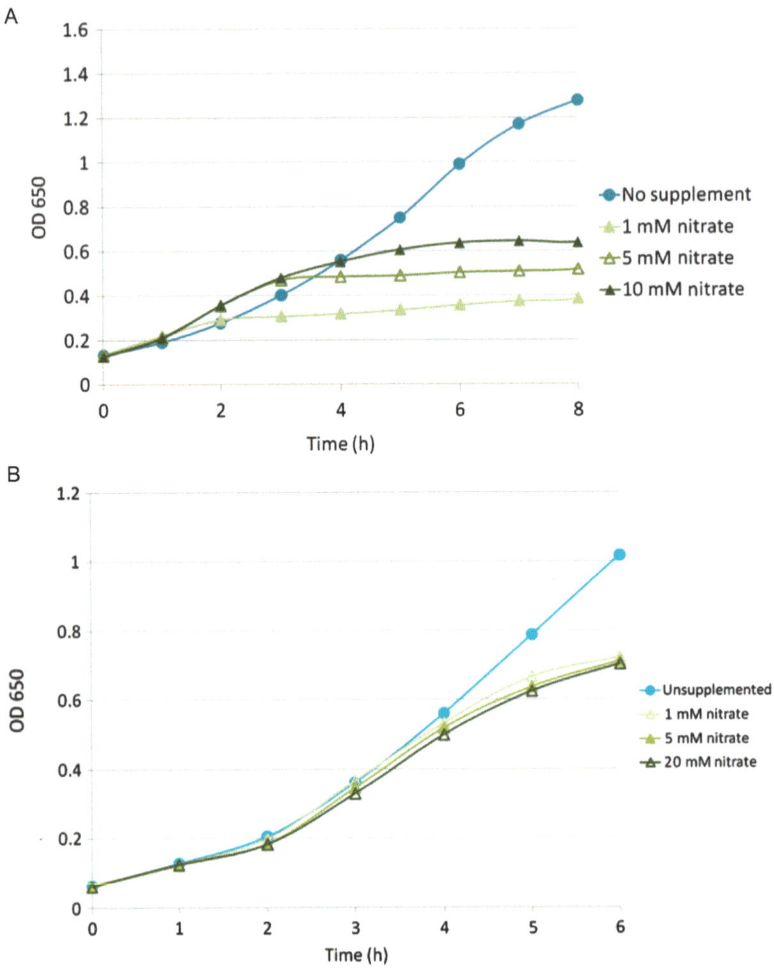

Fig. 4 Central role of the nitrate reductase NarG as a primary source of nitrosative stress. Bacteria were grown anaerobically in minimal salts medium augmented with 5% (v/v) nutrient broth and various concentrations of nitrate. (A) The strain used in these experiments is defective in all potential sources and sinks of NO: NirBD, NrfAB, Hmp, NorVW, Hcp, and YtfE. Note that the inhibition of growth is inversely dependent upon the concentration of nitrate added. (B) Almost complete relief of nitrosative stress by further deletion of the *narG* operon.

the periplasmic enzyme is preferentially synthesised in a carbon-sufficient culture when little nitrate is available. FNR activates expression of both systems. Complementary regulation depends upon the nitrate-responsive two-component regulatory system, NarX–NarL, which when activated by excess nitrate represses *nap* gene expression but induces *narGHJI* expression.

As much less NO is produced by the *Salmonella* periplasmic nitrate reductase than by the membrane-associated NarG (Rowley et al., 2012), different conclusions will be reached concerning a central role for NarG depending on the ratio of nitrate to carbon source used for growth.

Both *S. enterica* and *E. coli* have alternative nitrite reductases, the cytoplasmic NirBD and periplasmic NrfAB, both of which reduce nitrite to ammonia. Regulation of both of these pathways is critically dependent upon growth conditions. As with the two nitrate reductases, nitrate-activated NarX NarL activates expression from the *nirB* promoter but represses *nrfA* expression (Bell, Cole, & Busby, 1990; Constantinidou et al., 2006; Page, Griffiths, & Cole, 1990; Potter et al., 1999; Wang & Gunsalus, 2000). Nucleoid-associated proteins such as IHF and Fis regulate the two promoters differentially, but only the *nrfA* promoter is repressed by the NO-sensitive repressor, NsrR. There is also a second two-component system, NarQ and NarP: NarQ has a higher affinity for nitrate than NarX (Wang & Gunsalus, 2000; Wang, Tseng, & Gunsalus, 1999). The two pathways also respond differently to the available carbon source: glucose repression of the *nir* operon is mediated by the transcription factor, Cra (Tyson, Busby, & Cole, 1997). At the most recent count, the *nirB* and the *nrfA* promoters are known to be regulated by at least seven and six transcription factors, respectively (Browning, Cole, & Busby, 2008; Browning, Lee, Spiro, & Busby, 2010). Thus growth conditions critically determine whether there is sufficient NirB or NrfA to contribute significantly to NO production.

A role was proposed for NarX–NarL in regulating synthesis of the flavorubredoxin NO reductase, NorV. This result was likely to be an indirect consequence of NarL-dependent induction of NarG synthesis, which in turn led to the production of sufficient NO to activate NorR and hence to induction of *norV* and *norW* (Justino, Almeida, Goncalves, Teixeira, & Saraiva, 2006).

10.2 Dangers in Deducing Survival Strategies From Laboratory Experiments Under Extreme Conditions

Anaerobic bacteria rapidly detoxify NO by two main mechanisms: enzyme-catalysed reduction; and chemical reactions with cellular components. This leads to the proposal that a steady state is established in which the rate of NO production is essentially the same as its rate of removal. The same is true in plants, animals, and aerobic microorganisms, except that NO is removed by oxidation rather than by reduction. Although rate data for NO production and removal are available for various organisms, few measurements of actual concentrations of NO can be found in the literature. There are claims that the concentration of NO in the μM range occur in vivo (Hausladen et al., 2001), but supporting data from direct measurements are hard to find. What is clear, however, is that NO concentrations in the bacterial cytoplasm are typically in the nanomolar, not the micromolar range (Jarboe et al., 2008; Wang et al., 2016). The low concentrations of NO released by homolytic decomposition of GSNO is sufficient to inactivate NsrR repression without activating the GSNO global response (Jarboe et al., 2008). The effects of NO on different strains of *N. gonorrhoeae* vary depending on whether they are able to maintain the NO concentration in the range 0–10 nM or in the much higher concentration of 50–100 nM (Cardinale & Clark, 2005). Rarely have concentrations above 1 μM been reported in any natural environment. Proteins that regulate gene expression in response to NO or that metabolise it in vivo must therefore have dissociation constants in the low nM range. As most of these are metalloproteins, experiments with NO concentrations 100–1000 times greater than those that occur naturally risk generating data based upon nonphysiological chemical reactions with cellular components. Proteins with roles in NO sensing or metabolism are also potentially prone to chemical inactivation or destruction. Three examples of where this has probably misled data interpretation will be given.

Oxygen sensing by the FNR protein depends on a [4Fe–4S] iron–sulphur centre, which is inactivated on exposure to oxygen. FNR functions as a transcription activator at many promoters and also as a repressor at others that include the *hmp* promoter. The [4Fe–4S] centre also binds and is inactivated by NO. Before the discovery of the NO-sensing NsrR, experiments in which bacteria were incubated with 50 μM NO resulted in inactivation of FNR and hence release of *hmp* expression from FNR repression. This led to FNR being assigned an NO-sensing function (Cruz-Ramos et al., 2002; Poole et al., 1996). However, subsequent experiments clearly showed that FNR remains active at concentrations of NO that inactivate NsrR repression, even at the *hcp–hcr* promoter where FNR is essential to activate transcription (Constantinidou et al., 2006; Vine, Purewal, & Cole, 2011; Wang et al., 2016). One might argue that the requirement for FNR to activate *hcp–hcr* transcription is relaxed under conditions of severe nitrosative stress because NsrR repression is relieved. If so, inactivation of FNR would not be detected. This possibility could easily be eliminated and any damage to FNR could be assessed by comparing transcription of FNR-activated and FNR-repressed promoters in an *nsrR* mutant in

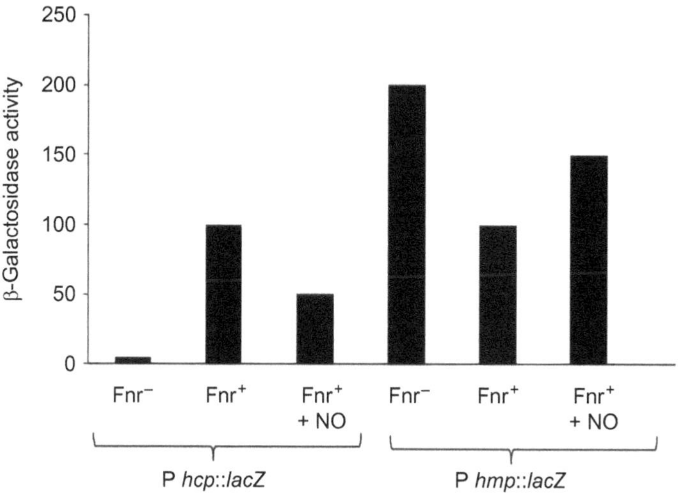

Fig. 5 Cartoon of hypothetical data for the effect of NO damage to FNR on transcription from of an FNR-activated and an FNR-repressed promoter. Idealised data from an experiment in which expression from the FNR-dependent *hcp* promoter decreases and the FNR-repressed hmp promoter increases in response to inactivation of FNR on exposure to NO. In this imaginary experiment, the cartoon implies that FNR is 50% less effective as a transcription activator or repressor. For many reasons, such idealised data are unlikely to be obtained. However, in a real experiment, it can confidently be predicted that the ratio of expression from P*hmp*:P*hcp* will increase with increasing FNR damage.

which effects of NsrR have been eliminated (Fig. 5). Results of such an experiment have not been reported.

The most credible role of FNR repression at the *hmp* promoter is to ensure that Hmp is produced mainly during aerobic growth when its NO oxygenase activity becomes critical for survival. Note, however, that there is precedence for genes required for an essentially aerobic function to be regulated by FNR. During anaerobic or oxygen-limited growth of *Shigella flexneri*, FNR activates synthesis of components of the type three secretion system in preparation for invasion of an aerobic host (Marteyn et al., 2010). Anaerobic induction of Hmp synthesis in response to NO might therefore be an example of the maxim "be prepared for trouble ahead", which in this example would be the need to be able to neutralise the nitrosative burst induced by invasion of a mammalian host.

Other transcription factors exploit a transition metal ligand or a haem group to sense NO. Examples include the widely distributed HcpR proteins that appear to use haem to bind NO, and the iron-sensing Fur proteins. It has been proposed that Fur enables pathogenic *Neisseria* to respond to an NO donor, but in these experiments the concentration of NO used was equivalent to 15 μM, far higher than those reported to occur naturally (compare D'Autréaux et al., 2002 with Barth & Clark, 2008). Even higher concentrations of the NO donor were required to relieve Fur repression completely. This is more consistent with chemical inactivation of Fur than a physiological response. Nevertheless, other groups working with other bacteria have also concluded not only that Fur regulates the nitrosative stress responses, but also binds to the relevant promoter regions (Crawford & Goldberg, 1998; Crawford et al., 2016; Gilberthorpe et al., 2007; Mukhopadhyay et al., 2004; Moore et al., 2004; Stamler, Lamas, & Fang, 2001). In each of these studies, the concentrations of NO or the chemicals used to generate NO were well above the physiologically relevant range. However, they cannot be dismissed because nitrosative damage to Fur or deletion of the *fur* gene clearly decreases the ability of *S. enterica*, *Staphylococcus aureus*, and other pathogens to survive in macrophages or in vivo where aerobic NO removal by Hmp is the key protection mechanism (Qu et al., 2009; Richardson, Dunman, & Fang, 2006; Troxell et al., 2011).

One of the most puzzling reports concluded that the *E. coli* OxyR protein, which is known to regulate part of the oxidative stress response, is also directly involved in regulating the response to nitrosative stress (Seth et al., 2012). The same group had repeatedly previously proposed roles for OxyR in the nitrosative stress response (Hausladen, Privalle, Keng, DeAngelo, & Stamler, 1996; Kim et al., 2002). The key result in the more recent paper was that deletion of *oxyR* resulted in increased protein S-nitrosylation. This increase coincided with OxyR-dependent transcription of a regulon that was distinct from the previously characterised OxyR regulon. The data confirmed that NO production is largely dependent upon NarG, but surprisingly anaerobic growth of the *oxyR* mutant in the presence of nitrate was severely impaired. It was then shown that OxyR is itself S-nitrosylated either by an external source of NO or during anaerobic growth with nitrate. The authors concluded that alternative posttranslational modifications enable OxyR to serve as a master regulator of some genes in the nitrosative and oxidative stress responses. But is there an alternative possible explanation for these surprising results?

Anaerobic growth by nitrate or nitrite respiration is totally dependent upon a functional FNR protein, and FNR is extremely sensitive to oxidative damage. Many of the data presented by Seth et al. (2012) could be secondary consequences of damage to FNR in the OxyR mutant, for example, during aerobic growth of the inoculum. If so, cultures of an *oxyR* mutant would be predicted to grow normally on nitrate if the inoculum is first adapted to anaerobic growth by glucose fermentation. The greater ability of dense cultures than dilute inocula to consume oxygen and ROS leads to two further predictions. First, the ability of an aerobic inoculum of an *oxyR* mutant to adapt to anaerobic growth in the presence of nitrate might be dependent upon the inoculum cell density. Secondly, an *oxyR* mutant would be able to grow by nitrate respiration if the inoculum is first grown anaerobically, for example with glycerol as carbon source and fumarate as electron acceptor. The more general point is that any repressor protein that requires an –SH group or transition metal centre (haem; iron–sulphur centre; nickel; or another type of metal centre) for function is vulnerable to inactivation by chemical attack by high concentrations of NO. It would therefore be reassuring to determine whether Fur, OxyR, SoxR–SoxS, PerR, and NmlR contribute significantly to resistance to nitrosative stress under conditions encountered by anaerobic bacteria in their natural environments.

11. CONCLUSIONS

Hopefully this review has shown that there is a vast but incomplete literature on how bacteria protect themselves against nitrosative stress. A major gap is knowledge of the concentrations of cytoplasmic NO and GSNO encountered by bacteria under conditions in their natural environment. The importance of Hmp both in vivo and in protecting bacteria in aerobic environments is totally convincing. Many other papers could have been cited to justify this statement. In contrast, no such dogmatic statements can be made about the anaerobic response to nitrosative stress, especially for bacteria other than laboratory strains of *E. coli*. Important roles have been established for the transcription factors NorV, NsrR, and the various Crp and FNR homologues that specifically bind NO with high affinity. Less so the widely assumed roles of OxyR, FNR, SoxR–SoxS, MelR, and even NmlR in sensing NO. It has been argued that in most of these cases changes in levels of transcription are due to secondary consequences of chemical damage to other transcription factors. This damage is far more extensive under the extreme conditions of laboratory experiments than would be encountered by bacteria in their natural oxygen-limited environments.

A pathway has been proposed for a previously unrecognised role for the YtfE protein working in partnership with the Hcp to repair nitrosative damage to iron–sulphur protein without the accumulation of toxic levels of NO in the cytoplasm. Experiments designed to detect direct interaction between

YtfE and Hcp were unsuccessful (Wang et al., 2016), and one or more proteins other than Hcr provide the electrons required to reduce NO to N_2O. The *hcr* gene is not found in most bacteria that encode Hcp: are FrdX and other redox proteins with a ferredoxin domain the electron donors? One of the greatest challenges is to establish the primary physiological role of YtfE by reconstituting its catalytic activity, if any, with purified components in vitro. Throughout the bacterial kingdom the expression of many genes of unknown function is induced by nitrosative stress. Their functions remain unknown.

Finally there is the question whether multiple defence mechanisms in some bacteria are complementary, or redundant. A specific example is *E. coli* in which we now know that NorV, NrfA, and Hcp (as well as Hmp!) can all reduce NO. Microbial physiologists will be wary of concluding that such a range of activities indicate redundancy. More likely is that the periplasmic NrfAB decreases the flux of exogenous NO into the cytoplasm, and that the NsrR-regulated Hcp limits cytoplasmic damage by the NO that enters the cytoplasm by diffusion or is generated by NarG. Should Hcp be unable to cope, synthesis of NorV and NorW is induced to eliminate the resulting stress (Chismon, Browning, Farrant, & Busby, 2010). In short, in enteric bacteria, NrfAB, Hcp, Hmp, and NorV are complimentary, not mutually redundant. Can we now find parallels in other bacterial groups?

REFERENCES

Abel, E., & Schmid, H. (1928). Kinetics of nitrous acid. 111. Kinetics of nitrous acid decomposition. *Physiological Chemistry*, *134*, 279–300.

Alvarez, L., Bricio, C., Blesa, A., Hidalgo, A., & Berenguer, J. (2014). Transferable denitrification capability of *Thermus thermophilus*. *Applied and Environmental Microbiology*, *80*, 19–28.

Alvarez, L., Bricio, C., Hidalgo, A., & Berenguer, J. (2014). Parallel pathways for nitrite reduction during anaerobic growth in *Thermus thermophilus*. *Journal of Bacteriology*, *196*, 1350–1358.

Arai, H., Hayashi, M., Kuroi, A., Ishii, M., & Igarashi, Y. (2005). Transcriptional regulation of the flavohemoglobin gene for aerobic nitric oxide detoxification by the second nitric oxide-responsive regulator of *Pseudomonas aeruginosa*. *Journal of Bacteriology*, *187*, 3960–3968.

Arendsen, A. F., Hadden, J., Card, G., McAlpine, J. S., Bailey, S., Zaitsev, V., et al. (1998). The "prismane" protein resolved: X-ray structure at 1.7 Å and multiple spectroscopy of two novel 4Fe clusters. *Journal of Biological Inorganic Chemistry*, *3*, 81–95.

Arnelle, D. R., & Stamler, J. S. (1995). NO^+, $NO^·$, and NO^- donation by S-nitrosothiols: Implications for regulation of physiological functions by S-nitrosylation and acceleration of disulphide formation. *Archives of Biochemistry and Biophysics*, *318*, 279–285.

Averill, B. A., & Tiedje, J. M. (1982). The chemical mechanism of microbial denitrification—A hypothesis. *FEBS Letters*, *138*, 8–11.

Balasiny, B., Rolfe, M. D., Vine, C. E., Bradley, C., Green, J., & Cole, J. A. (2018). Release of nitric oxide by the *Escherichia coli* YtfE (RIC) protein and its reduction by the hybrid cluster protein in an integrated pathway to minimise cytoplasmic nitrosative stress. *Microbiology*. In press.

Bang, I. S., Liu, L. M., Vazquez-Torres, A., Crouch, M. L., Stamler, J. S., & Fang, F. C. (2006). Maintenance of nitric oxide and redox homeostasis by the *Salmonella* flavohemoglobin Hmp. *Journal of Biological Chemistry, 281*, 28039–28047.

Barth, K., & Clark, V. L. (2008). Differences in nitric oxide steady states between arginine, hypoxanthine, uracil auxotrophs (AHU) and non-AHU strains of *Neisseria gonorrhoeae* during anaerobic respiration in the presence of nitrite. *Canadian Journal of Microbiology, 54*, 639–646.

Beaumont, H. J. E., Lens, S. I., Reijnders, W. N. M., Westerhoff, H. V., & van Spanning, R. J. M. (2004). Expression of nitrite reductase in *Nitrosomonas europaea* involves NsrR, a novel nitrite-sensitive transcription repressor. *Molecular Microbiology, 54*, 148–158.

Beckman, J. S., Ye, Y. Z., Anderson, P. G., Chen, J., Accavitti, M. A., Tarpey, M. M., et al. (1994). Extensive nitration of protein tyrosines in human atherosclerosis detected by immunohistochemistry. *Biological Chemistry Hoppe-Seyler, 375*, 81–88.

Bell, A. I., Cole, J. A., & Busby, S. J. W. (1990). Molecular genetic analysis of an FNR-dependent anaerobically-inducible *Escherichia coli* promoter. *Molecular Microbiology, 4*, 1753–1763.

Bodenmiller, D. M., & Spiro, S. (2006). The *yjbE* (*nsrR*) gene of *Escherichia coli* encodes a nitric-oxide sensitive transcriptional regulator. *Journal of Bacteriology, 188*, 874–881.

Braun, C., & Zumft, W. G. (1991). Marker exchange of the structural genes for nitric oxide reductase blocks the denitrification pathway of *Pseudomonas stutzeri* at nitric oxide. *Journal of Biological Chemistry, 266*, 22785–22788.

Browning, D. F., Cole, J. A., & Busby, S. J. (2008). Regulation by nucleoid-associated proteins at the *Escherichia coli nir* operon promoter. *Journal of Bacteriology, 190*, 7258–7267.

Browning, D. F., Lee, D. J., Spiro, S., & Busby, S. J. (2010). Down-regulation of the *Escherichia coli* K-12 *nrf* promoter by binding of the NsrR nitric oxide-sensing transcription repressor to an upstream site. *Journal of Bacteriology, 192*, 3824–3828.

Cabello, P., Pino, C., Olmo-Mira, M. F., Castillo, F., Roldan, M. D., & Moreno-Vivian, C. (2004). Hydroxylamine assimilation by *Rhodobacter capsulatus* E1F1, requirement of the *hcp* gene (hybrid cluster protein) located in the nitrate assimilation *nas* gene region for hydroxylamine reduction. *Journal of Biological Chemistry, 279*, 45485–45494.

Cadby, I. T., Faulkner, M., Cheneby, J., Long, J., van Helden, J., Dolla, A., & Cole, J. A. (2017). Coordinated response of the *Desulfovibrio desulfuricans* 27774 transcriptome to nitrate, nitrite and nitric oxide. *Scientific Reports*, 7, 16228.

Cadby, I. T., Ibrahim, S. A., Faulkner, M., Lee, D. J., Browning, D., Busby, S. J., et al. (2016). Regulation, sensory domains and roles of two *Desulfovibrio desulfuricans* ATCC27774 Crp family transcription factors, HcpR1 and HcpR2, in response to nitrosative stress. *Molecular Microbiology, 102*, 1120–1137.

Calmels, S., Oshima, H., & Bartsch, H. (1988). Nitrosamine formation by denitrifying and non-denitrifying bacteria: Implication of nitrite reductase and nitrate reductase in nitrosation catalysis. *Journal of General Microbiology, 134*, 221–226.

Cardinale, J. A., & Clark, V. L. (2005). Determinants of nitric oxide steady-state levels during anaerobic respiration by *Neisseria gonorrhoeae*. *Molecular Microbiology, 58*, 177–188.

Castiglione, N., Rinaldo, S., Giardina, G., & Cutruzzolà, F. (2009). The transcription factor DNR from *Pseudomonas aeruginosa* specifically requires nitric oxide and haem for the activation of a target promoter in *Escherichia coli*. *Microbiology, 155*, 2838–2844.

Cava, F., Laptenko, O., Borukhov, S., Shahlafi, Z., Blas-Galindo, E., Gomez-Puertas, P., et al. (2007). Control of the respiratory metabolism of *Thermus therophilus* by the nitrate respiration conjugative element *nce*. *Molecular Microbiology, 64*, 630–646.

Cava, F., Zafra, O., da Costa, M. S., & Berenguer, J. (2008). The role of the nitrate respiration element of *Thermus thermophilus* in the control and activity of the denitrification apparatus. *Environmental Microbiology, 10*, 522–533.

Chen, L., Liu, M. Y., LeGall, J., Fareira, P., Santos, H., & Xavier, A. V. (1993). Rubredoxin oxidase, a new flavo-hemo-protein, is the site of oxygen reduction to water by the "strict anaerobe" *Desulfovibrio gigas*. *Biochemical and Biophysical Research Communications, 193*, 100–105.

Chismon, D. L., Browning, D. F., Farrant, G. K., & Busby, S. J. (2010). Unusual organization, complexity and redundancy at the *Escherichia coli hcp-hcr* operon promoter. *Biochemical Journal, 430*, 61–68.

Constantinidou, C., Hobman, J. L., Griffiths, L., Patel, M. D., Penn, C. W., Cole, J. A., et al. (2006). A reassessment of the FNR regulon and transcriptomic analysis of the effects of nitrate, nitrite, NarXL, and NarQP as *Escherichia coli* adapts from aerobic to anaerobic growth. *Journal of Biological Chemistry, 281*, 4802–4815.

Corker, H., & Poole, R. K. (2003). Nitric oxide formation by *Escherichia coli*. Dependence on nitrite reductase, the NO-sensing FNR, and flavohemoglobin Hmp. *Journal of Biological Chemistry, 278*, 31584–31592.

Crack, J. C., Munnoch, J., Dodd, E. L., Knowles, F., Al Bassam, M. M., Kamali, S., et al. (2015). NsrR from *Streptomyces coelicolor* is a nitric oxide-sensing [4Fe-4S] cluster protein with a specialized regulatory function. *Journal of Biological Chemistry, 290*, 12689–12704.

Crack, J. C., Stapleton, M. R., Green, J., Thomson, A. J., & Le Brun, N. E. (2013). Mechanism of [4Fe-4S]$(Cys)_4$ cluster nitrosylation is conserved among NO-responsive regulators. *Journal of Biological Chemistry, 288*, 11492–11502.

Crack, J. C., Svistunenko, D. A., Munnoch, J., Thomson, A. J., Hutchings, M. I., & Le Brun, N. E. (2016). Differentiated, promoter-specific response of [4Fe-4S] NsrR DNA binding to reaction with nitric oxide. *The Journal of Biological Chemistry, 291*, 8663–8672.

Crawford, M. J., & Goldberg, D. E. (1998). Regulation of the *Salmonella typhimurium* flavohaemoglobin gene. A new pathway for bacterial gene expression in response to nitric oxide. *Journal of Biological Chemistry, 273*, 34028–34032.

Crawford, M. J., & Goldberg, D. E. (2006). Regulation of the *Salmonella typhimurium* flavohemoglobin gene. *Journal of Biological Chemistry, 281*, 3752.

Crawford, M. A., Henard, C. A., Tapscott, T., Porwollik, S., McClelland, M., & Vásquez-Torres, S. (2016). DksA-dependent transcriptional regulation in *Salmonella* experiencing nitrosative stress. *Frontiers in Microbiology, 7*, 444.

Cruz-Ramos, H., Crack, J., Wu, G., Hughes, M. N., Scott, C., Thomson, A. J., et al. (2002). NO sensing by FNR: Regulation of the *Escherichia coli* NO-detoxifying flavohemoglobin, HMP. *EMBO Journal, 21*, 3235–3244.

Cutruzzolà, F., Arcovito, A., Giardina, G., della Longa, S., D'Angelo, P., & Rinaldo, S. (2014). Distal-proximal crosstalk in the heme binding pocket of the NO sensor DNR. *Biometals, 27*, 763–773.

Das, A., Coulter, E. D., Kurtz, D. M., & Lungdahl, L. G. (2001). Five-gene cluster in *Clostridium thermoaceticum* consisting of two divergent operons encoding rubredoxin oxidoreductase-rubredoxin and rubrerythrin-type A flavoprotein-high-molecular-weight rubredoxin. *Journal of Bacteriology, 183*, 1560–1567.

D'Autréaux, B., Touati, D., Bersch, B., Latour, J. M., & Michaud-Soret, I. (2002). Direct inhibition by nitric oxide of the transcriptional ferric uptake regulation protein via nitrosylation of the iron. *Proceedings of the National Academy of Sciences of the United States of America, 99*, 16619–16624.

Ebert, M., Schweyen, P., Bröring, M., Laass, S., Härtig, A., & Jahn, D. (2017). Heme and nitric oxide binding by the transcriptional regulator DnrF from the marine bacterium *Dinoroseobacter shibae* increases *napD* promoter affinity. *Journal of Biological Chemistry, 292*, 15468–15480.

Efromovich, S., Grainger, D., Bodenmiller, D., & Spiro, S. (2008). Genome-wide identification of binding sites for the nitric oxide-sensitive transcriptional regulator NsrR. *Methods in Enzymology*, *437*, 211–233.

Elvers, K. T., Turner, S. M., Wainright, L. M., Marsden, G., Hinds, J., Cole, J. A., et al. (2005). NssR, a member of the Crp-FNR superfamily from *Campylobacter jejuni*, regulates a nitrosative stress-responsive regulon that includes both a single-domain and a truncated haemoglobin. *Molecular Microbiology*, *57*, 735–750.

Elvers, K. T., Wu, G., Gilberthorpe, N. J., Poole, R. K., & Park, S. F. (2004). Role of an inducible single-domain haemoglobin in mediating resistance to nitric oxide and nitrosative stress in *Campylobacter jejuni* and *Campylobacter coli*. *Journal of Bacteriology*, *186*, 5332–5341.

Figueiredo, M. C. O., Lobo, S. A. L., Sousa, S. H., Pereira, F. P., Wall, J. D., Nobre, L. S., et al. (2013). Hybrid cluster proteins and flavodiiron proteins afford protection to *Desulfovibrio vulgaris* upon macrophage infection. *Journal of Bacteriology*, *195*, 2684–2690.

Filenko, N. A., Spiro, D., Browning, D. F., Squire, D., Overton, T. W., Cole, J. A., et al. (2007). The NsrR regulon of *Escherichia coli* K-12 includes genes encoding the hybrid cluster protein and the periplasmic, respiratory nitrite reductase. *Journal of Bacteriology*, *189*, 4410–4417.

Flatley, J., Barrett, J., Pullan, S. T., Hughes, M. N., Green, J., & Poole, R. K. (2005). Transcriptional responses of *Escherichia coli* to S-nitrosoglutathione under defined chemostat conditions reveal major changes in methionine biosynthesis. *The Journal of Biological Chemistry*, *280*, 10065–10072.

Gardner, P. R., Costantino, G., Szabo, C., & Salzman, A. L. (1997). Nitric oxide sensitivity of the aconitases. *Journal of Biological Chemistry*, *272*, 25071–25076.

Gardner, A. M., & Gardner, P. R. (2002). Flavohemoglobin detoxifies nitric oxide in aerobic, but not anaerobic, cultures of *Escherichia coli*. Evidence for a novel inducible nitric oxide-scavenging activity. *Journal of Biological Chemistry*, *277*, 8166–8171.

Gardner, P. R., Gardner, A. M., Martin, L. A., & Salzman, A. L. (1998). Nitric oxide dioxygenase: An enzyme function for flavohemoglobin. *Proceedings of the National Academy of Sciences of the United States of America*, *95*, 10378–10383.

Gardner, A. M., Gessner, C. R., & Gardner, P. R. (2003). Regulation of the nitric oxide reduction operon (*norRVW*) in *Escherichia coli*. Role of NorR and sigma54 in the nitric oxide stress response. *Journal of Biological Chemistry*, *278*, 10081–10086.

Gardner, A. M., Helmick, R. A., & Gardner, P. R. (2002). Flavorubredoxin, an inducible catalyst for nitric oxidereduction and detoxification in *Escherichia coli*. *Journal of Biological Chemistry*, *277*, 8172–8177.

Gilberthorpe, N. J., Lee, M. E., Stevanin, T. M., Read, R. C., & Poole, R. K. (2007). NsrR: A key regulator circumventing *Salmonella enterica* serovar Typhimurium and nitrosative stress *in vivo* and in IFN-γ-stimulated J774.2 macrophages. *Microbiology*, *153*, 1756–1771.

Gilberthorpe, N. J., & Poole, R. K. (2008). Nitric oxide homeostasis in *Salmonella typhimurium*: Roles of respiratory nitrate reductase and flavohemoglobin. *Journal of Biological Chemistry*, *283*, 11146–11154.

Gomes, C. M., Giuffre, A., Forte, E., Vicente, J. B., Saraiva, L. M., Brunori, M., et al. (2002). A novel type of nitric-oxide reductase. *Escherichia coli* flavorubredooxin. *Journal of Biological Chemistry*, *277*, 25273–25276.

Haas, D., Camper, M., & Zimmermann, A. (1992). Anaerobic control in *Pseudomonas aeruginosa*. In E. Galli, S. Silver, & B. Witholt (Eds.), *Pseudomonas: Molecular biology and biotechnology* (pp. 177–187). Washington, DC: American Society for Microbiology.

Harrington, J. C., Wong, S. M. S., Rosadini, C. V., Garifulin, O., Boyartchuk, V., & Akerley, B. J. (2009). Resistance of *Haemophilus influenzae* to reactive nitrogen donors and gamma interferon-stimulated macrophages requires the formate-dependent nitrite reductase regulator-activated *ytfE* gene. *Infection and Immunity*, 77, 1945–1958.

Hausladen, A., Gow, A., & Stamler, J. S. (2001). Flavohemoglobin denitrosylase catalyzes the reaction of a nitroxyl equivalent with molecular oxygen. *Proceedings of the National Academy of Sciences of the United States of America*, *98*, 10108–10112.

Hausladen, A., Privalle, C. T., Keng, T., DeAngelo, J., & Stamler, J. S. (1996). Nitrosative stress: Activation of the transcription factor OxyR. *Cell*, *86*, 719–729.

Heo, J., Wolfe, M. T., Staples, C. R., & Ludden, P. W. (2002). Converting the NiFeS carbon monoxide dehydrogenase to a hydrogenase and a hydroxylamine reductase. *Journal of Bacteriology*, *184*, 5894–5897.

Hernández-Urzúa, E., Zamorano-Sánchez, D. S., Ponce-Coria, J., Morett, E., Grogan, S., Poole, R. K., et al. (2007). Multiple regulators of the Flavohaemoglobin (*hmp*) gene of *Salmonella enterica* serovar Typhimurium include RamA, a transcriptional regulator conferring the multidrug resistance phenotype. *Archives of Microbiology*, *187*, 67–77.

Husain, M., Jones-Carson, J., Liu, L., Song, M., Saah, J. R., Troxell, B., et al. (2014). Ferric uptake regulator-dependent antinitrosative defenses in *Salmonella enterica* serovar Typhimurium pathogenesis. *Infection and Immunity*, *82*, 333–340.

Hyduke, D. R., Jarboe, L. R., Tran, L. M., Chou, K. J., & Liao, J. C. (2007). Integrated network analysis identified nitric oxide response networks and dihydroxyacid dehydratase as a crucial target in *Escherichia coli*. *Proceedings of the National Academy of Sciences of the United States of America*, *104*, 8484–8489.

Isabella, V. M., Lapek, J. D., Jr., Kennedy, E. M., & Clark, V. L. (2009). Functional analysis of NsrR, a nitric oxide-sensing Rrf2 repressor in *Neisseria gonorrhoeae*. *Molecular Microbiology*, *71*, 227–239.

Jarboe, L. R., Hyduke, D. R., Tran, L. M., Chou, K. J., & Liao, J. C. (2008). Determination of the *Escherichia coli* S-nitrosoglutathione response network using integrated biochemical and systems analysis. *Journal of Biological Chemistry*, *283*, 5148–5157.

Johnston, S., Lin, S., Lee, P., Caffrey, S. M., Wildschut, J., Voordouw, J. K., et al. (2009). A genomic island of the sulphate-reducing bacterium *Desulfovibrio vulgaris* Hildenborough promotes survival under stress conditions while decreasing the efficiency of anaerobic growth. *Environmental Microbiology*, *11*, 981–991.

Justino, M. C., Almeida, C. C., Goncalves, V. L., Teixeira, M., & Saraiva, L. M. (2006). *Escherichia coli* YtfE is a di-iron protein with an important function in assembly of iron-sulphur clusters. *FEMS Microbiology Letters*, *257*, 278–284.

Justino, M. C., Almeida, C. C., Goncalves, V. L., Teixeira, M., & Saraiva, L. M. (2007). *Escherichia coli* di-iron YtfE protein is necessary for the repair of stress-damaged iron-sulphur clusters. *Journal of Biological Chemistry*, *282*, 10352–10359.

Justino, M. C., Baptista, J. M., & Saraiva, L. M. (2009). Di-iron proteins of the Ric family are involved in iron-sulfur cluster repair. *Biometals*, *22*, 99–108.

Justino, M. C., Ecobichon, C., Fernandes, A. F., Boneca, I. G., & Saraiva, L. M. (2012). *Helicobacter pylori* has an unprecedented nitric oxide detoxifying system. *Antioxidants & Redox Signaling*, *17*, 1190–1200.

Justino, M. C., Vicente, J. B., Teixeira, M., & Saraiva, L. M. (2005). New genes implicated in the protection of anaerobically grown *Escherichia coli* against nitric oxide. *Journal of Biological Chemistry*, *280*, 2636–2643.

Karlinsey, J. E., Bang, I.-S., Becker, L. A., Frawley, E. R., Porwollik, S., Robbins, H. F., et al. (2012). The NsrR regulon in nitrosative stress resistance of *Salmonella enterica* serovar Typhimurium. *Molecular Microbiology*, *85*, 1179–1193.

Kim, S. O., Merchant, K., Nudelman, R., Beyer, W. F., Jr., Keng, T., DeAngelo, J., et al. (2002). OxyR: A molecular code for redox-related signaling. *Cell*, *109*, 383–396.

Kim, S. O., Orii, Y., Lloyd, D., Hughes, M. N., & Poole, R. K. (1999). Anoxic function for the *Escherichia coli* flavohemoglobin (Hmp): Reversible binding of nitric oxide and reduction to nitrous oxide. *FEBS Letters*, *445*, 389–394.

Kommineni, S., Lama, A., Popescu, B., & Nakano, M. M. (2012). Global transcriptional control by NsrR in *Bacillus subtilis*. *Journal of Bacteriology*, *194*, 1679–1688.

Lee, Y.-Y., Shearer, N., & Spiro, S. (2006). Transcription factor NNR from *Paracoccus denitrificans* is a sensor of both nitric oxide and oxygen: Isolation of *nnr** alleles encoding effector-independent proteins and evidence for a haem-based sensing mechanism. *Microbiology*, *152*, 1461–1470.

Levine, J. S., Augustsson, T. R., Anderson, I. C., Hoell, J. M., Jr., & Brewer, D. A. (1984). Tropospheric sources of NO_x: Lightning and biology. *Atmospheric Environment*, *18*, 1797–1804.

Lewis, R. S., & Deen, W. M. (1994). Kinetics of the reaction of nitric oxide with oxygen in aqueous solutions. *Chemical Research in Toxicology*, *7*, 568–574.

Lin, H. Y., Bledsoe, P. J., & Stewart, V. (2007). Activation of *yeaR-yoaG* operon transcription by the nitrate-responsive regulator NarL is independent of oxygen-responsive regulator FNR in *Escherichia coli* K-12. *Journal of Bacteriology*, *189*, 7539–7548.

Liu, L., Hausladen, A., Zeng, M., Que, L., Heifman, J., & Stamler, J. S. (2001). A metabolic enzyme for S-nitrosothiol conserved from bacteria to humans. *Nature*, *410*, 490–494.

Lo, F.-C., Hsieh, C.-C., Maestre-Reyna, M., Chen, C.-Y., Ko, T.-P., Horng, Y.-C., et al. (2016). Crystal structure analysis of the repair of iron centers protein YtfE and its interaction with NO. *Chemistry*, *22*, 9768–9776.

Lobato, L., Bouzhir-Sima, L., Yamashita, T., Wilson, M. T., Vos, M. H., & Liebl, U. (2014). Dynamics of the heme-binding bacterial gas-sensing dissimilative nitrate respiration regulator (DNR) and activation barriers for ligand binding and escape. *Journal of Biological Chemistry*, *289*, 26514–26524.

Mania, D., Heylen, K., van Spanning, R. J. M., & Frostegård, Å. (2015). The nitrate ammonifying and *nosZ*-carrying bacterium *Bacillus vireti* is a potent source and sink for nitric and nitrous oxide under high nitrate conditions. *Environmental Microbiology*, *16*, 3196–3210.

Marteyn, B., West, N., Browning, D., Cole, J., Shaw, J., Palm, F., et al. (2010). Modulation of *Shigella* virulence in response to available oxygen in vivo. *Nature*, *465*, 355–358.

Membrillo-Hernández, J., Coopamah, M. D., Channa, A., Hughes, M. N., & Poole, R. K. (1998). A novel mechanism for upregulation of the *Escherichia coli* K-12 *hmp* (flavohaemoglobin) gene by the 'NO releaser', *S*-nitrosoglutathione: Nitrosation of homocysteine and modulation of MetR binding to the *glyA-hmp* intergenic region. *Molecular Microbiology*, *29*, 1101–1112.

Metheringham, R., & Cole, J. (1997). A reassessment of the genetic determinants, the effect of growth conditions and the availability of an electron donor on the nitrosating activity of *Escherichia coli* K-12. *Microbiology*, *143*, 2647–2656.

Miao, Z., & King, S. B. (2016). Recent advances in the chemical biology of nitroxyl (HNO) detection and generation. *Nitric Oxide*, *57*, 1–14.

Mills, P. C., Rowley, G., Spiro, S., Hinton, J. C. D., & Richardson, D. J. (2008). A combination of cytochrome *c* nitrite reductase (NrfA) and flavorubredoxin (NorV) protects *Salmonella enterica* serovar Typhimurium against killing by NO in anoxic environments. *Microbiology*, *154*, 1218–1228.

Moore, C. M., Nakano, M. M., Wang, T., Ye, R. W., & Helmann, J. D. (2004). Response of *Bacillus subtilis* to nitric oxide and the nitrosating agent sodium nitroprusside. *Journal of Bacteriology*, *186*, 4655–4664.

Mukhopadhyay, P., Zheng, M., Bedzyk, L. A., LaRossa, R. A., & Storz, G. (2004). Prominent roles of the NorR and Fur regulators in the *Escherichia coli* transcriptional response to reactive nitrogen species. *Proceedings of the National Academy of Sciences of the United States of America*, *101*, 745–750.

Nakano, M. M., Geng, H., Nakano, S., & Kobayashi, K. (2006). The nitric oxide-responsive regulator NsrR controls ResDE-dependent gene expression. *Journal of Bacteriology*, *188*, 5878–5887.

Nobre, L. S., Garcia-Serres, R., Todorovic, S., Hildebrandt, P., Teixeira, M., Latour, J.-M., et al. (2014). *Escherichia coli* RIC is able to donate iron to iron-sulfur clusters. *PLoS One*, *9*, e95222. https://doi.org/10.1371/journal.pone.0095222.

Ott, L. E., Pickering, K. E., Stenchikov, G. L., Allen, D. J., DeCaria, A. J., Ridley, B., et al. (2010). Production of lightning NO_x and its vertical distribution calculated from three-dimensional cloud-scale chemical transport model simulations. *Journal of Geophysical Research-Atmospheres, 115*(D4), D04301.
Overeijnder, M. L., Hagen, W. R., & Hagedoorn, P. L. (2009). A thermostable hybrid cluster protein from *Pyrococcus furiosus*: Effect of the loss of a three helix bundle subdomain. *Journal of Biological Inorganic Chemistry, 14*, 703–710.
Overton, T. W., Whitehead, R., Li, Y., Snyder, L. A., Saunders, N. J., Smith, H., et al. (2006). Coordinated regulation of the *Neisseria gonorrhoeae* truncated denitrification pathway by the nitric oxide-sensitive repressor, NsrR, and nitrite-insensitive NarQ-NarP. *Journal of Biological Chemistry, 281*, 33115–33126.
Pacher, P., Beckman, J. S., & Liaudet, L. (2007). Nitric oxide and peroxynitrite in health and disease. *Physiological Reviews, 87*, 315–424.
Page, L., Griffiths, L., & Cole, J. A. (1990). Different physiological roles of two independent pathways for nitrite reduction to ammonia by enteric bacteria. *Archives of Microbiology, 154*, 349–354.
Partridge, J. D., Bodenmiller, D. M., Humphrys, M. S., & Spiro, S. (2009). NsrR targets in the *Escherichia coli* genome: New insights into DNA sequence requirements for binding and a role for NsrR in the regulation of motility. *Molecular Microbiology, 73*, 680–694.
Payne, W. J., Riley, P. S., & Cox, C. D. (1971). Separate nitrite, nitric oxide, and nitrous oxide reducing fractions from *Pseudomonas perfectomarinus*. *Journal of Bacteriology, 106*, 356–361.
Pierik, A. J., Wolbert, R. B. G., Mutsaers, P. H. A., Hagen, W. R., & Veeger, C. (1992). Purification and biochemical characterization of a putative [6Fe-6S] prismane-cluster-containing protein from *Desulfovibrio vulgaris* (Hildenborough). *European Journal of Biochemistry, 206*, 697–704.
Poock, S. R., Leach, E. R., Moir, J. W., Cole, J. A., & Richardson, D. J. (2002). Respiratory detoxification of nitric oxide by the cytochrome c nitrite reductase of *Escherichia coli*. *Journal of Biological Chemistry, 277*, 23664–23669.
Poole, R. K., Anjum, M. F., Membrillo-Hernández, J., Kim, S. O., Hughes, M. N., & Stewart, V. (1996). Nitric oxide, nitrite, and FNR regulation of *hmp* (flavohemoglobin) gene expression in *Escherichia coli* K-12. *Journal of Bacteriology, 178*, 5487–5492.
Potter, L., Millington, P., Thomas, G., & Cole, J. (1999). Competition between *Escherichia coli* strains expressing either a periplasmic or a membrane-bound nitrate reductase: Does Nap confer a selective advantage during nitrate-limited growth? *Biochemical Journal, 344*, 77–84.
Pullan, S. T., Gidley, M. D., Jones, R. A., Barrett, J., Stevanin, T. M., Read, R. C., et al. (2007). Nitric oxide in chemostat-cultured *Escherichia coli* is sensed by FNR and other global regulators: Unaltered methionine biosynthesis indicates lack of S nitrosation. *Journal of Bacteriology, 189*, 1845–1855.
Qu, W., Zhou, Y., Shao, C., Sun, Y., Zhang, Q., Chen, C., et al. (2009). *Helicobacter pylori* proteins response to nitric oxide stress. *The Journal of Microbiology, 47*, 486–493.
Ralt, D., Wishnok, J. S., Fitts, R., & Tannenbaum, S. R. (1988). Bacterial catalysis of nitrosation: Involvement of the *nar* operon of *Escherichia coli*. *Journal of Bacteriology, 170*, 359–364.
Ren, B., Zhang, N., Yang, J., & Ding, H. (2008). Nitric oxide-induced bacteriostasis and modification of iron-sulphur proteins in *Escherichia coli*. *Molecular Microbiology, 70*, 953–964.
Richardson, A. R., Dunman, P. M., & Fang, F. C. (2006). The nitrosative stress response of *Staphylococcus aureus* is required for resistance to innate immunity. *Molecular Microbiology, 61*, 927–939.

Rodionov, D. A., Dubchak, I. L., Arkin, A. P., Alm, E. J., & Gelfand, M. S. (2005). Dissimilatory metabolism of nitrogen oxides in bacteria: Comparative reconstruction of transcriptional networks. *PLoS Computational Biology, 1*, 415–431.

Rodrigues, R., Vicente, J. B., Félix, R., Oliviera, S., Teixeira, M., & Rodrigues-Pousada, C. (2006). *Desulfovibrio gigas* flavodiiron protein affords protection against nitrosative stress *in vivo*. *Journal of Bacteriology, 188*, 2745–2751.

Rogstam, A., Larsson, J. T., Kjelgaard, P., & von Wachenfeldt, C. (2007). Mechanisms of adaptation to nitrosative stress in bacillus subtilis. *Journal of Bacteriology, 189*, 3063–3071.

Rowley, G., Hensen, D., Felgate, H., Arkenberg, A., Appia-Ayme, C., Prior, H. C., et al. (2012). Resolving the contributions of the membrane-bound and periplasmic nitrate reductase systems to nitric oxide and nitrous oxide production in *Salmonella enterica* serotype Typhimurium. *Biochemical Journal, 441*, 755–762.

Saul, R. L., Kkabir, S. H., Cohen, Z., Bruce, W. R., & Archer, M. C. (1981). Reevaluation of nitrate and nitrite levels in the human intestine. *Cancer Research, 41*, 2280–2283.

Schumann, U., & Huntrieser, H. (2007). The global lightning-induced nitrogen oxides source. *Atmospheric Chemistry and Physics, 7*, 3823–3907.

Schwartz, C. J., Giel, J. L., Patschkowski, T., Luther, C., Ruzicka, F. J., Beinert, H., et al. (2001). IscR, an Fe-S cluster-containing transcription factor, represses expression of *Escherichia coli* genes encoding Fe-S cluster assembly proteins. *Proceedings of the National Academy of Sciences of the United States of America, 98*, 14895–14900.

Seth, D., Hausladen, A., Wang, J.-Y., & Stamler, J. S. (2012). Endogenous protein S-nitrosylation in *E. coli*: Regulation by OxyR. *Science, 336*, 470–473.

Shepherd, M., Achard, M. E. S., Idris, A., Totsika, M., Phan, M.-D., Peters, K. M., et al. (2016). The cytochrome *bd*-I respiratory oxidase augments survival of multidrug-resistant *Escherichia coli* during infection. *Scientific Reports, 6*, 35285.

Shimizu, T., Tsutsuki, H., Matsumoto, A., Nakaya, H., & Noda, M. (2012). The nitric oxide reductase of enteroheamorrhagic *Escherichia coli* plays an important role for the survival within macrophages. *Molecular Microbiology, 85*, 492–512.

Silaghi-Dumitrescu, R., Coulter, E. D., Das, A., Ljungdahl, L. G., Jameson, G. N. L., Huynh, B. H., et al. (2003). A flavodiiron protein and high molecular weight rubredoxin from *Moorella thermoacetica* with nitric oxide reductase activity. *Biochemistry, 42*, 2806–2815.

Smith, M. S. (1983). Nitrous oxide production by *Escherichia coli* is correlated with nitrate reductase activity. *Applied and Environmental Microbiology, 45*, 679–689.

Spiro, S. (2007). Regulators of bacterial responses to nitrosative stress. *FEMS Microbiology Reviews, 31*, 193–211.

Squire, D. J. P., Xu, M., Cole, J. A., Busby, S. J. W., & Browning, D. F. (2009). Competition between NarL-dependent activation and Fis-dependent repression controls expression from the *Escherichia coli yeaR* and *ogt* promoters. *Biochemical Journal, 420*, 249–257.

Stamler, J. S., Lamas, S., & Fang, F. C. (2001). Nitrosylation: The prototypic redox-based signalling mechanism. *Cell, 106*, 675–683.

Stevanin, T. M., Read, R. C., & Poole, R. K. (2007). The *hmp* gene encoding the NO-inducible flavohaemoglobin in *Escherichia coli* confers a protective advantage in resisting killing within macrophages, but not in vitro: Links with swarming motility. *Gene, 398*, 62–68.

Stroeher, U. H., Kidd, S. P., Stafford, S. L., Jennings, M. P., Paton, J. C., & McEwan, A. G. (2007). A pneumococcal MerR-like regulator and S-nitrosoglutathione reductase are required for systemic virulence. *The Journal of Infectious Diseases, 196*, 1820–1826.

Strube, K., De Vries, S., & Cramm, R. (2007). Formation of a dinitrosyl iron complex by NorA, a nitric oxide-binding di-iron protein from *Ralstonia eutropha* H16. *Journal of Biological Chemistry, 282*, 20292–20300.

Tinajero-Trejo, M., Vreugdenhil, A., Sedelnikova, S. E., Davidge, K. S., & Poole, R. K. (2013). Nitric oxide reactivities of the two globins of the foodborne pathogen *Campylobacter jejuni*: Roles in protection from nitrosative stress and analysis of potential reductants. *Nitric Oxide, 34*, 65–75.

Tosques, I. E., Shi, J., & Shapleigh, J. P. (1996). Cloning and characterization of *nnrR*, whose product is required for the expression of proteins involved in nitric oxide metabolism in *Rhodobacter sphaeroides* 2.4.3. *Journal of Bacteriology, 178*, 4958–4964.

Troxell, B., Sikes, M. L., Fink, R. C., Vazquez-Torres, A., Jones-Carson, J., & Hassan, H. M. (2011). Fur negatively regulates *hns* and is required for the expression of HilA and virulence in *Salmonella enterica* serovar Typhimurium. *Journal of Bacteriology, 193*, 497–505.

Tucker, N. P., D'Autréaux, B., Studholm, D. J., Spiro, S., & Dixon, R. (2004). DNA binding activity of the *Escherichia coli* nitric oxide sensor NorR suggests a conserved target sequence in diverse proteobacteria. *Journal of Bacteriology, 186*, 6656–6660.

Tucker, N. P., Hicks, M. G., Clarke, T. A., Crack, J. C., Chandra, G., Le Brun, J. E., et al. (2008). The transcriptional repressor protein NsrR senses nitric oxide directly via a [2Fe-2S] cluster. *PLoS One, 3*, e3623.

Tyson, K., Busby, S., & Cole, J. (1997). Catabolite regulation of two *Escherichia coli* operons encoding nitrite reductases: Role of the Cra protein. *Archives of Microbiology, 168*, 240–244.

van den Berg, W. A. M., Hagen, W. R., & van Dongen, W. M. A. M. (2000). The hybrid-cluster protein ('prismane protein') from *Escherichia coli*. *European Journal of Biochemistry, 267*, 666–676.

Van Spanning, R. J., De Boer, A. P., Reijnders, W. N., Spiro, S., Westerhoff, H. V., Stouthamer, A. H., et al. (1995). Nitrite and nitric oxide reduction in *Paracoccus denitrificans* is under the control of NNR, a regulatory protein that belongs to the FNR family of transcriptional activators. *FEBS Letters, 360*, 151–154.

Van Spanning, R. J., Houben, E., Reijnders, W. N., Spiro, S., Westerhoff, H. V., & Saunders, N. (1999). Nitric oxide is a signal for NNR-mediated transcription activation in *Paracoccus denitrificans*. *Journal of Bacteriology, 181*, 4129–4132.

van Wonderen, J. H., Burlat, B., Richardson, D. J., Cheesman, M. R., & Butt, J. N. (2008). The nitric oxide reductase activity of cytochrome *c* nitrite reductase from *Escherichia coli*. *Journal of Biological Chemistry, 283*, 9587–9594.

Vasil'eva, S. V., Stupakova, M. V., Lobysheva, I. I., Mikoyan, V. D., & Vanin, A. F. (2001). Activation of the *Escherichia coli* SoxRS-regulon by nitric oxide and its physiological donors. *Biochemistry, 66*, 984–988.

Vasudevan, S. G., Armarego, W. L. F., Shaw, D. C., Lilley, P. E., & Poole, R. K. (1991). Isolation and nucleotide sequence of the *hmp* gene that encodes a hemoglobin-like protein in *Escherichia coli* K-12. *Molecular & General Genetics, 226*, 49–58.

Vine, C. E., & Cole, J. A. (2011). Nitrosative stress in *Escherichia coli*: Reduction of nitric oxide. *Biochemical Society Transactions, 39*, 213–215.

Vine, C. E., Justino, M. C., Saraiva, L., & Cole, J. A. (2010). A spontaneous 124-gene deletion during construction of a *ytfE* (*ric*) mutant accounts for growth defective phenotypes: Confirmation that a *ytfE* mutation results in loss of repair of iron centres in proteins damaged by oxidative or nitrosative stress. *Journal of Microbiological Methods, 81*, 77–79.

Vine, C. E., Purewal, S. K., & Cole, J. A. (2011). NsrR-dependent method for detecting nitric oxide accumulation in the *Escherichia coli* cytoplasm and enzymes involved in NO production. *FEMS Microbiology Letters, 325*, 118–124.

Vollack, K. U., Härtig, E., Körner, H., & Zumft, W. G. (1999). Multiple transcription factors of the FNR family in denitrifying *Pseudomonas stutzeri*: Characterization of four *FNR*-like genes, regulatory responses and cognate metabolic processes. *Molecular Microbiology, 31*, 1681–1694.

Vollack, K.-U., & Zumft, W. G. (2001). Nitric oxide signaling and transcriptional control of denitrification genes in *Pseudomonas stutzeri*. *Journal of Bacteriology*, *183*, 2516–2526.

Wang, H., & Gunsalus, R. P. (2000). The *nrfA* and *nirB* nitrite reductase operons in Escherichia coli are expressed differently in response to nitrate than to nitrite. *Journal of Bacteriology*, *182*, 5813–5822.

Wang, H., Tseng, C.-P., & Gunsalus, R. P. (1999). The *napF* and *narG* nitrate reductase operons in *Escherichia coli* are differentially expressed in response to submicromolar concentrations of nitrate but not nitrite. *Journal of Bacteriology*, *181*, 5303–5308.

Wang, J., Vine, C. E., Balasiny, B. K., Rizk, J., Bradley, C. L., Tinajero-Trejo, M., et al. (2016). The roles of the hybrid cluster protein, Hcp, and its reductase, Hcr, in high affinity nitric oxide reduction that protects anaerobic cultures of *Escherichia coli* against nitrosative stress. *Molecular Microbiology*, *100*, 877–892.

Weiss, B. (2006). Evidence for mutagenesis by nitric oxide during nitrate metabolism in *Escherichia coli*. *Journal of Bacteriology*, *188*, 829–833.

Whitehead, R. N., Overton, T. W., Snyder, L. A. S., McGowan, S. J., Smith, H., Cole, J. A., et al. (2007). The small FNR regulon of *Neisseria gonorrhoeae*: Comparison with the larger *Escherichia coli* FNR regulon and interaction with the NarQ-NarP regulon. *BMC Genomics*, *8*, 35.

Wildschut, J. D., Lang, R. M., Voordouw, J. K., & Voordouw, G. (2006). Rubredoxin: oxygen oxidoreductase enhances survival of *Desulfovibrio vulgaris* Hildenborough under microaerophilic conditions. *Journal of Bacteriology*, *188*, 6253–6260.

Wolfe, M. T., Heo, J., Garavelli, J. S., & Ludden, P. W. (2002). Hydroxylamine reductase activity of the hybrid cluster protein from *Escherichia coli*. *Journal of Bacteriology*, *184*, 5898–5902.

Yukl, E. T., Elbaz, M. A., Nakano, M. M., & Moënne-Loccoz, P. (2008). Transcription factor NsrR from *Bacillus subtilis* senses nitric oxide with a 4Fe-4S cluster. *Biochemistry*, *47*, 13084–13092.

Yurkiw, M. A., Voordouw, J., & Voordouw, G. (2012). Contribution of rubredoxin: Oxygen oxidoreductases and hybrid cluster proteins of *Desulfovibrio vulgaris* Hildenborough to survival under oxygen and nitrite stress. *Environmental Microbiology*, *14*, 2711–2725.

FSC
www.fsc.org
MIX
Papier | Fördert
gute Waldnutzung
FSC® C083411

Kolibri 360 GmbH
Ferdinand-Jühlke-Straße 7
99095 Erfurt, Deutschland
produktsicherheit@kolibri360.de